METHODS OF STATISTICAL PHYSICS

This graduate-level textbook on thermal physics covers classical thermodynamics, statistical mechanics, and their applications. It describes theoretical methods to calculate thermodynamic properties, such as the equation of state, specific heat, Helmholtz potential, magnetic susceptibility, and phase transitions of macroscopic systems.

In addition to the more standard material covered, this book also describes more powerful techniques, which are not found elsewhere, to determine the correlation effects on which the thermodynamic properties are based. Particular emphasis is given to the cluster variation method, and a novel formulation is developed for its expression in terms of correlation functions. Applications of this method to topics such as the three-dimensional Ising model, BCS superconductivity, the Heisenberg ferromagnet, the ground state energy of the Anderson model, antiferromagnetism within the Hubbard model, and propagation of short range order, are extensively discussed. Important identities relating different correlation functions of the Ising model are also derived.

Although a basic knowledge of quantum mechanics is required, the mathematical formulation is accessible, and the correlation functions can be evaluated either numerically or analytically in the form of infinite series. Based on courses in statistical mechanics and condensed matter theory taught by the author in the United States and Japan, this book is entirely self-contained and all essential mathematical details are included. It will constitute an ideal companion text for graduate students studying courses on the theory of complex analysis, classical mechanics, classical electrodynamics, and quantum mechanics. Supplementary material is also available on the internet at http://uk.cambridge.org/resources/0521580560/

TOMOYASU TANAKA obtained his Doctor of Science degree in physics in 1953 from the Kyushu University, Fukuoka, Japan. Since then he has divided his time between the United States and Japan, and is currently Professor Emeritus of Physics and Astronomy at Ohio University (Athens, USA) and also at Chubu University (Kasugai, Japan). He is the author of over 70 research papers on the two-time Green's function theory of the Heisenberg ferromagnet, exact linear identities of the Ising model correlation functions, the theory of super-ionic conduction, and the theory of metal hydrides. Professor Tanaka has also worked extensively on developing the cluster variation method for calculating various many-body correlation functions.

METHODS OF
STATISTICAL PHYSICS

TOMOYASU TANAKA

CAMBRIDGE
UNIVERSITY PRESS

CAMBRIDGE UNIVERSITY PRESS
Cambridge, New York, Melbourne, Madrid, Cape Town, Singapore,
São Paulo, Delhi, Dubai, Tokyo

Cambridge University Press
The Edinburgh Building, Cambridge CB2 8RU, UK

Published in the United States of America by Cambridge University Press, New York

www.cambridge.org
Information on this title: www.cambridge.org/9780521589581

First published 2002

A catalogue record for this publication is available from the British Library

Library of Congress Cataloguing in Publication data

Tanaka, Tomoyasu, 1919–
Methods of statistical physics / Tomoyasu Tanaka.
p.   cm.
Includes bibliographical references and index.
ISBN 0 521 58056 0 – ISBN 0 521 58958 4 (pb.)
1. Statistical physics.   I. Title.
QC174.8 .T36   2002
530.13 – dc21   2001035650

ISBN 978-0-521-58056-4 Hardback
ISBN 978-0-521-58958-1 Paperback

Transferred to digital printing 2010

To the late Professor Akira Harasima

Contents

Preface

This book may be used as a textbook for the first or second year graduate student who is studying concurrently such topics as theory of complex analysis, classical mechanics, classical electrodynamics, and quantum mechanics.

In a textbook on statistical mechanics, it is common practice to deal with two important areas of the subject: mathematical formulation of the distribution laws of statistical mechanics, and demonstrations of the applicability of statistical mechanics.

The first area is more mathematical, and even philosophical, especially if we attempt to lay out the theoretical foundation of the approach to a thermodynamic equilibrium through a succession of irreversible processes. In this book, however, this area is treated rather routinely, just enough to make the book self-contained.[†]

The second area covers the applications of statistical mechanics to many thermodynamic systems of interest in physics. Historically, statistical mechanics was regarded as the only method of theoretical physics which is capable of analyzing the thermodynamic behaviors of dilute gases; this system has a disordered structure and statistical analysis was regarded almost as a necessity.

Emphasis had been gradually shifted to the imperfect gases, to the gas–liquid condensation phenomenon, and then to the liquid state, the motivation being to be able to deal with correlation effects. Theories concerning rubber elasticity and high polymer physics were natural extensions of the trend. Along a somewhat separate track, starting with the free electron theory of metals, energy band theories of both metals and semiconductors, the Heisenberg–Ising theories of ferromagnetism, the Bloch–Bethe–Dyson theories of ferromagnetic spin waves, and eventually the Bardeen–Cooper–Schrieffer theory of super-conductivity, the so-called solid state physics, has made remarkable progress. Many new and powerful theories, such as

---

[†] The reader is referred to the following books for extensive discussions of the subject: R. C. Tolman, *The Principles of Statistical Mechanics*, Oxford, 1938, and D. ter Haar, *Elements of Statistical Mechanics*, Rinehart and Co., New York, 1956; and for a more careful derivation of the distribution laws, E. Schrödinger, *Statistical Thermodynamics*, Cambridge, 1952.

the diagrammatic methods and the methods of the Green's functions, have been developed as applications of statistical mechanics. One of the most important themes of interest in present day applications of statistical mechanics would be to find the strong correlation effects among various modes of excitations.

In this book the main emphasis will be placed on the various methods of accurately calculating the correlation effects, i.e., the thermodynamical average of a product of many dynamical operators, if possible to successively higher orders of accuracy. Fortunately a highly developed method which is capable of accomplishing this goal is available. The method is called the cluster variation method and was invented by Ryoichi Kikuchi (1951) and substantially reformulated by Tohru Morita (1957), who has established an entirely rigorous statistical mechanics foundation upon which the method is based. The method has since been developed and expanded to include quantum mechanical systems, mainly by three groups; the Kikuchi group, the Morita group, and the group led by the present author, and more recently by many other individual investigators, of course. The method was a theme of special research in 1951; however, after a commemorative publication,[†] the method is now regarded as one of the more standardized and even rather effective methods of actually calculating various many-body correlation functions, and hence it is thought of as textbook material of graduate level.

Chapter 6, entitled 'The cluster variation method', will constitute the centerpiece of the book in which the basic variational principle is stated and proved. An exact *cumulant expansion* is introduced which enables us to evaluate the Helmholtz potential at any degree of accuracy by increasing the number of *cumulant functions* retained in the variational Helmholtz potential. The mathematical formulation employed in this method is tractable and quite adaptable to numerical evaluation by computer once the cumulant expansion is truncated at some point. In Sec. 6.10 a four-site approximation and in Appendix 3 a tetrahedron-plus-octahedron approximation are presented in which up to six-body correlation functions are evaluated by the cluster variation method. The number of variational parameters in the calculation is only ten in this case, so that the numerical analysis by any computer is not very time consuming (Aggarwal and Tanaka, 1977). In the advent of much faster computers in recent years, much higher approximations can be carried out with relative ease and a shorter cpu time.

Chapter 7 deals with the infinite series representations of the correlation functions. During the history of the development of statistical mechanics there was a certain period of time during which a great deal of effort was devoted to the calculation of the exact infinite series for some physical properties, such as the partition function, the high temperature paramagnetic susceptibility, the low temperature

---

[†] *Progress in Theoretical Physics Supplement no. 115* 'Foundation and applications of cluster variation method and path probability method' (1994).

spontaneous magnetization, and both the high and low temperature specific heat for the ferromagnetic Ising model in the three-dimensional lattices by fitting different diagrams to a given lattice structure. The method was called the combinatorial formulation. It was hoped that these exact infinite series might lead to an understanding of the nature of mathematical singularities of the physical properties near the second-order phase transition. G. Baker, Jr. and his collaborators (1961 and in the following years) found a rather effective method called *Padé approximants*, and succeeded in locating the second-order phase transition point as well as the nature of the mathematical singularities in the physical properties near the transition temperature.

Contrary to the prevailing belief that the cluster variation type formulations would give only undesirable classical critical-point exponents at the second-order phase transition, it is demonstrated in Sec. 7.5 and in the rest of Chapter 7 that the infinite series solutions obtained by the cluster variation method (Aggarwal & Tanaka, 1977) yield exactly the same series expansions as obtained by much more elaborate combinatorial formulations available in the literature. This means that the most accurate critical-point exponents can be reproduced by the cluster variation method; a fact which is not widely known. The cluster variation method in this approximation yielded exact infinite series expansions for ten correlation functions simultaneously.

Chapter 8, entitled 'The extended mean-field approximation', is also rather unique. One of the most remarkable accomplishments in the history of statistical mechanics is the theory of superconductivity by Bardeen, Cooper, & Schrieffer (1957). The degree of approximation of the BCS theory, however, is equivalent to the mean-field approximation. Another more striking example in which the mean-field theory yields an exact result is the famous Dyson (1956) theory of spin-wave interaction which led to the T^4 term of the low temperature series expansion of the spontaneous magnetization. The difficult part of the formulation is not in its statistical formulation, but rather in the solution of a two-spin-wave eigenvalue problem. Even in Dyson's papers the separation between the statistical formulation and the solution of the two-spin-wave eigenvalue problem was not clarified, hence there were some misconceptions for some time. The *Wentzel theorem* (Wentzel, 1960) gave crystal-clear criteria for a certain type of Hamiltonian for which the mean-field approximation yields an exact result. It is shown in Chapter 8 that both the BCS reduced Hamiltonian and the spin-wave Hamiltonian for the Heisenberg ferromagnet satisfy the Wentzel criteria, and hence the mean-field approximation gives exact results for those Hamiltonians. For this reason the content of Chapter 8 is pedagogical.

Chapter 9 deals with some of the exact identities for different correlation functions of the two-dimensional Ising model. Almost 100 Ising spin correlation functions may be calculated exactly if two or three known correlation functions are fed into these identities. It is shown that the method is applicable to the three-dimensional Ising model, and some 18 exact identities are developed for the diamond lattice (Appendix 5). When a large number of correlation functions are introduced there arises a problem of naming them such that there is no confusion in assigning two different numbers to the same correlation function appearing at two different locations in the lattice. The so-called *vertex number representation* is introduced in order to identify a given cluster figure on a given two-dimensional lattice.

In Chapter 10 an example of oscillatory behavior of the radial distribution (or pair correlation) function, up to the seventh-neighbor distance, which shows at least the first three peaks of oscillation, is found by means of the cluster variation method in which up to five-body correlation effects are taken into account. The formulation is applied to the order–disorder phase transition in the super-ionic conductor AgI. It is shown that the entropy change of the first-order phase transition thus calculated agrees rather well with the observed latent heat of phase transition. Historically, the radial distribution function in a classical monatomic liquid, within the framework of a continuum theory, is calculated only in the three-body (super-position) approximation, and only the first peak of the oscillatory behavior is found. The model demonstrated in this chapter suggests that the theory of the radial distribution function could be substantially improved if the lattice gas model is employed and with applications of the cluster variation method.

Chapter 11 gives a brief introduction of the Pfaffian formulation applied to the reformulation of the famous Onsager partition function for the two-dimensional Ising model. The subject matter is rather profound, and detailed treatments of the subject in excellent book form have been published (Green & Hurst, 1964; McCoy & Wu, 1973).

Not included are the diagrammatic method of many-body problem, the Green's function theories, and the linear response theory of transport coefficients. There are many excellent textbooks available on those topics.

The book starts with an elementary and rather brief introduction of classical thermodynamics and the ensemble theories of statistical mechanics in order to make the text self-contained. The book is not intended as a philosophical or fundamental principles approach, but rather serves more as a recipe book for statistical mechanics practitioners as well as research motivated graduate students.

Tomoyasu Tanaka

Acknowledgements

I should like to acknowledge the late Professor Akira Harasima, who, through his kind course instruction, his books, and his advice regarding my thesis, was of invaluable help whilst I was a student at Kyushu Imperial University during World War II and a young instructor at the post-war Kyushu University. Professor Harasima was one of the pioneer physicists in the area of theories of monatomic liquids and surface tension during the pre-war period and one of the most famous authors on the subjects of classical mechanics, quantum mechanics, properties of matter, and statistical mechanics of surface tension. All his physics textbooks are written so painstakingly and are so easy to understand that every student can follow the books as naturally and easily as Professor Harasima's lectures in the classroom. Even in the present day physics community in Japan, many of Professor Harasima's textbooks are best sellers more than a decade after he died in 1986. The present author has tried to follow the Harasima style, as it may be called, as much as possible in writing the *Methods of Statistical Mechanics*.

The author is also greatly indebted to Professor Tohru Morita for his kind leadership in the study of statistical mechanics during the four-year period 1962–6. The two of us, working closely together, burned the enjoyable late-night oil in a small office in the Kean Hall at the Catholic University of America, Washington, D.C. It was during this period that the cluster variation method was given full blessing.

1

The laws of thermodynamics

1.1 The thermodynamic system and processes

A physical system containing a large number of atoms or molecules is called the *thermodynamic system* if macroscopic properties, such as the temperature, pressure, mass density, heat capacity, etc., are the properties of main interest. The number of atoms or molecules contained, and hence the volume of the system, must be sufficiently large so that the conditions on the surfaces of the system do not affect the macroscopic properties significantly. From the theoretical point of view, the size of the system must be infinitely large, and the mathematical limit in which the volume, and proportionately the number of atoms or molecules, of the system are taken to infinity is often called the *thermodynamic limit*.

The *thermodynamic process* is a process in which some of the macroscopic properties of the system change in the course of time, such as the flow of matter or heat and/or the change in the volume of the system. It is stated that the system is in *thermal equilibrium* if there is no thermodynamic process going on in the system, even though there would always be microscopic molecular motions taking place. The system in thermal equilibrium must be uniform in density, temperature, and other macroscopic properties.

1.2 The zeroth law of thermodynamics

If two thermodynamic systems, A and B, each of which is in thermal equilibrium independently, are brought into thermal contact, one of two things will take place: either (1) a flow of heat from one system to the other or (2) no thermodynamic process will result. In the latter case the two systems are said to be in thermal equilibrium with respect to each other.

The zeroth law of thermodynamics *If two systems are in thermal equilibrium with each other, there is a physical property which is common to the two systems. This common property is called the temperature.*

Let the condition of thermodynamic equilibrium between two physical systems A and B be symbolically represented by

$$A \Leftrightarrow B. \tag{1.1}$$

Then, experimental observations confirm the statement

$$\text{if} \quad A \Leftrightarrow C \quad \text{and} \quad B \Leftrightarrow C, \quad \text{then} \quad A \Leftrightarrow B. \tag{1.2}$$

Based on preceding observations, some of the physical properties of the system C can be used as a measure of the temperature, such as the volume of a fixed amount of the chemical element mercury under some standard atmospheric pressure. The zeroth law of thermodynamics is the assurance of the existence of a property called the *temperature*.

1.3 The thermal equation of state

Let us consider a situation in which two systems A and B are in thermal equilibrium. In particular, we identify A as the thermometer and B as a system which is homogeneous and isotropic. In order to maintain equilibrium between the two, the volume V of B does not have to have a fixed value. The volume can be changed by altering the hydrostatic pressure p of B, yet maintaining the equilibrium condition in thermal contact with the system A. This situation may be expressed by the following equality:

$$f_B(p, V) = \theta_A, \tag{1.3}$$

where θ_A is an *empirical temperature* determined by the thermometer A.

The thermometer A itself does not have to be homogeneous and isotropic; however, let A also be such a system. Then,

$$f_B(p, V) = f_A(p_A, V_A). \tag{1.4}$$

For the sake of simplicity, let p_A be a constant. Usually p_A is chosen to be one atmospheric pressure. Then f_A becomes a function only of the volume V. Let us take this function to be

$$f_A(p_A, V_A) = 100 \left[\frac{V_A - V_0}{V_{100} - V_0} \right]_A, \tag{1.5}$$

where V_0 and V_{100} are the volumes of A at the freezing and boiling temperatures of water, respectively, under one atmospheric pressure. This means

$$\theta = 100 \frac{V_A - V_0}{V_{100} - V_0}. \tag{1.6}$$

If B is an arbitrary substance, (1.3) may be written as

$$f(p, V) = \theta. \tag{1.7}$$

In the above, the volume of the system A is used as the thermometer; however, the pressure p could have been used instead of the volume. In this case the volume of system A must be kept constant. Other choices for the thermometer include the resistivity of a metal. The temperature θ introduced in this way is still an empirical temperature. An equation of the form (1.7) describes the relationship between the pressure, volume, and temperature θ and is called the *thermal equation of state*. In order to determine the functional form of $f(p, V)$, some elaborate measurements are needed. To find a relationship between small changes in p, V and θ, however, is somewhat easier. When (1.7) is solved for p, we can write

$$p = p(\theta, V). \tag{1.8}$$

Differentiating this equation, we find

$$dp = \left(\frac{\partial p}{\partial \theta} \right)_V d\theta + \left(\frac{\partial p}{\partial V} \right)_\theta dV. \tag{1.9}$$

If the pressure p is kept constant, i.e., $dp = 0$, the so-called *isobaric process*,

$$\left(\frac{\partial p}{\partial \theta} \right)_V d\theta + \left(\frac{\partial p}{\partial V} \right)_\theta dV = 0. \tag{1.10}$$

In this relation, one of the two changes, either $d\theta$ or dV, can have an arbitrary value; however, the ratio $dV/d\theta$ is determined under the condition $dp = 0$. Hence the notation $(\partial V / \partial \theta)_p$ is appropriate. Then,

$$\left(\frac{\partial p}{\partial \theta} \right)_V + \left(\frac{\partial p}{\partial V} \right)_\theta \left(\frac{\partial V}{\partial \theta} \right)_p = 0. \tag{1.11}$$

$(\partial p / \partial \theta)_V$ is the rate of change of p with θ under the condition of constant volume, the so-called *isochoric process*. Since V is kept constant, p is a function only of θ. Therefore

$$\left(\frac{\partial p}{\partial \theta} \right)_V = \frac{1}{\left(\frac{\partial \theta}{\partial p} \right)_V}. \tag{1.12}$$

Hence (1.11) is rewritten as

$$\left(\frac{\partial p}{\partial V}\right)_\theta \left(\frac{\partial V}{\partial \theta}\right)_p \left(\frac{\partial \theta}{\partial p}\right)_V = -1. \tag{1.13}$$

This form of equation appears very often in the formulation of thermodynamics. In general, if a relation $f(x, y, z) = 0$ exists, then the following relations hold:

$$\left(\frac{\partial x}{\partial y}\right)_z = \frac{1}{\left(\frac{\partial y}{\partial x}\right)_z}, \quad \left(\frac{\partial x}{\partial y}\right)_z \left(\frac{\partial y}{\partial z}\right)_x \left(\frac{\partial z}{\partial x}\right)_y = -1. \tag{1.14}$$

The quantity

$$\beta = \frac{1}{V}\left(\frac{\partial V}{\partial \theta}\right)_p \tag{1.15}$$

is called the *volume expansivity*. In general, β is almost constant over some range of temperature as long as the range is not large. Another quantity

$$K = -V\left(\frac{\partial p}{\partial V}\right)_\theta \tag{1.16}$$

is called the *isothermal bulk modulus*. The reciprocal of this quantity,

$$\kappa = -\frac{1}{V}\left(\frac{\partial V}{\partial p}\right)_\theta, \tag{1.17}$$

is called the *isothermal compressibility*. Equation (1.9) is expressed in terms of these quantities as

$$dp = \beta K \, d\theta - \frac{K}{V} dV. \tag{1.18}$$

1.4 The classical ideal gas

According to laboratory experiments, many gases have the common feature that the pressure, p, is inversely proportional to the volume, V; i.e., the product pV is constant when the temperature of the gas is kept constant. This property is called the *Boyle–Marriot law*,

$$pV = F(\theta), \tag{1.19}$$

where $F(\theta)$ is a function only of the temperature θ. Many real gases, such as oxygen, nitrogen, hydrogen, argon, and neon, show small deviations from this behavior; however, the law is obeyed increasingly more closely as the density of the gas is lowered.

Thermodynamics is a branch of physics in which thermal properties of physical systems are studied from a macroscopic point of view. The formulation of the theories does not rely upon the existence of a system which has idealized properties. It is, nevertheless, convenient to utilize an idealized system for the sake of theoretical formulation. The *classical ideal gas* is an example of such a system.

Definition *The ideal gas obeys the Boyle–Marriot law at any density and temperature.*

Let us now construct a thermometer by using the ideal gas. For this purpose, we take a fixed amount of the gas and measure the volume change due to a change of temperature, θ_p, while the pressure of the gas is kept constant. So,

$$\theta_p = 100 \frac{V - V_0}{V_{100} - V_0}, \tag{1.20}$$

where V_0 and V_{100} are the volumes of the gas at the freezing and boiling temperatures, respectively, of water under the standard pressure. This scale is called the *constant-pressure gas thermometer*.

It is also possible to define a temperature scale by measuring the pressure of the gas while the volume of the gas is kept constant. This temperature scale is defined by

$$\theta_V = 100 \frac{p - p_0}{p_{100} - p_0}, \tag{1.21}$$

where p_0 and p_{100} are the pressures of the gas at the freezing and boiling temperatures, respectively, of water under the standard pressure. This scale is called the *constant-volume gas thermometer*.

These two temperature scales have the same values at the two fixed points of water by definition; however, they also have the same values in between the two fixed temperature points.

From (1.20) and (1.21),

$$\theta_p = 100 \frac{pV - pV_0}{pV_{100} - pV_0}, \qquad \theta_V = 100 \frac{pV - p_0 V}{p_{100} V - p_0 V}, \tag{1.22}$$

and, since $pV_0 = p_0 V$ and $pV_{100} = p_{100} V$,

$$\theta_p = \theta_V, \tag{1.23}$$

and hence we may set $\theta_p = \theta_V = \theta$ and simply define

$$pV_0 = p_0 V = (pV)_0, \qquad pV_{100} = p_{100} V = (pV)_{100}, \tag{1.24}$$

and

$$\theta = 100 \frac{pV - (pV)_0}{(pV)_{100} - (pV)_0}. \tag{1.25}$$

When this equation is solved for pV, we find that

$$pV = \frac{(pV)_{100} - (pV)_0}{100} \left[\theta + \frac{100(pV)_0}{(pV)_{100} - (pV)_0} \right]. \tag{1.26}$$

If we define

$$\frac{(pV)_{100} - (pV)_0}{100} = R',$$

$$\theta + \frac{100(pV)_0}{(pV)_{100} - (pV)_0} = \Theta, \tag{1.27}$$

(1.26) can then be written in the following form:

$$pV = R'\Theta. \tag{1.28}$$

Θ is called the *ideal gas temperature*. It will be shown later in this chapter that this temperature becomes identical with the *thermodynamic temperature scale*.

The difference between θ and Θ is given by

$$\Theta_0 = 100 \frac{(pV)_0}{(pV)_{100} - (pV)_0}. \tag{1.29}$$

According to laboratory experiments, the value of this quantity depends only weakly upon the type of gas, whether oxygen, nitrogen, or hydrogen, and in particular it approaches a common value, Θ_0, in the limit as the density of the gas becomes very small:

$$\Theta_0 = 273.15. \tag{1.30}$$

We can calculate the volume expansivity β for the ideal gas at the freezing point of water $\theta = 0$:

$$\beta = \frac{1}{V_0} \left(\frac{\partial V}{\partial \Theta} \right)_p = \frac{1}{V_0} \frac{R'}{p} = \frac{R'}{R'\Theta_0} = \frac{1}{\Theta_0}. \tag{1.31}$$

When the value $\Theta_0 = 273.15$ is introduced, we find

$$\beta = 0.0036610. \tag{1.32}$$

This value may be favorably compared with experimental measurements.

1.5 The quasistatic and reversible processes

The *quasistatic process* is defined as a thermodynamic process which takes place unlimitedly slowly. In the theoretical formulation of thermodynamics it is customary to consider a sample of gas contained in a cylinder with a frictionless piston. The walls of the cylinder are made up of a diathermal, i.e., a perfectly heat conducting metal, and the cylinder is immersed in a heat bath at some temperature. In order to cause any heat transfer between the heat bath and the gas in the cylinder there must be a temperature difference; and similarly there must be a pressure difference between the gas inside the cylinder and the applied pressure to the piston in order to cause any motion of the piston in and out of the cylinder. We may consider an ideal situation in which the temperature difference and the pressure difference are adjusted to be infinitesimally small and the motion of the piston is controlled to be unlimitedly slow. In this ideal situation any change or process of heat transfer along with any mechanical work upon the gas by the piston can be regarded as reversible, i.e., the direction of the process can be changed in either direction, by compression or expansion. Any gadgets which might be employed during the course of the process are assumed to be brought back to the original condition at the end of the process. Any process designed in this way is called a quasistatic process or a *reversible process* in which the system maintains an equilibrium condition at any stage of the process.

In this way the thermodynamic system, a sample of gas in this case, can make some finite change from an initial state P_1 to a final state P_2 by a succession of quasistatic processes. In the following we often state that a thermodynamic system undergoes a finite change from the initial state P_1 to the final state P_2 by reversible processes.

1.6 The first law of thermodynamics

Let us consider a situation in which a macroscopic system has changed state from one equilibrium state P_1 to another equilibrium state P_2, after undergoing a succession of reversible processes. Here the processes mean that a quantity of heat energy Q has cumulatively been absorbed by the system and an amount of mechanical work W has cumulatively been performed upon the system during these changes.

The first law of thermodynamics *There would be many different ways or routes to bring the system from state P_1 to the state P_2; however, it turns out that the sum*

$$W + Q \tag{1.33}$$

is independent of the ways or the routes as long as the two states P_1 and P_2 are fixed, even though the quantities W and Q may vary individually depending upon the different routes.

This is the fact which has been experimentally confirmed and constitutes the first law of thermodynamics. In (1.33) the quantities W and Q must be measured in the same units.

Consider, now, the case in which P_1 and P_2 are very close to each other and both W and Q are very small. Let these values be $d'W$ and $d'Q$. According to the first law of thermodynamics, the sum, $d'W + d'Q$, is independent of the path and depends only on the initial and final states, and hence is expressed as the difference of the values of a quantity called the *internal energy*, denoted by U, determined by the physical, or thermodynamic, state of the system, i.e.,

$$dU = U_2 - U_1 = d'W + d'Q. \tag{1.34}$$

Mathematically speaking, $d'W$ and $d'Q$ are not *exact differentials* of state functions since both $d'W$ and $d'Q$ depend upon the path; however, the sum, $d'W + d'Q$, is an exact differential of the state function U. This is the reason for using primes on those quantities. More discussions on the exact differential follow later in this chapter.

1.7 The heat capacity

We will consider one of the thermodynamical properties of a physical system, the *heat capacity*. The heat capacity is defined as the amount of heat which must be given to the system in order to raise its temperature by one degree. The *specific heat* is the heat capacity per unit mass or per mole of the substance.

From the first law of thermodynamics, the amount of heat $d'Q$ is given by

$$d'Q = dU - d'W = dU + pdV, \qquad d'W = -pdV. \tag{1.35}$$

These equations are not yet sufficient to find the heat capacity, unless dU and dV are given in terms of $d\Theta$, the change in ideal gas temperature. In order to find these relations, it should be noted that the thermodynamic state of a single-phase system is defined only when two variables are fixed. The relationship between U and Θ is provided by the *caloric equation of state*

$$U = U(\Theta, V), \tag{1.36}$$

and there is a thermal equation of state determining the relationship between p, V, and Θ:

$$p = p(\Theta, V). \tag{1.37}$$

In the above relations, we have chosen Θ and V as the independent variables to specify the thermodynamic state of the system. We could have equally chosen other sets, such as (Θ, p) or (p, V). Which of the sets is chosen depends upon the situation, and discussions of the most convenient set will be given in Chapter 2.

Let us choose the set (Θ, V) for the moment; then, one finds that

$$dU = \left(\frac{\partial U}{\partial \Theta}\right)_V d\Theta + \left(\frac{\partial U}{\partial V}\right)_\Theta dV, \tag{1.38}$$

$$d'Q = \left(\frac{\partial U}{\partial \Theta}\right)_V d\Theta + \left[\left(\frac{\partial U}{\partial V}\right)_\Theta + p\right] dV, \tag{1.39}$$

and the heat capacity, C, is given by

$$C = \left(\frac{\partial U}{\partial \Theta}\right)_V + \left[\left(\frac{\partial U}{\partial V}\right)_\Theta + p\right]\left(\frac{dV}{d\Theta}\right)_{\text{process}}. \tag{1.40}$$

The notation $(dV/d\Theta)_{\text{process}}$ means that the quantity is not just a function only of Θ, and the process must be specified.

The *heat capacity at constant volume (isochoric)*, C_V, is found by setting $dV = 0$, i.e.,

$$C_V = \left(\frac{\partial U}{\partial \Theta}\right)_V. \tag{1.41}$$

The heat capacity for an arbitrary process is expressed as

$$C = C_V + \left[\left(\frac{\partial U}{\partial V}\right)_\Theta + p\right]\left(\frac{dV}{d\Theta}\right)_{\text{process}}. \tag{1.42}$$

The *heat capacity at constant pressure (isobaric)*, C_p, is given by

$$C_p = C_V + \left[\left(\frac{\partial U}{\partial V}\right)_\Theta + p\right]\left(\frac{\partial V}{\partial \Theta}\right)_p, \tag{1.43}$$

where $(\partial V/\partial \Theta)_p$ is found from the thermal equation of state, and $(\partial U/\partial V)_\Theta$ is from the caloric equation of state. This quantity may be rewritten as

$$\left(\frac{\partial U}{\partial V}\right)_\Theta = \frac{C_p - C_V}{\left(\frac{\partial V}{\partial \Theta}\right)_p} - p. \tag{1.44}$$

The denominator is expressed in terms of the volume expansivity, β, i.e.,

$$\beta = \frac{1}{V}\left(\frac{\partial V}{\partial \Theta}\right)_p, \tag{1.45}$$

and then

$$\left(\frac{\partial U}{\partial V}\right)_{\Theta} = \frac{C_p - C_V}{\beta V} - p. \qquad (1.46)$$

This equation expresses the volume dependence of the internal energy in terms of C_p, C_V, and β.

For many real gases, if the experimentally measured values of C_p, C_V, and β are introduced into the above equation, the right hand side becomes vanishingly small, especially if the state of the gas is sufficiently removed from the saturation point; an experimental fact which led to the definition of a classical ideal gas.

Definition *The thermal and caloric equations of state for the classical ideal gas are defined, respectively, by*

$$p = p(\Theta, V) = \frac{nR\Theta}{V}, \qquad U = U(\Theta), \qquad (1.47)$$

where n is the quantity of the gas measured in the number of moles of the gas and R is a constant.

It is worthwhile noting the fact that the definition of a mole can be given within the framework of thermodynamics, i.e., the amount of the gas is adjusted in such a way that the quantity pV/Θ becomes equal for all gases. Thermodynamics is a macroscopic physics, and hence the formulation of thermodynamics can be developed without taking any atomic structure of the working system into consideration.

One important property of the classical ideal gas follows immediately from the above definition of the equations of state and (1.46):

$$\left(\frac{\partial U}{\partial V}\right)_{\Theta} = 0, \quad C_p = C_V + \beta p V = C_V + nR. \qquad (1.48)$$

1.8 The isothermal and adiabatic processes

Let us now discuss some other properties of the ideal gas. There are two commonly employed processes in the formulation of thermodynamics.

One is the *isothermal process*. In this process, the physical system, such as an ideal gas, is brought into thermal contact with a heat reservoir of temperature Θ, and all the processes are performed at constant temperature. For an ideal gas,

$$pV = \text{constant}, \quad d\Theta = 0. \qquad (1.49)$$

The lines drawn in the $p - V$ plane are called the *isotherms*.

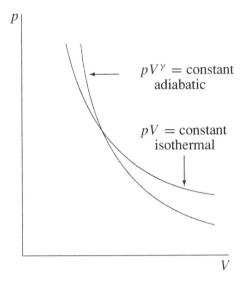

Fig. 1.1.

The other process is the *adiabatic* process. In this process, the physical system is isolated from any heat reservoir, and hence there is no heat transfer in and out of the system. A passage, or line, in the $p - V$ plane in an adiabatic process will now be found.

From (1.39), for an adiabatic process

$$\left(\frac{\partial U}{\partial \Theta}\right)_V = C_V, \quad \left(\frac{\partial U}{\partial V}\right)_\Theta = 0, \quad C_V d\Theta + p dV = 0, \tag{1.50}$$

which yields a differential equation

$$\frac{dV}{V} + \frac{C_V}{nR}\frac{d\Theta}{\Theta} = 0, \tag{1.51}$$

with the solution

$$V \Theta^{\frac{C_V}{nR}} = \text{constant.} \tag{1.52}$$

If this is combined with $\Theta = pV/nR$, we find

$$p^{\frac{C_V}{nR}} V^{\frac{nR+C_V}{nR}} = \text{constant,} \tag{1.53}$$

which yields

$$pV^\gamma = \text{constant,} \tag{1.54}$$

where $\gamma = C_p/C_V$.

1.9 The enthalpy

Let us go back to the first law of thermodynamics,

$$d'Q = dU + p\,dV, \tag{1.55}$$

and construct an equation in which the temperature Θ and pressure p are used as independent variables. In order to accomplish this, both dU and dV must be expressed in terms of $d\Theta$ and dp, i.e.,

$$dU = \left(\frac{\partial U}{\partial \Theta}\right)_p d\Theta + \left(\frac{\partial U}{\partial p}\right)_\Theta dp,$$

$$dV = \left(\frac{\partial V}{\partial \Theta}\right)_p d\Theta + \left(\frac{\partial V}{\partial p}\right)_\Theta dp. \tag{1.56}$$

Then,

$$d'Q = \left[\left(\frac{\partial U}{\partial \Theta}\right)_p + p\left(\frac{\partial V}{\partial \Theta}\right)_p\right] d\Theta + \left[\left(\frac{\partial U}{\partial p}\right)_\Theta + p\left(\frac{\partial V}{\partial p}\right)_\Theta\right] dp. \tag{1.57}$$

This suggests that the quantity H, called the *enthalpy* and defined by

$$H = U + pV, \tag{1.58}$$

is a convenient quantity when Θ and p are used as the independent variables:

$$d'Q = \left(\frac{\partial H}{\partial \Theta}\right)_p d\Theta + \left[\left(\frac{\partial H}{\partial p}\right)_\Theta - V\right] dp. \tag{1.59}$$

The heat capacity at constant pressure (isobaric), C_p, is found by setting $dp = 0$, i.e.,

$$C_p = \left(\frac{\partial H}{\partial \Theta}\right)_p. \tag{1.60}$$

The heat capacity for an arbitrary process is expressed as

$$C = C_p + \left[\left(\frac{\partial H}{\partial p}\right)_\Theta - V\right]\left(\frac{dp}{d\Theta}\right)_{\text{process}}. \tag{1.61}$$

1.10 The second law of thermodynamics

Let us examine closely the reversibility characteristics of the processes which take place in nature. There are three forms of statement concerning the reversibility vs. irreversibility argument. (Note that the terminologies *cyclic engine* and *cyclic process* are used. The cyclic engine is a physical system which performs a succession of processes and goes back to the state from which it started at the end of the processes. The physical conditions of the surroundings are assumed also to go back to the original state.)

The second law of thermodynamics is stated in the following three different forms.

Clausius's statement *It is impossible to operate a cyclic engine in such a way that it receives a quantity of heat from a body at a lower temperature and gives off the same quantity of heat to a body at a higher temperature without leaving any change in any physical system involved.*

Thomson's statement[†] *It is impossible to operate a cyclic engine in such a way that it converts heat energy from one heat bath completely into a mechanical work without leaving any change in any physical system involved.*

Ostwald's statement *It is impossible to construct a perpetual machine of the second kind.*

The *perpetual machine of the second kind* is a machine which negates Thomson's statement. For this reason, Ostwald's statement is equivalent to Thomson's statement.

If one of the statements mentioned above is accepted to be true, then other statements are proven to be true. All the above statements are, therefore, equivalent to one another.

In order to gain some idea as to what is meant by the proof of a theorem in the discussion of the second law of thermodynamics, the following theorem and its proof are instructive.

Theorem 1.1 *A cyclic process during which a quantity of heat is received from a high temperature body and the same quantity of heat is given off to a low temperature body is an irreversible process.*

Proof If this cyclic process is reversible it would then be possible to take away a quantity of heat from a body at a lower temperature and give off the same quantity of heat to a body at a higher temperature without leaving any changes in the surroundings. This reverse cycle would then violate Clausius's statement. For this reason, if the Clausius statement is true, then the statement of this theorem is also true.

[†] William Thomson, later Lord Kelvin, developed the second law of thermodynamics in 1850.

It will be left to the Exercises to prove that the following statements are all true if one of the preceding statements is accepted to be true:

generation of heat by friction is an irreversible process;
free expansion of an ideal gas into a vacuum is an irreversible process;
a phenomenon of flow of heat by heat conduction is an irreversible process.

The motion of a pendulum is usually treated as a reversible phenomenon in classical mechanics; however, if one takes into account the frictional effect of the air, then the motion of the pendulum must be treated as an irreversible process. Similarly, the motion of the Moon around the Earth is irreversible because of the tidal motion on the Earth. Furthermore, if any thermodynamic process, such as the flow of heat by thermal conduction or the free expansion of a gas into a vacuum, is involved, all the natural phenomena must be regarded as irreversible processes.

Another implication of the second law of thermodynamics is that the direction of the irreversible flow of heat can be used in defining the direction of the temperature scale. If two thermodynamic systems are not in thermal equilibrium, then a flow of heat takes place from the body at a higher temperature to the body at a lower temperature.

1.11 The Carnot cycle

The *Carnot cycle* is defined as a cyclic process which is operated under the following conditions.

Definition *The Carnot cycle is an engine capable of performing a reversible cycle which is operated between two heat reservoirs of empirical temperatures θ_2 (higher) and θ_1 (lower).*

The *heat reservoir* or *heat bath* is interpreted as having an infinitely large heat capacity and hence its temperature does not change even though there is heat transfer into or out of the heat bath. The terminologies heat reservoir (R) and heat bath may be used interchangeably in the text.

The Carnot cycle, C, receives a positive quantity of heat from the higher temperature reservoir and gives off a positive quantity of heat to the lower temperature reservoir. Since this is a reversible cycle, it is possible to operate the cycle in the reverse direction. Such a cycle is called the *reverse Carnot cycle*, $\bar{C}$. $\bar{C}$ undergoes a cyclic process during which it receives a positive quantity of heat from a lower temperature reservoir and gives off a positive amount of heat to the reservoir at a higher temperature.

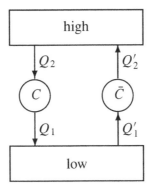

Fig. 1.2.

Theorem 1.2 *The Carnot cycle, C, performs positive work on the outside body while the reverse Carnot cycle, C̄, receives positive work from the outside body.*

Proof Let us consider a reverse Carnot cycle $\bar{C}$. This cycle, by definition, takes up a quantity of heat from the lower temperature heat reservoir and gives off a positive quantity of heat to the higher temperature heat reservoir. If, contrary to the assumption, the outside work is zero, the cycle would violate the Clausius statement. The quantity of work from the outside body cannot be negative, because this would mean that the cyclic engine could do positive work on the outside body, which could be converted into heat in the higher temperature heat bath. The net result is that the reverse cycle would have taken up a quantity of heat and transferred it to the higher temperature reservoir. This would violate the Clausius statement. For this reason the reverse Carnot cycle must receive a positive quantity of work from an outside body.

Next, let us consider a normal Carnot cycle working between two heat reservoirs. The amount of work performed on the outside body must be positive, because, if it were negative, the cycle would violate the Clausius statement by operating it in the reverse direction.

1.12 The thermodynamic temperature

In this section the *thermodynamic temperature* will be defined. To accomplish this we introduce an empirical temperature scale, which may be convenient for practical purposes, e.g., a mercury column thermometer scale. The only essential feature of the empirical temperature is that the scale is consistent with the idea of an irreversible heat conduction, i.e., the direction of the scale is defined in such a way

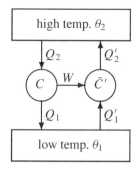

Fig. 1.3.

that a quantity of heat can flow irreversibly from a body at a higher temperature to a body at a lower temperature. Let us prepare two heat baths of temperatures θ_1 and θ_2, $\theta_1 < \theta_2$, and suppose that two Carnot cycles, C and C', are operated between the two reservoirs.

C receives heat Q_2 from θ_2, gives off heat Q_1 to θ_1, and performs mechanical work W on the outside.
C' receives heat Q'_2 from θ_2, gives off heat Q'_1 to θ_1, and performs mechanical work W on the outside.

Let us now reverse the direction of the Carnot cycle C', call it the reverse Carnot cycle $\bar{C}'$, and consider a combined system $C + \bar{C}'$.

$C + \bar{C}'$ receives heat $Q'_1 - Q_1$ from θ_1, gives off heat $Q'_2 - Q_2$ to θ_2 and performs no work on the outside.

The next step is to examine whether $Q'_1 - Q_1$ is positive or negative.

If $Q'_1 - Q_1 > 0$, then $Q'_2 - Q_2 > 0$, because of the first law. This would, however, violate the Clausius statement.
If $Q'_1 - Q_1 < 0$, then $Q'_2 - Q_2 < 0$. The combined cycle, however, becomes equivalent with irreversible conduction of heat, which is in contradiction with the performance of the reversible Carnot cycle.
The only possibility is, then, $Q'_1 - Q_1 = 0$, and $Q'_2 - Q_2 = 0$.

An important conclusion is that all the Carnot cycles have the same performance regardless of the physical nature of the individual engine, i.e.,

$$Q_1 = Q_1(\theta_1, \theta_2, W), \qquad Q_2 = Q_2(\theta_1, \theta_2, W). \qquad (1.62)$$

Furthermore, if the same cycle is repeated, the heat and work quantities are doubled, and in general

$$Q_1(\theta_1, \theta_2, nW) = nQ_1(\theta_1, \theta_2, W),$$
$$Q_2(\theta_1, \theta_2, nW) = nQ_2(\theta_1, \theta_2, W); \qquad (1.63)$$

and in turn

$$\frac{Q_1(\theta_1, \theta_2, nW)}{Q_2(\theta_1, \theta_2, nW)} = \frac{Q_1(\theta_1, \theta_2, W)}{Q_2(\theta_1, \theta_2, W)}. \tag{1.64}$$

This means that the ratio Q_2/Q_1 depends only on θ_1 and θ_2, not on the amount of work W. So,

$$\frac{Q_2}{Q_1} = f(\theta_1, \theta_2), \tag{1.65}$$

where f is a function which does not depend upon the type of the Carnot cycle. Let us suppose that two Carnot cycles are operated in a series combination as is shown in Fig. 1.4. Then, from the preceding argument,

$$\frac{Q_2}{Q_1} = f(\theta_1, \theta_2), \quad \frac{Q_1}{Q_0} = f(\theta_0, \theta_1). \tag{1.66}$$

If we look at the combination of the cycle C, heat bath R_1, and cycle C' as another Carnot cycle, we have

$$\frac{Q_2}{Q_0} = f(\theta_0, \theta_2). \tag{1.67}$$

This means

$$f(\theta_1, \theta_2) = \frac{f(\theta_0, \theta_2)}{f(\theta_0, \theta_1)}. \tag{1.68}$$

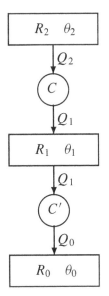

Fig. 1.4.

Since the left hand side of the equation does not depend upon θ_0, the right hand side of the equation is not allowed to contain θ_0. Therefore,

$$f(\theta_1, \theta_2) = \frac{\phi(\theta_2)}{\phi(\theta_1)}. \tag{1.69}$$

Equation (1.65) can be expressed as

$$\frac{Q_2}{Q_1} = \frac{\phi(\theta_2)}{\phi(\theta_1)}. \tag{1.70}$$

Based on the above finding, the thermodynamic temperature scale can be introduced according to the following equation:

$$\frac{Q_2}{Q_1} = \frac{T_2}{T_1}. \tag{1.71}$$

This equation, however, fixes only the ratio of the temperatures. In order to determine the scale of the temperature, two well defined fixed points are needed: one is the ice point and the other is the saturation vapor point of water under the standard atmospheric conditions, and the temperature difference between the two points is defined to be 100 degrees. When the Carnot cycle is operated between the heat baths kept at these temperatures, we have

$$\frac{T_0 + 100}{T_0} = \frac{Q_s}{Q_0}, \tag{1.72}$$

where T_0 is the ice point, Q_s is the quantity of heat received from the heat bath at the boiling point, and Q_0 is the quantity of heat given off to the heat source at the ice point.

In principle, it is possible to measure the ratio Q_s/Q_0, and the value must be independent of the physical systems used as the Carnot cycle. In this way the absolute scale of T_0 is found to be

$$T_0 = 273.15. \tag{1.73}$$

This scale of temperatue is called the *kelvin*,[†] or the *absolute temperature*, and is denoted by K. More recently, it has been agreed to use the triple point of water as the only fixed point, which has been defined to be 273.16 K.

From the practical point of view, the efficiency of an engine is an important quantity which is defined by

$$\eta = \frac{Q_2 - Q_1}{Q_2}. \tag{1.74}$$

[†] After Lord Kelvin (William Thomson (1854)).

The efficiency of the Carnot cycle is independent of the physical system used as the Carnot cycle, and is expressed as

$$\eta = \frac{T_2 - T_1}{T_2}. \tag{1.75}$$

1.13 The Carnot cycle of an ideal gas

It will be established in this section that the temperature provided by the ideal gas thermometer is identical to the thermodynamic temperature. Let us consider a Carnot cycle using an ideal gas operated between the heat reservoirs kept at T_2 and T_1. The ideal gas is a substance for which the P, V, Θ relation (among other properties) is given by

$$pV = nR\Theta, \tag{1.76}$$

where p is the hydrostatic pressure under which the gas is kept in a cylinder of volume V and in thermal equilibrium conditions at the ideal gas temperature Θ. R is a universal constant, independent of the type of gas, and n is the amount of gas measured in units of moles. Θ is the ideal gas absolute temperature. Any real gas behaves very much like an ideal gas as long as the mass density is sufficiently small.

Let us assume that the Carnot cycle is made up of the following four stages:

Stage (i)
    The ideal gas is initially prepared at state $A(p_0, V_0, \Theta_1)$. The gas is then isolated from the heat bath and compressed adiabatically until the temperature of the gas reaches Θ_2. At the end of this process the gas is in state $B(p_1, V_1, \Theta_2)$.

Stage (ii)
    The gas is brought into thermal contact with a heat bath at temperature Θ_2 and it is now allowed to expand while the temperature of the gas is kept at Θ_2 until the gas reaches the state $C(p_2, V_2, \Theta_2)$. This process is an isothermal expansion.

Stage (iii)
    The gas is again isolated from the heat bath and allowed to expand adiabatically until it reaches state $D(p_3, V_3, \Theta_1)$.

Stage (iv)
    The gas is brought into thermal contact with the heat bath at temperature Θ_1 and then compressed isothermally until it is brought back to its initial state $A(p_0, V_0, \Theta_1)$.

All the foregoing processes are assumed to be performed quasistatically and hence reversibly. It was shown in Sec. 1.8 that the pressure and volume of the ideal gas change according to the law $pV^\gamma = \text{constant}$ during an adiabatic process. Here, $\gamma = C_p/C_V$.

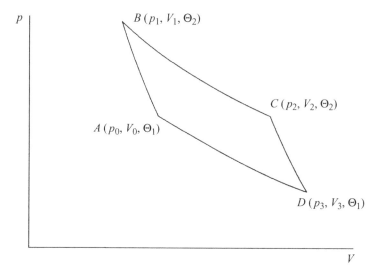

Fig. 1.5. A Carnot cycle in the p–V plane.

Let us now examine the energy balance in each of the preceding processes. γ is assumed to be constant for the gas under consideration.

Process (i): $A \rightarrow B$

There is no transfer of heat during this process. The mechanical work performed upon the gas is

$$W_1 = \int_{V_1}^{V_0} p \, dV,$$

$$pV^\gamma = p_0 V_0^\gamma = p_1 V_1^\gamma = k,$$

$$\int_{V_1}^{V_0} p \, dV = \int_{V_1}^{V_0} \frac{k}{V^\gamma} \, dV$$

$$= \frac{k}{\gamma - 1} \left[\frac{1}{V_1^{\gamma-1}} - \frac{1}{V_0^{\gamma-1}} \right]$$

$$= \frac{1}{\gamma - 1} \left[\frac{p_1 V_1^\gamma}{V_1^{\gamma-1}} - \frac{p_0 V_0^\gamma}{V_0^{\gamma-1}} \right]$$

$$= \frac{1}{\gamma - 1} (p_1 V_1 - p_0 V_0). \tag{1.77}$$

Since

$$p_0 V_0 = R' \Theta_1, \qquad p_1 V_1 = R' \Theta_2, \tag{1.78}$$

the work performed upon the gas during stage (i) is given by

$$W_1 = \frac{R'}{\gamma - 1}(\Theta_2 - \Theta_1). \tag{1.79}$$

Process (ii): $B \rightarrow C$

Since the internal energy of an ideal gas is independent of the volume and depends only on the temperature, the work performed by the gas on the outside body must be equal to the amount of heat taken in during the isothermal process. The work performed on the outside body during this process is given by

$$W_2 = \int_{V_1}^{V_2} p\,dV = \int_{V_1}^{V_2} \frac{R'\Theta_2}{V}\,dV$$

$$= R'\Theta_2 \log \frac{V_2}{V_1}$$

$$= Q_2, \tag{1.80}$$

where Q_2 is the quantity of heat received from the heat bath Θ_2.

Process (iii): $C \rightarrow D$

The process is similar to process (i). The work performed on the outside is given by

$$W_3 = \frac{R'}{\gamma - 1}(\Theta_2 - \Theta_1). \tag{1.81}$$

Process (iv): $D \rightarrow A$

As in process (ii), the work performed upon the gas is

$$W_4 = R'\Theta_1 \log \frac{V_3}{V_0} = Q_1, \tag{1.82}$$

where Q_1 is the heat quantity transferred to the heat bath Θ_1.

Throughout the entire process, the heat quantity Q_2 from the high temperature bath, Q_1 given off to the low temperature bath, and the total work on the outside are given by

$$Q_2 = R'\Theta_2 \log \frac{V_2}{V_1},$$

$$Q_1 = R'\Theta_1 \log \frac{V_3}{V_0},$$

$$W = R'\left(\Theta_2 \log \frac{V_2}{V_1} - \Theta_1 \log \frac{V_3}{V_0}\right). \tag{1.83}$$

Separately, we have

$$\Theta_1 V_0^{\gamma-1} = \Theta_2 V_1^{\gamma-1}, \qquad \Theta_1 V_3^{\gamma-1} = \Theta_2 V_2^{\gamma-1}, \tag{1.84}$$

and hence

$$\left(\frac{V_0}{V_1}\right)^{\gamma-1} = \left(\frac{V_3}{V_2}\right)^{\gamma-1}; \quad \frac{V_2}{V_1} = \frac{V_3}{V_0}. \tag{1.85}$$

Finally, from (1.80),

$$Q_2 : Q_1 : W = \Theta_2 : \Theta_1 : \Theta_2 - \Theta_1. \tag{1.86}$$

From this, we can conclude that not only W but both Θ_1 and Θ_2 are positive. If we compare the above equation with (1.69) we find that

$$\frac{T_2}{T_1} = \frac{\Theta_2}{\Theta_1}, \tag{1.87}$$

where T_1 and T_2 are the thermodynamic temperatures of the heat baths Θ_1 and Θ_2. Further, if the ice point and the boiling point are chosen to be the fixed reference points, we can confirm that

$$T_0 = \Theta_0 \quad \text{and} \quad T = \Theta. \tag{1.88}$$

1.14 The Clausius inequality

Let us consider a cyclic engine, C, which can exchange heat, during one cycle, with n different reservoirs, $R_1, R_2, \ldots, R_n$, at thermodynamic temperatures, $T_1, T_2, \ldots, T_n$. C receives heat quantities, $Q_1, Q_2, \ldots, Q_n$ from these reservoirs, respectively. The reservoirs are not necessarily external with respect to the cycle C; they may be part of the engine C.

Performance of C could be either reversible or irreversible depending upon the situation. After completion of one cycle, another heat reservoir R at temperature T is prepared, and also an additional n cyclic reversible cycles, $C_1, C_2, \ldots, C_n$, are prepared. These reversible cycles perform the following processes, i.e., $C_i, i = 1, 2, \ldots, n$, receives heat quantity Q_i'' from the reservoir R_i and Q_i' from the reservoir R.

We note that the sign of the heat quantity is now defined to be positive when the cycle receives the heat quantity. Because of this change, (1.87) for a Carnot cycle is now expressed as

$$\frac{Q_1}{T_1} + \frac{Q_2}{T_2} = 0 \quad \text{(Carnot cycle)}. \tag{1.89}$$

When applying (1.89) to the cycles C_i, we see that

$$\frac{Q_i'}{T} + \frac{Q_i''}{T_i} = 0, \quad i = 1, 2, \ldots, n. \tag{1.90}$$

Now, all the additional cycles C_i are adjusted in such a way that all the heat reservoirs return to the original conditions, i.e., the heat quantities satisfy the

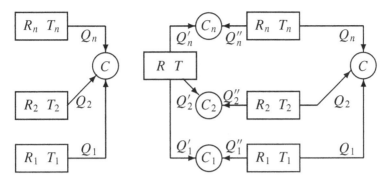

Fig. 1.6. The Clausius inequality.

conditions:

$$Q_i + Q_i'' = 0, \quad i = 1, 2, \ldots, n. \tag{1.91}$$

When (1.90) and (1.91) are combined, we find

$$\frac{1}{T} \sum_{i=1,\ldots,n} Q_i' = \sum_{i=1,\ldots,n} \frac{Q_i}{T_i}. \tag{1.92}$$

If $\sum_{i=1,\ldots,n}(Q_i/T_i) = 0$ then $\sum_{i=1,\ldots,n} Q_i' = 0.$

Hence R returns to the initial state. Because of the first law of thermodynamics there is no work performed on the outside by C_i. This means that the cycle C is reversible.

If $\sum_{i=1,\ldots,n}(Q_i/T_i) > 0$ then $\sum_{i=1,\ldots,n} Q_i' > 0.$

There is a net flow of heat out of the reservoir R and, according to the first law of thermodynamics, this must be converted into some work performed on the outside. This will violate Thomson's statement and means that this cannot occur in nature.

If $\sum_{i=1,\ldots,n}(Q_i/T_i) < 0$ then $\sum_{i=1,\ldots,n} Q_i' < 0.$

According to the first law of thermodynamics, there must be some work performed on the entire system by some external agent and some heat is transferred to R. This is an irreversible process, because its reverse process would violate Thomson's statement. Since all the cycles, except C, are reversible the irreversibility must be due to the performance of C.

In summary,

if $\sum_{i=1,\ldots,n}(Q_i/T_i) = 0$, the cycle C is reversible;
if $\sum_{i=1,\ldots,n}(Q_i/T_i) < 0$, the cycle C is irreversible;
the case $\sum_{i=1,\ldots,n}(Q_i/T_i) > 0$, does not occur in nature.

In the limit as the number of heat reservoirs goes to infinity while the magnitude of each quantity of heat transferred becomes infinitesimally small, the preceding summation is replaced by an integration:

$$\lim_{n \to \infty} \sum_{i=1,\dots,n} \frac{Q_i}{T_i} = \oint \frac{\mathrm{d}'Q}{T}. \tag{1.93}$$

$\oint(\mathrm{d}'Q/T) < 0$ implies the cycle C is irreversible;
$\oint(\mathrm{d}'Q/T) = 0$ implies the cycle C is reversible.

These two conditions are called the *Clausius inequality*.

1.15 The entropy

Let us asssume that an engine performs a cycle starting from the state P_1, through a path K, reaches the state P_2, and returns to the initial state P_1 through a path K''. If the entire cycle is a succession of infinitesimally small reversible (quasistatic) processes, the following condition is satisfied:

$$\int_{P_1 K P_2} \frac{\mathrm{d}'Q}{T} + \int_{P_2 K'' P_1} \frac{\mathrm{d}'Q}{T} = 0. \tag{1.94}$$

If, however, there is another path K' from P_1 to P_2, along which the processes are reversible, a similar condition is satisfied:

$$\int_{P_1 K' P_2} \frac{\mathrm{d}'Q}{T} + \int_{P_2 K'' P_1} \frac{\mathrm{d}'Q}{T} = 0. \tag{1.95}$$

Since the second terms of (1.94) and (1.95) are the same, we should have

$$\int_{P_1 K P_2} \frac{\mathrm{d}'Q}{T} = \int_{P_1 K' P_2} \frac{\mathrm{d}'Q}{T}. \tag{1.96}$$

The paths K and K' are arbitrary as long as they are reversible paths. Under such conditions it becomes unnecessary to specify the paths, and the quantity

$$\int_{P_1 \to P_2} \frac{\mathrm{d}'Q}{T} \tag{1.97}$$

becomes independent of the paths, as long as they are reversible processes.

Let us now define a quantity called S defined by

$$S = \int_{P_0 \to P \text{ (rev)}} \frac{\mathrm{d}'Q}{T}, \tag{1.98}$$

where P_0 is some standard state. The choice of P_0 is arbitrary; however, once it is chosen, it should be fixed for the rest of the formulation.

S is a quantity which depends upon the thermodynamic state of the system, and is called the *entropy* of the system at the state P.

Carnot's theorem *S is a state function defined by* $S = \int_{P_0 \to P(\text{rev})} \frac{d'Q}{T}$

Let S_1 and S_2 be the entropies at the states P_1 and P_2, respectively; then,

$$S_1 = \int_{P_0(\text{rev})}^{P_1} \frac{d'Q}{T}, \quad S_2 = \int_{P_0(\text{rev})}^{P_2} \frac{d'Q}{T}, \quad (1.99)$$

and we have

$$S_2 - S_1 = \int_{P_1(\text{rev})}^{P_2} \frac{d'Q}{T}. \quad (1.100)$$

If the two states P_1 and P_2 are close to each other, then

$$dS = \left(\frac{d'Q}{T}\right)_{(\text{rev})}. \quad (1.101)$$

Let us next consider a path, from P_2 back to P_1, along which the process contains some irreversible steps. Many things must be specified in order to characterize an irreversible process, such as some complicated velocity pattern in the case of the occurrence of vortices, a temperature distribution pattern in the case of conduction of heat, etc. On the contrary, a reversible process can be characterized by a small number of parameters. Since the combined process is an irreversible one, the Clausius inequality applies:

$$\int_{P_1(\text{rev})}^{P_2} \frac{d'Q}{T} + \int_{P_2(\text{irr})}^{P_1} \frac{d'Q}{T} < 0, \quad (1.102)$$

where

$$\int_{P_1(\text{rev})}^{P_2} = S_2 - S_1. \quad (1.103)$$

We then have

$$\int_{P_2(\text{irr})}^{P_1} \frac{d'Q}{T} + S_2 - S_1 < 0, \quad (1.104)$$

or

$$\int_{P_1(\text{irr})}^{P_2} \frac{d'Q}{T} < S_2 - S_1. \quad (1.105)$$

If, however, the process $P_1 \to P_2$ approaches reversibility, then

$$\int_{P_1(\text{rev})}^{P_2} \frac{d'Q}{T} = S_2 - S_1. \quad (1.106)$$

For an infinitesimal irreversible process

$$\left(\frac{d'Q}{T}\right)_{irr} < dS. \tag{1.107}$$

If the system is entirely isolated from the rest of the world, for an infinitesimal irreversible process,

$$0 < S_2 - S_1, \quad 0 < dS. \tag{1.108}$$

This is called the *principle of irreversible increase of the entropy in an isolated system* or *Carnot's law*.

Free expansion of an ideal gas into a vacuum is an example of irreversible increase of the entropy [Exercise 1.3].

In the above discussion, the entropy is defined as a function which depends upon the thermodynamic state of the system in thermal equilibrium; however, the entropy can be defined even in a nonequilibrium state. The system may be divided into many small subsystems in such a way that each of the subsystems are regarded as in thermodynamic equilibrium. Then the entropy, S, of the total system can be defined as given by

$$S = \sum_i S_i.$$

Now, for each subsystem, the principle of irreversible increase of the entropy can be applied if the total system is isolated from the rest of the universe. In an isolated system the entropy increases steadily toward a maximum value consistent with the equilibrium conditions.

1.16 General integrating factors

The concept of entropy has been introduced in many different formulations. It was originally introduced by Clausius, who employed a cyclic process performed on the ideal gas. In 1909 Carathéodory proposed an axiomatic definition of the entropy based on a general set of assumptions without utilizing the ideal gas as the working system. Since the Carathéodory formulation, it has become customary to introduce the concept of entropy without resort to the ideal gas. Max Planck further improved Carathéodory's formulation in a form which is conceptually simpler. H. Hausen (1935) further improved Planck's formulation, and the *Hausen formulation* will be followed in the remainder of this chapter.

Let us first consider a differential form in two dimensions:

$$P(x, y)dx + Q(x, y)dy, \tag{1.109}$$

which can be expressed as an *exact differential* if multiplied by an appropriate function $\mu_1(x, y)$:

$$dz = \mu_1(x, y)P(x, y)dx + \mu_1(x, y)Q(x, y)dy. \tag{1.110}$$

Such a function $\mu_1(x, y)$ is called the *integrating factor*.

The necessary and sufficient condition for the integrating factor is the condition of differentiability, expressed as

$$\frac{\partial \mu_1 P}{\partial y} = \frac{\partial \mu_1 Q}{\partial x}, \tag{1.111}$$

or

$$P\frac{\partial \mu_1}{\partial y} - Q\frac{\partial \mu_1}{\partial x} + \mu_1\left(\frac{\partial P}{\partial y} - \frac{\partial Q}{\partial x}\right) = 0. \tag{1.112}$$

Let us next consider the problem of finding a general solution of the above equation in the form $\mu = \mu_1 v$ when a particular solution μ_1 is known. This means that v must satisfy the following equation:

$$v\left[P\frac{\partial \mu_1}{\partial y} - Q\frac{\partial \mu_1}{\partial x} + \mu_1\left(\frac{\partial P}{\partial y} - \frac{\partial Q}{\partial x}\right)\right] + \mu_1\left[P\frac{\partial v}{\partial y} - Q\frac{\partial v}{\partial x}\right] = 0. \tag{1.113}$$

Since μ_1 is a particular solution, the first terms vanish, and hence v must be a solution of the equation

$$P\frac{\partial v}{\partial y} - Q\frac{\partial v}{\partial x} = 0. \tag{1.114}$$

It is easily seen that this equation is satisfied by $v = z(x, y)$, i.e.,

$$\frac{\partial z}{\partial x} = \mu_1 P, \qquad \frac{\partial z}{\partial y} = \mu_1 Q,$$

$$P\frac{\partial z}{\partial y} - Q\frac{\partial z}{\partial x} = P\mu_1 Q - Q\mu_1 P = 0. \tag{1.115}$$

One also finds that any function $\Psi(z)$ of z can be chosen to be v because of the condition

$$P\Psi'(z)\frac{\partial z}{\partial y} - Q\Psi'(z)\frac{\partial z}{\partial x} = 0. \tag{1.116}$$

We may therefore conclude that the most general integrating factor can be found in the form

$$\mu = \mu_1 \Psi(z). \tag{1.117}$$

1.17 The integrating factor and cyclic processes

For the sake of simplicity, let us consider a thermodynamic system of which the state is specified by one of the sets of two variables, $(p, V), (p, \theta)$, or (V, θ). This is possible because a thermodynamic state of a one-component system is characterized by $U = U(p, V)$ and $p = p(V, \theta)$.

Let us consider a quasistatic infinitesimal process during which there is a transfer of heat q to the system and a mechanical work w performed upon the system. According to the first law of thermodynamics, these quantities are related to an increment dU of the internal energy U by

$$q = dU - w. \tag{1.118}$$

When the mechanical work is due to a volume change under the influence of hydrostatic pressure p, the relation is expressed as

$$q = dU + pdV. \tag{1.119}$$

Both q and w are infinitesimal quantities; however, they are not exact differentials of thermodynamic quantities, i.e., there are no functions of the physical state such that q and w are expressed as

$$q = df(V, \theta), \quad w = dg(V, \theta). \tag{1.120}$$

In order to define the entropy without resort to the ideal gas it is necessary to find an integrating factor which does not depend upon any specific system used as a working system and to make the quantity of heat, q, an exact differential expressed by $d\sigma$. Let this integrating factor be called τ and make the quantity σ be a property which is completely determined by the thermodynamic state of the physical system, i.e.,

$$d\sigma = \frac{q}{\tau}. \tag{1.121}$$

The quantity σ defined in this way should be a function only of the thermodynamic state of the system and should be independent of the processes by which the system has been brought to the present state, i.e.,

$$\sigma = \sigma(p, \theta). \tag{1.122}$$

For a system for which the state is completely defined in terms of the volume V, the hydrostatic pressure p, and an empirical temperature θ, the above relation is expressed as

$$d\sigma = \frac{(dU + pdV)}{\tau}. \tag{1.123}$$

In order for $d\sigma$ to be an exact differential,

$$\frac{\partial^2 \sigma}{\partial V \partial \theta} = \frac{\partial^2 \sigma}{\partial \theta \partial V}. \tag{1.124}$$

This equation, when the differentials are carried out, leads to

$$\frac{\partial \sigma}{\partial V} = \frac{1}{\tau} \left[\left(\frac{\partial U}{\partial V} \right)_\theta + p \right],$$

$$\frac{\partial}{\partial \theta} \left(\frac{\partial \sigma}{\partial V} \right) = \left[\left(\frac{\partial U}{\partial V} \right)_\theta + p \right] \frac{-1}{\tau^2} \left(\frac{\partial \tau}{\partial \theta} \right)_V + \frac{1}{\tau} \left[\frac{\partial}{\partial \theta} \left(\frac{\partial U}{\partial V} \right) + \left(\frac{\partial p}{\partial \theta} \right)_V \right],$$

$$\frac{\partial \sigma}{\partial \theta} = \frac{1}{\tau} \left(\frac{\partial U}{\partial \theta} \right)_V,$$

$$\frac{\partial}{\partial V} \left(\frac{\partial \sigma}{\partial \theta} \right) = \frac{-1}{\tau^2} \left(\frac{\partial U}{\partial \theta} \right)_V \left(\frac{\partial \tau}{\partial V} \right)_\theta + \frac{1}{\tau} \left[\frac{\partial}{\partial V} \left(\frac{\partial U}{\partial \theta} \right) \right], \tag{1.125}$$

and then

$$\left[\left(\frac{\partial U}{\partial V} \right)_\theta + p \right] \left(\frac{\partial \ln \tau}{\partial \theta} \right)_V - \left(\frac{\partial U}{\partial \theta} \right)_V \left(\frac{\partial \ln \tau}{\partial V} \right)_\theta = \left(\frac{\partial p}{\partial \theta} \right). \tag{1.126}$$

The integrating factor, τ, will be found by solving this equation, if p and U are functions of V and θ, i.e., by employing the thermal and caloric equations of state of the system.

Let one particular solution of this equation be τ_0 and the corresponding σ be σ_0, then, from the general properties of the integrating factor we discussed earlier,

$$\tau = \frac{\tau_0}{\psi(\sigma_0)} \tag{1.127}$$

will be another integrating factor, where $\psi(\sigma_0)$ is an arbitrary function of σ_0.

Since the integrating factor τ_0 is a solution of (1.126), it will, naturally, depend upon the thermal and caloric equations of the particular system that we employ.

As will be demonstrated in Sec. 1.20, there exists a universal integrating factor which does not depend upon any particular physical system employed.

If $d\sigma$ is an exact differential, then the total sum of $d\sigma$ over a cyclic process will be zero. Inversely, any differential quantity whose total sum over a cyclic process is zero is an exact differential. For this reason we shall define a general integrating factor by the condition that the total sum of the exact differential over any reversible cyclic process must be equal to zero instead of finding the integrating factor as a solution of the differential equation mentioned above.

1.18 Hausen's cycle

Let us consider a system made up of n physical subsystems and assume that all the subsystems are in thermal equilibrium individually and independently of each other. Initially it is assumed that the physical state of each of the subsystems involved is determined by two state variables, V_i and θ_i, and a mechanical work performed on each subsystem by some outside body is expressed as $w_i = -p_i dV_i$.[†]

Because of this assumption and the discussion developed in the previous section, it is possible to define an integrating factor τ for each of the subsystems independently. One can see, then, that each of the integrating factors, say τ_i, and the corresponding state function σ_i depend upon the particular properties of the ith system.

In this way, the physical state of each subsystem will be defined from now on by two variables σ_i and τ_i instead of the original variables V_i and θ_i:

$$\sigma_i = \sigma_i(V_i, \theta_i), \quad \tau_i = \tau_i(V, \theta_i). \tag{1.128}$$

Process (i) Isothermal within each subsystem; adiabatic
for the combined system

All the n subsystems are brought into thermal equilibrium at a common temperature θ. Let them exchange heat quantities with respect to one another, although the entire process is done adiabatically as a whole. Let these heat quantities be $q_1, q_2, \ldots, q_n$. Because of the condition that the entire system is totally isolated,

$$q_1 + q_2 + \cdots + q_n = 0. \tag{1.129}$$

Let each of the state functions be $\sigma_1, \sigma_2, \ldots, \sigma_n$, and then

$$\tau_1 d\sigma_1 + \tau_2 d\sigma_2 + \cdots + \tau_n d\sigma_n = 0,$$
$$\tau_i(\sigma_i, \theta) \Rightarrow \tau_i(\sigma_i', \theta). \tag{1.130}$$

Process (ii) Adiabatic for each subsystem individually

Each subsystem undergoes a reversible adiabatic process, such as an adiabatic expansion or compression. The process of each individual subsystem is adjusted in such a way that the final temperature of all the subsystems is brought to the same temperature, say θ'. The σ of each system does not change because of the adiabatic process. The τ of each system will change, however,

$$\tau_i(\sigma_1', \theta) \Rightarrow \tau_i'(\sigma_1', \theta'). \tag{1.131}$$

[†] This assumption will be removed later.

Process (iii) Isothermal, within each subsystem; adiabatic
for the combined system

All the n subsystems are brought into thermal equilibrium at the common temperature θ'. Let them exchange heat quantities with respect to one another, although the total combined system undergoes an adiabatic process as a whole, i.e.,

$$\tau_1' d\sigma_1' + \tau_2' d\sigma_2' + \cdots + \tau_n' d\sigma_n' = 0. \tag{1.132}$$

Since the quantities of heat exchange for the $n - 1$ systems can be controlled arbitrarily, they may be adjusted in the following way:

$$d\sigma_1' = -d\sigma_1, \quad d\sigma_2' = -d\sigma_2, \ldots, d\sigma_{n-1}' = -d\sigma_{n-1}. \tag{1.133}$$

The quantity of heat exchange of the nth system cannot be arbitrary but is adjusted according to (1.132) because of the adiabatic condition of Process (iii).

Process (iv) Adiabatic for each system individually

Each subsystem undergoes a reversible adiabatic process, such as an adiabatic expansion or compression. The process of each individual subsystem is adjusted in such a way that the final temperature of all the subsystems is brought back to the same initial value, θ. The σ's of $n - 1$ subsystems are brought back to the initial values $\sigma_1, \sigma_2, \ldots, \sigma_{n-1}$ because of the conditions (1.133). This means that those $n - 1$ systems have returned to their initial state. At this point it is not trivially clear whether the nth system has returned to its initial state or not. If the nth system had not returned to its initial state, then one has to conclude that the whole system has not undergone a reversible cycle. As it turns out, however, the nth system has indeed returned to its initial state.

1.19 Employment of the second law of thermodynamics

The question of whether the nth system has returned to its initial state or not has an important implication for the existence of a universal relationship among the integrating factors of the individual physical systems.

Without loss of generality, we can assume that the integrating factor of the nth system has a positive value at the temperature θ. If the value were negative we could trace the following discussion in reverse order, or one can argue that if τ_n is an integrating factor, then $-\tau_n$ is also an integrating factor. The assumption of a positive value of the integrating factor is therefore acceptable.

Now assume that the value of σ_n at the end of Hausen's cycle is different from its initial value by the amount $-d\sigma_n$ (i.e., decreased). Then transfer a quantity of heat $q_n = \tau_n d\sigma_n$ to the system from an outside heat bath of temperature θ. In this way, all the n systems have been brought back to their initial state. During this last process the combined system received a heat quantity equal to q_n, all the systems have returned to their original states, and hence the amount of the external work (the work performed on the external body) must be, according to the first law of thermodynamics, equal to $-w$ (w is the work performed on the system). The entire (combined) system is, then, a perpetual machine of the second kind, the existence of which is negated by the second law of thermodynamics. The perpetual motion machine of the second kind, if it existed, could convert a quantity of heat entirely into mechanical work while in thermal contact with a heat bath of a single temperature. (See Thomson's statement or Ostwald's statement of the second law of thermodynamics, Sec. 1.10). In conclusion, σ_n must have been brought back to its original value. If the change in σ_n is $d\sigma_n$ (i.e., increased) the order of the four processes can be reversed and then $d\sigma_n$ will be negative. In the end, the n working systems all go back to their initial states in Hausen's cycle.

1.20 The universal integrating factor

In the Hausen cycle all the n subsystems undergo reversible cycles and the following relations hold:

$$d\sigma_1' = -d\sigma_1, d\sigma_2' = -d\sigma, \ldots, d\sigma_n' = -d\sigma_n, \qquad (1.134)$$

and hence the following relation of general validity will be obtained:

$$\left(\frac{\tau_1}{\tau_n} - \frac{\tau_1'}{\tau_n'}\right) d\sigma_1 + \left(\frac{\tau_2}{\tau_n} - \frac{\tau_2'}{\tau_n'}\right) d\sigma_2 + \cdots + \left(\frac{\tau_{n-1}}{\tau_n} - \frac{\tau_{n-1}'}{\tau_n'}\right) d\sigma_{n-1} = 0. \quad (1.135)$$

Now, since the values of $d\sigma_1, d\sigma_2, \ldots, d\sigma_{n-1}$ can be independently chosen, in order for the above equation to be valid, the following equalities should hold:

$$\frac{\tau_1}{\tau_1'} = \frac{\tau_2}{\tau_2'} = \cdots = \frac{\tau_{n-1}}{\tau_{n-1}'} = \frac{\tau_n}{\tau_{n'}} = f(\theta', \theta). \qquad (1.136)$$

Now let us consider a situation in which θ' is fixed while θ is arbitrarily changed. Then the ratio (τ/τ') will be the same for all the n subsystems. This means that the ratio (τ/τ') is independent of the particular properties of the subsystems and that the ratio must be a function only of the common temperature θ. This common function is now called $f(\theta)$.

As we saw earlier, we may employ the variables σ and θ in place of V and θ for every system to define the physical state of every system. This in turn implies that the integrating factor is also a function of σ and θ. Since $\sigma_1' = \sigma_1, \sigma_2' = \sigma_2, \ldots, \sigma_n' = \sigma_n$

in the present situation, the following relations will hold:

$$\tau_1 = \tau_1(\sigma_1, \theta), \tau_2 = \tau_2(\sigma_2, \theta), \ldots, \tau_n = \tau_n(\sigma_n, \theta),$$
$$\tau_1' = \tau_1'(\sigma_1, \theta'), \tau_2' = \tau_2'(\sigma_2, \theta'), \ldots, \tau_n' = \tau_n'(\sigma_n, \theta'). \tag{1.137}$$

Since θ' is fixed as a reference, τ_1' is a function of σ_1, τ_2' is a function of $\sigma_2, \ldots$, and τ_n' is a function of σ_n,

$$\tau_1' = \psi_1(\sigma_1), \quad \tau_2' = \psi_2(\sigma_2), \ldots, \tau_n' = \psi_n(\sigma_n), \tag{1.138}$$

and hence (1.136) is expressed as

$$\frac{\tau_1}{\psi_1(\sigma_1)} = \frac{\tau_2}{\psi_2(\sigma_2)} = \cdots = \frac{\tau_n}{\psi_n(\sigma_n)} = f(\theta). \tag{1.139}$$

Thus $f(\theta)$ becomes the integrating factor common to all the n subsystems.

It will be shown in the following that the same $f(\theta)$ is the universal integrating factor even for a physical system whose state is not characterized by two variables and the mechanical work is not represented by $-pdV$.

Suppose that the nth subsystem in Hausen's cycle does not satisfy the conditions just mentioned. There will be a common integrating factor for the remaining $n - 1$ subsystems as argued above. It is therefore possible to express (1.129) and (1.132) as

$$f(\theta)[d\sigma_1 + d\sigma_2 + \cdots + d\sigma_{n-1}] + q_n = 0, \tag{1.140}$$
$$f(\theta')[d\sigma_1' + d\sigma_2' + \cdots + d\sigma_{n-1}'] + q_n' = 0; \tag{1.141}$$

and furthermore, because of the conditions (1.134), the above two equations can be combined into

$$\frac{q_n}{f(\theta)} + \frac{q_n'}{f(\theta')} = 0. \tag{1.142}$$

By employing the second law of thermodynamics, we can conclude that at the end of Hausen's cycle the nth system must have been brought back to its initial state, and hence for the combined system as a whole the sum of the quantities

$$\sum \left[\frac{q_n}{f(\theta)} \right] = 0. \tag{1.143}$$

The $f(\theta)$ may be determined by the state properties of some suitably chosen substance. For instance, by utilizing a substance whose state is characterized by two variables, V and θ, then the internal energy is determined as a function of those variables. In (1.126), τ is replaced by $f(\theta)$ and we solve the differential equation.

In this way, by utilizing a universal integrating factor $f(\theta) = T$ the heat quantity is expressed as

$$q = T\,d\sigma. \tag{1.144}$$

Exercises

1.1 Derive the following relation:

$$\frac{\kappa_T}{\kappa_{ad}} = \frac{C_p}{C_V} = \gamma, \tag{E1.1}$$

where κ_T and κ_{ad} are the isothermal and adiabatic compressibilities, respectively.

[**Hint**]

From (1.39), (1.41), and (1.44) we have

$$d'Q = C_V\,dT + \frac{C_p - C_V}{\left(\frac{\partial V}{\partial T}\right)_p}dV. \tag{E1.2}$$

In an adiabatic process, $d'Q = 0$,

$$C_V\,dT + \frac{C_p - C_V}{\left(\frac{\partial V}{\partial T}\right)_p}dV = 0. \tag{E1.3}$$

If we take p and V as the independent variables,

$$dT = \left(\frac{\partial T}{\partial V}\right)_p dV + \left(\frac{\partial T}{\partial p}\right)_V dp, \tag{E1.4}$$

then

$$\left[C_V\left(\frac{\partial T}{\partial V}\right)_p + \frac{C_p - C_V}{\left(\frac{\partial V}{\partial T}\right)_p}\right]dV + C_V\left(\frac{\partial T}{\partial p}\right)_V dp = 0. \tag{E1.5}$$

Since

$$\left(\frac{\partial T}{\partial V}\right)_p = \frac{1}{\left(\frac{\partial V}{\partial T}\right)_p}, \tag{E1.6}$$

we have

$$C_p\,dV + C_V\left(\frac{\partial T}{\partial p}\right)_V\left(\frac{\partial V}{\partial T}\right)_p dp = 0, \tag{E1.7}$$

and

$$\kappa_{ad} = -\frac{1}{V}\left(\frac{dV}{dp}\right)_{ad}; \tag{E1.8}$$

therefore

$$-C_p\kappa_{ad} + C_V\left(\frac{\partial T}{\partial p}\right)_V\left(\frac{\partial V}{\partial T}\right)_p = 0. \tag{E1.9}$$

Separately, since

$$\left(\frac{\partial T}{\partial p}\right)_V \left(\frac{\partial V}{\partial T}\right)_p = -\left(\frac{\partial V}{\partial p}\right)_T = V\kappa_T, \qquad \text{(E1.10)}$$

finally we have

$$\frac{\kappa_T}{\kappa_{\text{ad}}} = \frac{C_p}{C_V} = \gamma. \qquad \text{(E1.11)}$$

1.2 Prove that generation of heat by friction is an irreversible process.

[**Hint**]

Suppose that the phenomenon is reversible. Then the heat developed by friction can be converted into the mechanical work which was used for generating the heat. This reversible cycle would, then, violate the second law (negation of existence of the perpetual machine of the second kind).

1.3 Prove that the free expansion of an ideal gas into a vacuum is irreversible.

[**Hint**]

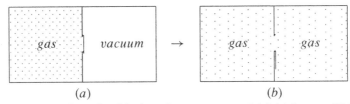

(a)   (b)

Fig. 1.7. Free expansion of an ideal gas into a vacuum: (a) initial state; (b) final state.

Suppose that this phenomenon is reversible. Let the cyclic engine which brings the expanded gas into the original state, i.e., the gas is confined in the left half of the container, be called the cycle C. This cycle C does not leave any change in the surroundings. Let us consider another process in which the same gas confined in the left half of the cylinder is in thermal contact with a heat reservoir at temperature θ, and the gas is expanded quasistatically while the temperature is kept the same as that of the reservoir until the gas reaches the final state, Fig. 1.7(b). During the process the gas performs mechanical work W, which is equal to the amount of heat transferred from the reservoir to the gas. Now, we can utilize the cycle C and bring the gas to the original state, reversibly. The end result is that a quantity of heat W is transferred from the heat reservoir, and the same amount of mechanical work is performed by the gas upon the outside body without leaving any change in the surroundings. This cyclic process will violate Thomson's statement.

1.4 Prove that it is justified to assume that two different Carnot cycles can perform the same amount of work on the outside body when operated between the same set of two heat reservoirs.

1.5 Prove that the process in which a solid body at temperature T_1 is brought into thermal contact with a heat reservoir at temperature T_2 is an irreversible process by comparing the entropy change and the integral $\int_{T_1}^{T_2}(\mathrm{d}'Q/T)$.

[Hint]

The amount of heat transferred from the reservoir to the solid body is $C(T_2 - T_1)$, and hence

$$\frac{1}{T_2} \int_{T_1}^{T_2} d'Q = \frac{C(T_2 - T_1)}{T_2} = \frac{C}{T_2} \int_{T_1}^{T_2} dT. \qquad (E1.12)$$

In order to calculate the entropy change, it is necessary to find a quasistatic path along which T_2 is reached starting from T_1. The quantity of heat necessary to change the temperature of the solid body by an infinitesimal magnitude dT reversibly is given by $C\,dT$, where C is the heat capacity of the solid body. Then the entropy change is found by

$$S_2 - S_1 = \int_{T_1}^{T_2} \frac{C\,dT}{T} = C \log \frac{T_2}{T_1}. \qquad (E1.13)$$

It is easy to compare the two results directly:

$$C \int_{T_1}^{T_2} \frac{dT}{T} - C \int_{T_1}^{T_2} \frac{d'Q}{T_2} = C \int_{T_1}^{T_2} \left(\frac{1}{T} - \frac{1}{T_2} \right) dT. \qquad (E1.14)$$

It is interesting to note that this is always positive regardless of whether $T_1 < T_2$ or $T_1 > T_2$.

1.6 Calculate the entropy change when two bodies at temperatures T_1 and T_2 are brought into thermal contact.

[Hint]

Assume that the two bodies are made up of the same material, have the same mass, and that C is their heat capacity. The entropy changes of the two bodies are given by

$$\Delta S_1 = \int_{T_1}^{T_f} \frac{C}{T} dT = C \log \frac{T_f}{T_1}, \quad \Delta S_2 = \int_{T_2}^{T_f} \frac{C}{T} dT = C \log \frac{T_f}{T_2}, \quad (E1.15)$$

and hence,

$$\Delta S = \Delta S_1 + \Delta S_2 = C \left(\log \frac{T_f}{T_1} + \log \frac{T_f}{T_2} \right). \qquad (E1.16)$$

Since the final temperature is $T_f = \frac{1}{2}(T_1 + T_2)$,

$$\Delta S = 2C \log \frac{1}{2} \left[\left(\frac{T_1}{T_2} \right)^{\frac{1}{2}} + \left(\frac{T_2}{T_1} \right)^{\frac{1}{2}} \right]. \qquad (E1.17)$$

ΔS is positive unless $T_1 = T_2$.

1.7 Assume that a uniform bar of length L initially has a uniform temperature gradient, and that the ends are kept at temperatures T_A and T_B. The bar is, then, detached from any heat reservoirs and irreversible heat flow takes place. After a while the temperature gradient will disappear and the bar will attain a thermal equilibrium with the final temperature $T_f = \frac{1}{2}(T_A + T_B)$. Show that the entropy change is given by

$$\Delta S = -CL \int_0^{\frac{1}{2}} \left[\log\left(1 - \frac{aL}{T_f}\xi\right) + \log\left(1 + \frac{aL}{T_f}\xi\right) \right] d\xi, \qquad \text{(E1.18)}$$

where $a = (T_B - T_A)/L$, and C is the heat capacity of the bar. Show further that the integrand is negative for a positive ξ, and hence ΔS is positive.

2

Thermodynamic relations

2.1 Thermodynamic potentials

After having discussed the second law of thermodynamics in Chapter 1, we can now use the absolute temperature, T, and entropy, S, as the thermodynamical variables to describe any thermodynamic relations. In this chapter we find some thermodynamic relations which are valid regardless of the available physical systems. These relations contain some derivatives of thermodynamic variables with respect to some other thermodynamic variables. From the mathematical operation point of view, the derivative means a comparison of function values at two adjacent points (states) which are infinitesimally close to each other. These two adjacent states must be connected by a quasistatic process in order to calculate the change of the quantity. In the following derivation, therefore, we should only use relations which are valid for the quasistatic processes.

According to the first law of thermodynamics,

$$d'Q = dU - d'W. \tag{2.1}$$

Further, because the process is quasistatic,

$$d'W = -pdV, \quad d'Q = TdS. \tag{2.2}$$

Consequently,

$$TdS = dU + pdV, \quad \text{or} \quad dU = TdS - pdV. \tag{2.3}$$

The second form of (2.3) indicates that the total differential of U is expressed in terms of S and V, the independent variables. Hence,

$$T = \left(\frac{\partial U}{\partial S}\right)_V, \quad p = -\left(\frac{\partial U}{\partial V}\right)_S. \tag{2.4}$$

The second relation has a special physical interpretation, i.e., the change in the internal energy is equal to the amount of mechanical work in an adiabatic process,

38

since the entropy of the system does not change in the quasistatic adiabatic process. Alternatively,

$$dU = -(pdV)_{\text{ad}}.$$ (2.5)

The second form of (2.3) may be expressed as

$$dS = \frac{1}{T}dU + \frac{p}{T}dV,$$ (2.6)

indicating that the independent variables are changed to U and V. This will lead to the relations

$$\frac{1}{T} = \left(\frac{\partial S}{\partial U}\right)_V, \quad \frac{p}{T} = \left(\frac{\partial S}{\partial V}\right)_U.$$ (2.7)

This set of relations has an especially important implication in the formulation of statistical mechanics in Chapter 3. It means that if the entropy is known as a function of the internal energy and the volume, then the thermodynamical (reciprocal) temperature and pressure of the system will be found from the partial derivatives of the entropy S with respect to the internal energy U and volume V, respectively.

Similarly, we may consider the enthalpy H which was introduced in Sec. 1.9. Differentiating (1.58), we find

$$dH = dU + pdV + Vdp,$$ (2.8)

and, if combined with the second form of (2.3),

$$dH = TdS + Vdp.$$ (2.9)

This is the total differential equation in which S and V are the independent variables. Hence,

$$T = \left(\frac{\partial H}{\partial S}\right)_p, \quad V = \left(\frac{\partial H}{\partial p}\right)_S.$$ (2.10)

Let us next consider the function

$$F = U - TS.$$ (2.11)

The newly introduced function is called the *Helmholtz potential*, and it plays a much more convenient role in thermodynamics than the functions U or H. The differential of this function is

$$dF = dU - TdS - SdT = -pdV - SdT.$$ (2.12)

The independent variables are now V and T, and we find

$$p = -\left(\frac{\partial F}{\partial V}\right)_T, \quad S = -\left(\frac{\partial F}{\partial T}\right)_V.$$ (2.13)

Equation (2.11) may be rewritten as

$$U = F + TS, \tag{2.14}$$

and, from (2.13), we obtain

$$U = F - T \left(\frac{\partial F}{\partial T} \right)_V = -T^2 \left[\frac{\partial \left(\frac{F}{T} \right)}{\partial T} \right]_V. \tag{2.15}$$

This is called the *Gibbs–Helmholtz equation*.

The fourth quantity to be discussed is the *Gibbs potential*, which is defined by

$$G = U - TS + pV = F + pV = H - TS. \tag{2.16}$$

Whichever form of the definition we take, the differential of G is given by

$$dG = -SdT + Vdp. \tag{2.17}$$

It is seen that the independent variables are changed into T and p, and the following relations are obtained:

$$S = \left(\frac{\partial G}{\partial T} \right)_p, \quad V = \left(\frac{\partial G}{\partial p} \right)_T. \tag{2.18}$$

There is another expression also called the *Gibbs–Helmholtz equation*:

$$H = G + TS = G - T \left(\frac{\partial G}{\partial T} \right)_p = -T^2 \left[\frac{\partial \left(\frac{G}{T} \right)}{\partial T} \right]_p. \tag{2.19}$$

The quantities U, S, H, F, and G are called the *thermodynamic potentials* depending upon the choice of the independent variables. In Table 2.1, we summarize the combinations of the independent variables and thermodynamic potentials.

Table 2.1. *Thermodynamic potentials and associated independent variables.*

Thermodynamic potentials	Independent variables
U	S, V
S	U, V
H	S, p
F	T, V
G	T, p

2.2 Maxwell relations

If we apply the theorem of differentiability of a function of two variables, $z = z(x, y)$, expressed as

$$\frac{\partial}{\partial x}\left(\frac{\partial z}{\partial y}\right) = \frac{\partial}{\partial y}\left(\frac{\partial z}{\partial x}\right), \tag{2.20}$$

to the thermodynamic potentials (all of which are functions of two independent variables), we obtain a set of thermodynamic relations as follows:

$$\frac{\partial}{\partial S}\left(\frac{\partial U}{\partial V}\right) = \frac{\partial}{\partial V}\left(\frac{\partial U}{\partial p}\right) \Rightarrow \left(\frac{\partial p}{\partial S}\right)_V = -\left(\frac{\partial T}{\partial V}\right)_S, \tag{2.21}$$

$$\frac{\partial}{\partial S}\left(\frac{\partial H}{\partial p}\right) = \frac{\partial}{\partial p}\left(\frac{\partial H}{\partial S}\right) \Rightarrow \left(\frac{\partial V}{\partial S}\right)_p = \left(\frac{\partial T}{\partial p}\right)_S, \tag{2.22}$$

$$\frac{\partial}{\partial V}\left(\frac{\partial F}{\partial T}\right) = \frac{\partial}{\partial T}\left(\frac{\partial F}{\partial V}\right) \Rightarrow \left(\frac{\partial S}{\partial V}\right)_T = \left(\frac{\partial p}{\partial T}\right)_V, \tag{2.23}$$

$$\frac{\partial}{\partial p}\left(\frac{\partial G}{\partial T}\right) = \frac{\partial}{\partial T}\left(\frac{\partial G}{\partial p}\right) \Rightarrow \left(\frac{\partial S}{\partial p}\right)_T = -\left(\frac{\partial V}{\partial T}\right)_p. \tag{2.24}$$

These are called the *Maxwell relations*.

Since the Maxwell relations are very useful, it is worthwhile memorizing a scheme which enables us to reproduce the relations at any time, and at any place. Many different schemes are known; one is shown in Fig. 2.1.

The simple and small diagram shown in Fig. 2.1 contains a lot of information:

$U = U(S, V)$,
$dU = T dS - p dV$, $T \to S$ reverse, $p \to V$ direct,
$F = F(T, V)$,
$dF = -S dT - p dV$, $S \to T$ direct, $p \to V$ direct,
$G = G(T, p)$,
$dG = -S dT + V dp$, $S \to T$ direct, $V \to p$ reverse,
$H = H(S, p)$,
$dH = T dS + V dp$, $T \to S$ reverse, $V \to p$ reverse.

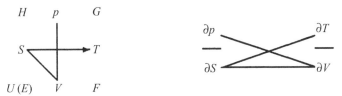

a negative sign appears

Fig. 2.1.

For instance, the term $T\,dS$ is given a positive sign wherever it appears in the above differential equations. The positive sign is given when the order T to S in the product is in reverse order compared with the order indicated by the arrow on the left hand side of Fig. 2.1. The term $S\,dT$, wherever it appears in the differential equations, is given a negative sign because the order S to T in the product is the same as the order indicated by the arrow in the same diagram. The same rule applies to the product of p and V. It is quite unfortunate that the sign of a product is not positive when the orders in the product and in the arrow are the same. This is due entirely to an *ad hoc* way of arranging four variables in the order $p - V - S - T$, which is easy to memorize.

There is a rule of sign for the Maxwell relation which states that when p and T are on the same side of the equation, both sides have the same sign, whereas there is a minus sign when p and T are not on the same side of the equation.

2.3 The open system

When the amount of material substance is changeable in the thermodynamic system, or if one is interested in the changes in thermodynamic potentials due to a change in the concentration, the system is called an *open system*.

All the foregoing differential equations are modified due to the addition of a new term, and we have

$$dU = T\,dS - p\,dV + \mu\,dN, \tag{2.25}$$

where μ is a new quantity called the *chemical potential*, and dN is the change in the amount of material substance usually measured in *moles*. Since dU is regarded as the total differential of the thermodynamic potential, U, the following relations hold:

$$T = \left(\frac{\partial U}{\partial S}\right)_{V,N}, \quad -p = \left(\frac{\partial U}{\partial V}\right)_{S,N}, \quad \mu = \left(\frac{\partial U}{\partial N}\right)_{S,V}. \tag{2.26}$$

The Helmholtz potential is

$$F = U - TS, \quad dF = dU - T\,dS - S\,dT, \tag{2.27}$$

and, substituting from (2.25),

$$dF = -S\,dT - p\,dV + \mu\,dN, \tag{2.28}$$

and hence

$$-S = \left(\frac{\partial F}{\partial T}\right)_{V,N}, \quad -p = \left(\frac{\partial F}{\partial V}\right)_{S,N}, \quad \mu = \left(\frac{\partial F}{\partial N}\right)_{T,V}. \tag{2.29}$$

This set of relations indicates that the Helmholtz potential is a thermodynamic potential described by independent variables T, V, and N.

The Gibbs potential is

$$G = F + pV, \quad dG = dF + pdV + Vdp, \tag{2.30}$$

and, substituting from (2.28),

$$dG = -SdT + Vdp + \mu dN. \tag{2.31}$$

This indicates that the Gibbs potential is a thermodynamic potential described by independent variables T, p, and N, and we obtain the following relations:

$$-S = \left(\frac{\partial G}{\partial T}\right)_{p,N}, \quad V = \left(\frac{\partial G}{\partial p}\right)_{T,N}, \quad \mu = \left(\frac{\partial G}{\partial N}\right)_{T,p}. \tag{2.32}$$

The last relation indicates that the chemical potential, μ, is the rate of change of the Gibbs potential with the change of the number of moles of the material substance, and hence the change in the Gibbs potential, when the temperature T and the pressure p are kept constant, is proportional to the change in the amount of the material substance, i.e.,

$$dG = \mu(T, p)dN, \quad \text{for} \quad dT = 0, \quad dp = 0. \tag{2.33}$$

This equation may be integrated with respect to the variable N to yield

$$G(T, p, N) = \int_0^N \mu(T, p)dN = \mu(T, p)\int_0^N dN = \mu(T, p)N. \tag{2.34}$$

This integration process may be called the charging-up process and may be described as a material substance carried into the system, bit by bit, from a material source (reservoir) while the temperature and pressure are kept unchanged.

In this way, we may establish that

$$G(T, p, N) = \mu(T, p)N = U - TS + pV. \tag{2.35}$$

The second equality in this equation leads to an interesting differential relation:

$$\mu dN + N d\mu = -SdT + Vdp + \mu dN, \tag{2.36}$$

and on eliminating μdN from the equation, we find

$$N d\mu = -SdT + Vdp, \quad \text{or} \quad N d\mu + SdT - Vdp = 0. \tag{2.37}$$

This equation is called the *Gibbs–Duhem relation*, and indicates the fact that when p and T are changed, the chemical potential μ changes accordingly; and so we have a relationship between these three parameters.

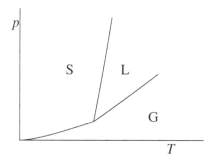

Fig. 2.2.

2.4 The Clausius–Clapeyron equation

Many simple systems (substances) can be found in the gas, liquid, and solid phases depending upon the temperature and pressure. Fig. 2.2 shows a typical $p - T$ phase diagram of such systems.

Along the boundary line, e.g., between the gas and liquid phases, the two phases can coexist at the same temperature and pressure in both phases. The transition from one phase to the other is associated with a discontinuity in the internal energy, $\Delta U = L$, or the entropy, ΔS. This is called the *first-order transition*. There is an end point of the gas–liquid transition called the *critical temperature* where the system can be changed from the gas phase to liquid phase with no discontinuity in the internal energy but with a singularity in the specific heat. This is called the *second-order transition*.

The slope of the phase boundary line, or the two-phase coexistence line, is given by the *Clausius–Clapeyron equation*

$$\frac{\mathrm{d}p}{\mathrm{d}T} = \frac{L}{T\,\Delta v}, \tag{2.38}$$

where L is the discontinuity in the internal energy, called the *latent heat*, and Δv is the accompanying change in the molar volume.

In order to derive this equation, let us consider a fixed amount, N moles, of a single-component system which is divided into two coexisting phases, phase I and phase II, at given pressure p and temperature T. Since the coexistence condition is characterized by the values of p and T, the most convenient thermodynamical potential is the Gibbs potential $G(p, T)$. This is a closed system, and

$$G(p, T) = N_{\mathrm{I}}\mu_{\mathrm{I}}(p, T) + N_{\mathrm{II}}\mu_{\mathrm{II}}(p, T),$$
$$N = N_{\mathrm{I}} + N_{\mathrm{II}}, \tag{2.39}$$

where N_{I} and N_{II} are the number of moles in the respective phases.

Since the equilibrium values of N_{I} and N_{II} are such that the Gibbs potential is a minimum with respect to virtual variations of N_{I} and N_{II}, they are determined by the conditions

$$\delta G = \delta N_I \mu_I + \delta N_{II} \mu_{II} = 0,$$
$$\delta N = \delta N_I + \delta N_{II} = 0. \tag{2.40}$$

If the second relation is substituted into the first, we find that

$$\mu_I = \mu_{II}. \tag{2.41}$$

This is the equilibrium condition under constant pressure and temperature across the phase boundary.

If we consider two points which are separated by an infinitesimal distance, δp and δT, along the coexistence line in each phase,

$$\delta \mu_I = v_I \delta p - s_I \delta T,$$
$$\delta \mu_{II} = v_{II} \delta p - s_{II} \delta T, \tag{2.42}$$

where v_I and v_{II} are the molar volumes and s_I and s_{II} are the molar entropies in the respective phases.

Since (2.41) holds at each point along the coexistence line, the changes also satisfy

$$\delta \mu_I = \delta \mu_{II}, \tag{2.43}$$

and hence, from (2.42),

$$\left(\frac{dp}{dT} \right)_{\text{coex}} = \frac{s_{II} - s_I}{v_{II} - v_I} = \frac{L}{T(v_{II} - v_I)}, \tag{2.44}$$

where

$$L = T(s_{II} - s_I) \tag{2.45}$$

is the latent heat of transition from phase I to phase II.

Water is an example of a system which can exist in many different solid phases depending upon the temperature and pressure (see Fig. 2.3).

It is well known that liquid water has a larger mass density than that of ordinary ice (phase I), and hence $s_{II} - s_I$ and $v_{II} - v_I$ have opposite signs, represented by the negative slope of $\frac{dp}{dT}$

$$\left(\frac{dT}{dp} \right)_{\text{melt}} < 0, \quad \left(\frac{dp}{dT} \right)_{\text{melt}} = \frac{1}{\frac{dT}{dp}} < 0. \tag{2.46}$$

If we look at the phase diagram of water (Fig. 2.3) there is another transition between phases VII and VIII (both solid phases). In this case the slope of dT/dp is zero, which means that the volume change at the transition is zero. This transition is therefore characterized as the first-order phase transition without volume

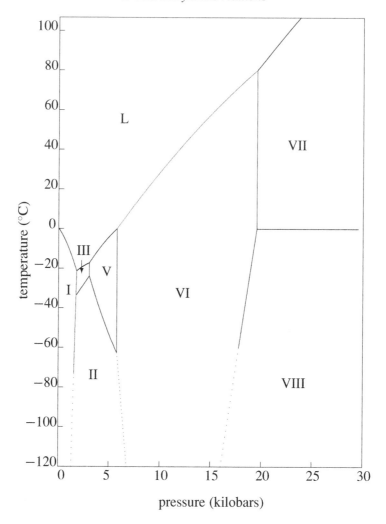

Fig. 2.3. The phase diagram of ice. For more details, see Fletcher (1970).

change. A statistical mechanics interpretation of this transition will be presented in Chapter 9.

2.5 The van der Waals equation

In 1873 van der Waals presented, in his doctoral dissertation, a famous equation of state which bears his name. The equation accounts for deviations from the ideal gas law in a convenient way and is expressed as

$$\left(p + \frac{a}{V^2}\right)(V - b) = RT, \tag{2.47}$$

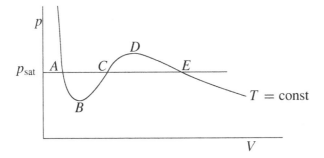

Fig. 2.4. The van der Waals equation of state.

where a and b are parameters characteristic of the given nonideal gas. Loosely speaking, a/V^2 represents the effect of the attractive interaction and b represents an excluded volume effect due to the repulsive interaction, respectively, between the constituent gas molecules.

At sufficiently low temperatures along an isotherm there appears a range of volume, from B to D in Fig. 2.4, over which the derivative of the pressure with respect to the volume becomes positive, $\partial p/\partial v > 0$. This positive derivative represents a mechanical instability, and the gas cannot exist in one homogeneous phase. Along the isotherm, E represents the saturation vapor pressure, and if the gas is further compressed past E a portion of the gas condenses into the liquid phase. The actual isotherm becomes a horizontal line, $E - C - A$. The van der Waals equation is, therefore, supplemented by a horizontal line representing the gas–liquid coexisting phases at the saturation pressure p_{sat} for the given temperature.

A common procedure used to find the saturation vapor pressure line, $A - C - E$, is known as the *Maxwell rule*.

When the saturation vapor pressure line, which is horizontal with a constant temperature and pressure, is drawn, it will intersect with the van der Waals isotherm of the same temperature at three points and a figure of eight $(A \rightarrow B \rightarrow C \rightarrow D \rightarrow E \rightarrow C \rightarrow A)$ will be constructed. The Maxwell rule stipulates that

$$\text{area } ABCA = \text{area } CDEC. \qquad (2.48)$$

In order to show that this is a thermodynamically acceptable procedure, the second law of thermodynamics may be conveniently invoked. The thermodynamic process along the loop $(A \rightarrow B \rightarrow C \rightarrow D \rightarrow E \rightarrow C \rightarrow A)$ is a reversible-cyclic process operated in thermal contact with one heat reservoir of the given temperature, and hence the total mechanical work performed during the cycle must be equal to zero according to Thomson's statement of the second law of thermodynamics:

$$\int_{ABCDECA} p\, dV = 0. \qquad (2.49)$$

Another proof of the Maxwell rule is found by integrating the following equation along the van der Waals isotherm:

$$dG = V\,dp - S\,dT, \quad dT = 0,$$

$$\int_{ABCDE} dG = G_{\mathrm{G}} - G_{\mathrm{L}} = n(\mu_{\mathrm{G}} - \mu_{\mathrm{L}}) = \int_{ABCDE} V\,dp, \tag{2.50}$$

where n is the number of moles of the system.

If the horizontal tie-line, ACE, is properly drawn between the coexisting liquid (L) and gas (G) phases,

$$\mu_{\mathrm{G}} = \mu_{\mathrm{L}} \tag{2.51}$$

must be satisfied, and hence

$$\int_{ABCDE} V\,dp = 0, \tag{2.52}$$

which also stipulates that the two areas are equal:

$$\text{area } ABCA = \text{area } CDEC. \tag{2.53}$$

2.6 The grand potential

Let us define a new thermodynamic potential Ω, which is defined by

$$\Omega = -pV = U - \mu N - TS. \tag{2.54}$$

If we take the differential of Ω, we obtain

$$d\Omega = dU - T\,dS - S\,dT - \mu\,dN - N\,d\mu = -N\,d\mu - p\,dV - S\,dT. \tag{2.55}$$

This equation shows that Ω is a thermodynamic potential, called the *grand potential*, for which μ, V, and T are the independent variables, and

$$N = -\left(\frac{\partial \Omega}{\partial \mu}\right)_{V,T}, \quad p = -\left(\frac{\partial \Omega}{\partial V}\right)_{T,\mu}, \quad S = -\left(\frac{\partial \Omega}{\partial T}\right)_{V,\mu}. \tag{2.56}$$

Exercises

2.1 Derive the following relation:

$$\left(\frac{\partial C_V}{\partial V}\right)_T = T\left(\frac{\partial^2 p}{\partial T^2}\right)_V. \tag{E2.1}$$

[**Hint**]

$$C_V = T\left(\frac{\partial S}{\partial T}\right)_T. \tag{E2.2}$$

Differentiating both sides of this equation with respect to V yields

$$\left(\frac{\partial C_V}{\partial V}\right)_T = T\frac{\partial^2 S}{\partial V \partial T} = T\frac{\partial}{\partial T}\left(\frac{\partial S}{\partial V}\right)_T. \tag{E2.3}$$

Using the Maxwell relation (2.23)

$$\left(\frac{\partial C_V}{\partial V}\right)_T = T\frac{\partial}{\partial T}\left(\frac{\partial p}{\partial T}\right)_V = T\frac{\partial^2 p}{\partial T^2}. \tag{E2.4}$$

Similarly

$$\left(\frac{\partial C_p}{\partial p}\right)_T = -T\left(\frac{\partial^2 V}{\partial T^2}\right)_p. \tag{E2.5}$$

2.2 Derive the following relation:

$$T\,dS = C_V dT + T\left(\frac{\partial p}{\partial T}\right)_V dV. \tag{E2.6}$$

[**Hint**]

$$T\,dS = T\left(\frac{\partial S}{\partial T}\right)_V dT + T\left(\frac{\partial S}{\partial V}\right)_T dV$$

$$= T C_V dT + T\left(\frac{\partial p}{\partial T}\right)_V. \tag{E2.7}$$

The second line follows by the Maxwell relation (2.23).

3

The ensemble theory

3.1 Microstate and macrostate

In the formulation of statistical mechanics we are concerned with the theoretical prediction of thermodynamic properties of a physical system which contains a large number of particles such as electrons, atoms, and molecules by some statistical average processes performed over an appropriately prepared *statistical sample*. There are many different ways to prepare the sample. J. Willard Gibbs (1901) coined the name *ensemble* for the statistical sample, and three different ensembles are introduced, i.e., the *microcanonical*, *canonical*, and *grand canonical* ensembles.

We can choose for the physical system any thermodynamical system such as a single-component gas, liquid, or solid, as well as a mixture of many components, as long as the system is in the condition of thermodynamical equilibrium. In order to establish a fundamental principle of statistical mechanics, however, we naturally choose as simple a system as possible, such as the one-component dilute gas made up of structureless monatomic molecules. We then extend the fomulation step by step to more complicated systems. In this chapter, formulation of the three Gibbs ensembles will be developed.

The microcanonical ensemble is a collection of identical replicas of a given physical system which is a gas made up of noninteracting structureless particles. Firstly, the system is assumed to be contained in a box of volume V, the number of particles is equal to N, and the total energy is given in a narrow range between E and $E + dE$. Let us first limit our discussion to the case of an ideal gas made up of identical, yet permanently distinguishable, particles.

Since the total energy of the system is the sum of the energies of individual particles, the total energy is given by

$$E(n) = \frac{\hbar^2}{2m} \left(\frac{2\pi}{L} \right)^2 \sum_i \left(n_{ix}^2 + n_{iy}^2 + n_{iz}^2 \right), \tag{3.1}$$

where L is the side length of the cube containing the gas, and n_{ix}, n_{iy}, and n_{iz} are integers characterizing the energy of the plane-wave state of the ith particle $(i = 1, 2, \ldots, N)$.

A *microstate* of the system is defined as the state characterized by one set of ordered arrangements of integers n_{ix}, n_{iy}, n_{iz} $(i = 1, \ldots, N)$. Another microstate, which is characterized by a different set of integers $n'_{ix}, n'_{iy}, n'_{iz}$ $(i = 1, \ldots, N)$, is regarded as a different microstate from the previous one, as long as this second set is different from the first, even if the second set is a rearrangement of the integers in the first set. According to this definition, interchanging the plane-wave states of two particles means creating a new microstate.

The *macrostate* of the gas will now be defined. In order to accomplish this, the almost continuous one-particle plane-wave energy spectrum is divided up into small energy intervals of variable width. The magnitude of the width is arbitrary as long as it is small enough to be assigned a definite energy value but contains a reasonably large number, g_i $(i = 1, 2, \ldots)$, of one-particle energy eigenstates. These energy intervals are numbered according to the increasing order of the energy, $\epsilon_1 < \epsilon_2 < \cdots < \epsilon_i < \cdots$. A macrostate of the gas is defined by the set of numbers of particles found in each of these energy intervals. We can set up the following list:

$$
\begin{array}{ccccc}
n_1 & n_2 & \cdots & n_i & \cdots \\
\epsilon_1 & \epsilon_2 & \cdots & \epsilon_i & \cdots \\
g_1 & g_2 & \cdots & g_i & \cdots ,
\end{array}
\tag{3.2}
$$

where n_i is the number of particles and g_i is the number of nearly degenerate one-particle energy eigenstates included in the ith energy interval ϵ_i. We are not concerned with which of the particles is in which of the energy intervals; i.e., the identification of individual particles is totally ignored.

The number of microstates included in a given macrostate is

$$
W_{\text{macro}} = \frac{N!}{n_1! n_2! \cdots n_i! \cdots} g_1^{n_1} g_2^{n_2} \cdots g_i^{n_i} \cdots .
\tag{3.3}
$$

The reason for the above expression is because $N!$ exchanges of the particles in different one-particle eigenstates among themselves can create new microstates, except $n_i!$ exchanges of particles within the degenerate energy state ϵ_i.

The set of numbers, n_i, of particles in these energy intervals satisfy the following two conditions:

$$
E = \sum_i n_i \epsilon_i, \qquad N = \sum_i n_i.
\tag{3.4}
$$

The total energy, E, given in (3.4) is the same as the total energy, $E(n)$, given by (3.1), except that n_i nearly degenerate energies occurring in (3.1) are grouped together in (3.4).

3.2 Assumption of equal *a priori* probabilities

Since the particles are assumed to be distinguishable, any set of values of $3N$ integers, as long as it is consistent with the total energy E appearing in (3.1), is acceptable in the definition of an *N-particle eigenstate*. We can see, therefore, that there are many *n*-particle eigenstates with total energy in the narrow range between E and $E + dE$.

The *microcanonical ensemble* is now constructed in such a way that every distinct N-particle state is represented with an equal probability in the ensemble. Each of these states is a microstate and has an equal statistical weight in the ensemble. This is called the assumption of *equal a priori probabilities*. There is no way to prove that one microstate is preferred to another, and hence the assumption is taken as a plausible postulate in the statistical mechanics formulation.

3.3 The number of microstates

Let us now find the total number of microstates for the given energy range between E and $E + dE$. The expression for the total energy can be rewritten as

$$
\begin{aligned}
n_{1x}^2 + n_{1y}^2 &+ n_{1z}^2 + n_{2x}^2 + n_{2y}^2 + n_{2z}^2 + \cdots + n_{Nx}^2 + n_{Ny}^2 + n_{Nz}^2 \\
&= R^2 \\
&= (L/2\pi)^2 (2mE/\hbar^2).
\end{aligned}
\tag{3.5}
$$

This is the equation of a sphere (the radius is equal to R) in the $3N$-dimensional space in which axes are integers, hence any small volume in this space is exactly equal to the number of microstates. The constant energy surface is a sphere in this $3N$-dimensional space. The number of microstates contained in a thin spherical shell between two energy surfaces E and $E + dE$ is equal to the volume given by

$$
\Omega(3N) R^{3N-1} dR,
\tag{3.6}
$$

where $\Omega(3N)$ is the angular factor for the spherical surface in the $3N$-dimensional space given by

$$
\Omega(3N) = 2(\pi)^{\frac{3N}{2}} / \Gamma(3N/2)
\tag{3.7}
$$

(see Appendix 1, Sec. A1.2). The number of microstates in the ensemble $W(V, N, E, \mathrm{d}E)$ is hence

$$W(V, N, E, \mathrm{d}E) = V^N (2\pi m/\hbar^2)^{\frac{3N}{2}} E^{\frac{3N}{2}-1} \mathrm{d}E / \Gamma(3N/2). \qquad (3.8)$$

The entropy of the system is defined by

$$S = k \log W(V, N, E, \mathrm{d}E). \qquad (3.9)$$

This is the most celebrated relation in the history of statistical mechanics proposed by L. Boltzmann. The proportionality constant k introduced in this equation is called the *Boltzmann constant*. It should be noted that S satisfies all the essential properties of the entropy introduced in the formulation of classical thermodynamics.

When the relation (3.8) is introduced into (3.9), and if the asymptotic form of the gamma function for a large argument is used, the entropy of a gas of distinguishable and noninteracting particles is found to be

$$S = kN \left[\log V + \tfrac{3}{2} \log(2\pi m/h^2) + \tfrac{3}{2} \log(2E/3N) + \tfrac{3}{2} \right]. \qquad (3.10)$$

Once the definition of entropy is accepted, we can return to the thermodynamic relation:

$$\mathrm{d}S = (1/T)\mathrm{d}E + (p/T)\mathrm{d}V - (\mu/T)\mathrm{d}N, \qquad (3.11)$$

and the reciprocal temperature β, the pressure p, and the chemical potential μ are found as

$$\beta = 1/kT = (1/k)(\partial S/\partial E) = \tfrac{3}{2}N/E, \quad E = \tfrac{3}{2}NkT,$$
$$p/T = (\partial S/\partial V) = kN/V, \quad pV = NkT,$$
$$\mu = kT \left[\log p - \tfrac{5}{2} \log T - \tfrac{3}{2} \log(2\pi m/h^2) - \tfrac{5}{2} \log k - \tfrac{5}{2} \right]. \qquad (3.12)$$

3.4 The most probable distribution

It was found in the previous section that there are a huge number of microstates of the gas made up of N distinguishable particles, with the energy between E and $E + \mathrm{d}E$; see (3.8). At this point, the definition of the microcanonical ensemble may be restated as the statistical ensemble in which every one of the microstates included in (3.8) is equally represented.

Any point in the $3N$-dimensional space given by (3.5) is called a representative point in the microcanonical ensemble and will represent the gas in a particular microstate. All the representative points in this ensemble have the same energy E. Some of these microstates may be far removed from the thermodynamic equilibrium conditions. For instance, a microstate in which one of the particles is given the total energy E and all other particles have no energy is also included in this ensemble as one of the extreme cases.

One of the most important objectives of statistical mechanics is to find some of the macrostates which contain a huge number of microstates and which would most appropriately represent the thermodynamic equilibrium conditions. For this purpose we consider a macrostate which is most likely to be found, i.e., the macrostate which contains the largest number of microstates.

The statistical mechanics formulation based on this assumption is called the method of the *most probable distribution*. Let us look at the number W_{macro}; the number of microstates contained in a macrostate defined by a set of particle numbers in different energy intervals:

$$W_{macro} = \frac{N!}{n_1!n_2!\cdots n_i!\cdots} g_1^{n_1} g_2^{n_2} \cdots g_i^{n_i} \cdots. \tag{3.13}$$

It should be noted that this W_{macro} is different from the W in (3.8).

Instead of trying to find a maximum of W_{macro}, let us find a maximum of the logarithm of W_{macro}:

$$\log W_{macro} = \log(N!) - \sum_i [\log(n_i!) - n_i \log g_i]$$

$$= (N \log N - N) - \sum_i n_i [\log n_i - 1 - \log g_i], \tag{3.14}$$

where Stirling's formula has been employed (see Appendix 1, Sec. A1.2),

$$\log(N!) = N \log N - N. \tag{3.15}$$

Because of the restricting conditions (3.4), we should consider the following expression:

$$\log W'_{macro} = (N \log N - N) - \sum_i n_i [\log n_i - 1 - \log g_i]$$

$$- \alpha \left(\sum_i n_i - N \right) - \beta \left(\sum_i n_i \epsilon_i - E \right), \tag{3.16}$$

where α and β are the Lagrange undetermined multipliers.

When $\log W'_{macro}$ is maximized with respect to n_i, we obtain

$$n_i = g_i \exp[-(\alpha + \beta \epsilon_i)], \qquad \exp(-\alpha) = N/q(\beta),$$

$$q(\beta) = \sum_i g_i \exp(-\beta \epsilon_i), \tag{3.17}$$

where $q(\beta)$ is called the *molecular partition function*. The probability that a gas particle is found in one of the free particle eigenstates is

$$P(\epsilon_i) = n_i/N = \frac{g_i}{q(\beta)} \exp(-\beta\epsilon_i). \tag{3.18}$$

This is called the *Maxwell–Boltzmann distribution*.

3.5 The Gibbs paradox

The entropy of the gas made up of distinguishable noninteracting particles is given by (3.10), i.e.,

$$S = k \log W = kN \left[\log V + \tfrac{3}{2}\log(2\pi mk/h^2) + \tfrac{3}{2}\log T + \tfrac{3}{2}\right]. \tag{3.19}$$

This has one unacceptable feature, i.e., the entropy is not given as an extensive quantity because of the $\log V$ term. If the amount of the gas is doubled without changing the temperature and pressure, i.e., N and V are doubled, the entropy S given by (3.19) will not be doubled. In order to trace back the source of this feature, let us look at the $\log W_{\text{macro}}$ term (3.14) again,

$$\log W_{\text{macro}} = (N \log N - N) - \sum_i n_i[\log n_i - 1 - \log g_i]. \tag{3.20}$$

When the Maxwell–Boltzmann distribution (3.18) is introduced into this equation, we find

$$\log W_{\text{macro}} = \beta E + N \log q(\beta), \tag{3.21}$$

where $q(\beta)$ is given by

$$q(\beta) = \sum_i g_i \exp(-\beta\epsilon_i) = V(2\pi m/\hbar^2)^{\frac{3}{2}}(kT)^{\frac{3}{2}}, \tag{3.22}$$

and

$$\begin{aligned} S &= k \log W_{\text{macro}} \\ &= kN \left[\log V + \tfrac{3}{2}\log(2\pi mk/\hbar^2) + \tfrac{3}{2}\log T + \tfrac{3}{2}\right]. \end{aligned} \tag{3.23}$$

Equation (3.21) can be rewritten as

$$\log W_{\text{macro}} = \tfrac{3}{2}N + \log[q(\beta)^N]. \tag{3.24}$$

Gibbs proposed to modify this equation to

$$\log W_{\text{macro}} = \tfrac{3}{2}N + \log[q(\beta)^N/N!]; \tag{3.25}$$

then the entropy (3.23) will be modified to

$$S = kN \left[\log(V/N) + \tfrac{3}{2}\log(2\pi mk/h^2) + \tfrac{3}{2}\log T + \tfrac{5}{2}\right], \tag{3.26}$$

which now has a correct dependence on the variables V and N. Gibbs did not give any microscopic explanation for the necessity to introduce the factor $N!$ in (3.25), and this has, henceforth, been called the *Gibbs paradox*.

At this point an important remark must be made. The entropy (3.23) calculated by the number of microstates represented by the most probable distribution is identical to the entropy calculated from the total number of microstates, (3.8) and (3.9), contained in the microcanonical ensemble. This is a rather surprising result, i.e., a result stating that 'a part is equal to the whole'; however, it also means that the majority of microstates are in close proximity to the most probable distribution. This superficially inconsistent result is due to the fact that two huge quantities are equal in the order of magnitude criterion and the quantities which are smaller in order of magnitude have no consequence.

Another feature worth noting is the temperature dependence of the entropy. The $\log T$ term in (3.26) does not represent a correct temperature dependence at low temperatures. This term will diverge to minus infinity as T approaches to zero. This indicates that the concept of a gas made up of distinguishable particles is only an approximation acceptable at high temperatures.

3.6 Resolution of the Gibbs paradox: quantum ideal gases

The unacceptable feature of the entropy expression for the classical ideal gas was corrected by artificial division by $N!$ in (3.25); however, the necessity for this correction had not been explained until the development of quantum mechanics and application of it to ideal gases. One should note the fact that the assumption of permanent distinguishability of identical particles is in contradiction with the principles of quantum mechanics, and therefore this assumption must be corrected.

The microstate of a gas made up of indistinguishable particles is defined differently from that comprising distinguishable particles. In this case a microstate is defined only by assigning the number, n_i, of particles in each of the one-particle energy eigenstates. There is no way of telling which particle is in which one-particle eigenstate. In order to find the number of physically realizable microstates, two characteristically different cases must be considered.

The first is the case of particles obeying *Fermi–Dirac statistics*. In this case, each one-particle eigenstate can accommodate only zero or one particle. g_i degenerate eigenstates, in the ith energy eigenvalue ϵ_i, are classified into two groups, i.e., n_i occupied states and $g_i - n_i$ unoccupied states. Hence the number of microstates is given by

$$W_{\text{F.D.}} = \prod_i \frac{g_i!}{n_i!(g_i - n_i)!}. \tag{3.27}$$

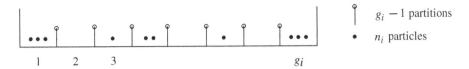

Fig. 3.1. Distribution of particles and partitions.

The second case is the system of particles obeying *Bose–Einstein statistics*. In this case, any one-particle eigenstate can be occupied by any number of particles, and the number of microstates is given by

$$W_{\text{B.E.}} = \prod_i \frac{(n_i + g_i - 1)!}{n_i!(gi - 1)!}.$$ (3.28)

The combinatorial expression (3.28) is most easily found by looking at the number of ways of arranging n_i indistinguishable particles and $g_i - 1$ partitions, mixed together, in a one-dimensional array; i.e., the total number of arrangements of $n_i + g_i - 1$ items in a row, of which n_i items are of one kind and the remaining $g_i - 1$ items are of the other kind. In such arrangements, if two partitions appear consecutively the gap between the two partitions can be regarded as a one-particle state which is vacant.

Let us, as an illustration, assume Fermi–Dirac statistics to find the most probable distribution of the macrostate and calculate the entropy.

Introducing the Lagrange undetermined multipliers, α and β, the modified variational function is given by

$$\log W'_{\text{F.D.}}$$
$$= \sum_i [g_i(\log g_i - 1) - n_i(\log n_i - 1) - (g_i - n_i)(\log(g_i - n_i) - 1)]$$
$$- \beta \left(\sum_i n_i \epsilon_i - E \right) - \alpha \left(\sum_i n_i - N \right).$$ (3.29)

When the variation with respect to n_i is taken, we obtain the *Fermi–Dirac distribution*,

$$n_i = \frac{g_i}{[\exp(\alpha + \beta \epsilon_i) + 1]},$$ (3.30)

where α and β are determined in such a way that

$$\sum_i n_i = N, \qquad \sum_i n_i \epsilon_i = E,$$ (3.31)

are satisfied.

In order to see if the Gibbs paradox has been resolved, let us take the limit of a dilute gas, i.e., $n_i \ll g_i$. It means that either

$$\exp(\alpha + \beta\epsilon_i) \gg 1 \quad \text{or} \quad \exp{-(\alpha + \beta\epsilon_i)} \ll 1 \tag{3.32}$$

is satisfied. In this limit the Fermi–Dirac distribution is reduced to the Maxwell–Boltzmann distribution,

$$n_i = g_i \exp{-(\alpha + \beta\epsilon_i)}, \qquad \exp(-\alpha) = N/q(\beta),$$
$$q(\beta) = \sum_i g_i \exp(-\beta\epsilon_i) = V(2\pi m/h^2)^{\frac{3}{2}}(kT)^{\frac{3}{2}}. \tag{3.33}$$

As long as (3.27) is used for counting the number of microstates, the entropy of the ideal gas at low densities, even if using the Maxwell–Boltzmann distribution (3.33), will be given by

$$S = k[\alpha N + \beta E + N], \tag{3.34}$$

and again, using (3.33), we obtain

$$S = kN\left[\log q(\beta) - \log N + \tfrac{5}{2}\right]$$
$$= kN\left[\log(V/N) + \tfrac{3}{2}\log(2\pi mk/h^2) + \tfrac{3}{2}\log T + \tfrac{5}{2}\right]. \tag{3.35}$$

This result is identical to (3.26), and it is seen that the Gibbs paradox has been resolved. The same is true assuming Bose–Einstein statistics.

3.7 Canonical ensemble

In the formulation of the microcanonical ensemble in Sec. 3.1 to Sec. 3.6, each of the microstates in the ensemble can be looked at as representing a replica of the system in the laboratory. The total number of the replicas, which is the same as the number of microstates, is $W(V, N, E, dE)$, see (3.8), a huge number proportional to the Nth power of some finite quantity.

From the thermodynamical point of view, the description of the physical state of the system in terms of its microstate contains too much information which cannot possibly be observed even from a microscopic point of view. For this reason, a description of the ensemble in terms of the macrostates has been introduced. The number of microstates contained in a particular macrostate is given by

$$W_{\text{macro}} = \frac{N!}{n_1! n_2! \cdots n_i! \cdots} g_1^{n_1} g_2^{n_2} \cdots g_i^{n_i} \cdots, \tag{3.36}$$

and there is a relationship between this number and $W(V, N, E, dE)$, the total number of microstates in the ensemble, given by

$$W(V, N, E, dE) = \sum_{\{n_i\}} W_{macro}(n_1, n_2, \ldots, n_i, \ldots),$$ (3.37)

where the summation is taken over all distinct sets of $\{n_i\}$ satisfying conditions (3.4).

One of the findings in the previous sections is that by looking at the most probable macrostate in the ensemble, the probability that a particle is found in one of the one-particle energy eigenstates, ϵ_i, is given by the Maxwell–Boltzmann distribution (3.18):

$$P(\epsilon_i) = \exp(-\beta\epsilon_i)/q(\beta), \quad \beta = 1/kT,$$ (3.38)

where $q(\beta)$ is the molecular partition function

$$q(\beta) = \sum_i \exp(-\beta\epsilon_i).$$ (3.39)

Here, we are interested in finding a statistical distribution law for the system in which the constituent particles interact strongly so that there is no way of defining the one-particle energy eigenstate.

The system is assumed to be made up of N strongly interacting particles, confined in a box of volume V and immersed in a large heat bath maintained at temperature T. It is, nevertheless, possible to define the energy eigenvalues $E_i, i = 1, 2, \ldots, \infty$, by solving the N-particle Schrödinger equation with appropriate boundary conditions. Because the system is immersed in the heat bath of temperature T, the system cannot be found in one of the possible energy eigenstates forever; instead the energy of the system would fluctuate rapidly around some possible average value $\overline{E}$.

In the statistical mechanics formulation in this kind of thermodynamic situation, one is interested in finding the average value of the energy, $\overline{E}$, the entropy S, and all other thermodynamic potentials of the system. In particular, one is interested in finding the statistical probability that the system will be found in one of the possible energy eigenvalues E_i. The task is most readily accomplished by means of the *canonical ensemble* formulation.

In the form presented by E. Schrödinger (1952), the heat bath surrounding the laboratory system under consideration is replaced by a collection of a huge number, $M - 1$, of identical replicas of the laboratory system. These $M - 1$ replicas are found in many different energy eigenvalues having the same energy level scheme as that of the laboratory system. Furthermore, it is assumed that the replicas interact with one another as well as with the laboratory system through short range interaction between constituent particles across the interfacial boundaries. The entire system, the laboratory system plus $M - 1$ replica systems, which may be called the super-system, is isolated from the rest of the universe, and is found in thermal equilibrium at temperature T.

Let us specify the way in which this super-system is prepared. Suppose that there are $M_1, M_2, \ldots, M_i, \ldots$ systems having the energy eigenvalues $E_i, E_2, \ldots, E_i, \ldots$, respectively, in the super-system. There are two conditions imposed:

$$E = \sum_i M_i E_i, \quad M = \sum_i M_i, \quad i = 1, 2, \ldots, \infty. \tag{3.40}$$

The interaction energy between the replica systems does not appear in the energy balance equation because it is proportional to the total interfacial area which is an order of magnitude smaller than any of the energy eigenvalues, which are a volume-proportional quantity. The interfacial interaction energy is negligible in the thermodynamic limit mentioned in Sec. 1.1. The summations over the energy eigenstates go to infinity, but many of the M_i values are zero, and in particular all of the M_i values for which $E < E_i$ are identically zero.

The super-system as a whole is assumed to be isolated from the rest of the universe and hence the energy E is a fixed quantity. The number M can be a huge number much larger than any macroscopically large number.

Since each of the M systems is a macroscopic system, they are distinguishable permanently from one another, and hence the total number of distinctly different ways of distributing the entire energy over M systems is given by

$$W = \frac{M!}{M_1! M_2! \cdots M_i! \cdots} g_1^{M_1} g_2^{M_2} \cdots g_i^{M_i} \cdots, \tag{3.41}$$

where g_i is the multiplicity of the respective energy eigenvalue.

This number W is also the total number of distinct ways of preparing the replicas of the super-system, not the replicas of the laboratory system, for a fixed set of $\{M_i\}$, and can be the size of the statistical ensemble. If all the different sets of $\{M_i\}$ which satisfy the conditions (3.40) are included, the size of the ensemble can be still larger.

It is clear that the formulation developed herein is parallel to the microcanonical ensemble formulation. In the following we may again use the method of the most probable distribution.

The M_i values satisfying the condition of the most probable distribution are now found in exactly the same way as in the microcanonical ensemble formulation:

$$M_i = \frac{M}{Z(\beta)} g_i \exp(-\beta M_i), \tag{3.42}$$

where $Z(\beta)$ is the normalization factor called the *canonical partition function*,

$$Z(\beta) = \sum_i g_i \exp(-\beta E_i). \tag{3.43}$$

The probability that a system randomly chosen in the super-system is found in the energy eigenvalue E_i is given by

$$P(E_i) = \frac{g_i}{Z(\beta)} \exp(-\beta E_i). \tag{3.44}$$

$P(E_i)$ is called the *canonical distribution*.

Since β is introduced as the Lagrange undetermined multiplier for conditions (3.40), the following equation must be satisfied:

$$\overline{E} = \sum_i g_i E_i \exp(-\beta E_i) \Big/ \sum_i g_i \exp(-\beta E_i)$$

$$= -\partial \log Z(\beta)/\partial \beta. \tag{3.45}$$

3.8 Thermodynamic relations

We have introduced β, the Lagrange multiplier for the total energy of the super-gas; however, neither the total energy nor the total number of systems in the super-gas is actually measured. For this reason, the physical interpretation of β is still not clear.

In this section we shall find another way of identifying the physical interpretation of β. Let us recall the fact that our system is made up of N strongly interacting particles, confined in a cube of volume V, and immersed in a heat bath made up of the rest of the super-gas. Because of thermal contact with the heat bath, the energy of the system can change from time to time in the neighborhood of an average value E which depends on the temperature of the heat bath. The probability of finding the system in one of the eigenstates, E_i, is given by the canonical distribution

$$P(E_i) = \exp(-\beta E_i)/Z(\beta), \tag{3.46}$$

where $Z(\beta)$ is the *canonical partition function* defined by

$$Z(\beta) = \exp(-\beta F) = \sum_i \exp(-\beta E_i), \tag{3.47}$$

where F is a new quantity to be identified later on.

Let us consider some small changes in the pertaining physical properties, i.e., the values of the parameters β and V. When the volume V of the container is changed infinitely slowly, the eigenvalues will change without inducing any transition from one eigenstate to another (the *adiabatic approximation* in quantum mechanics)

$$\delta E_i = \frac{\partial E_i}{\partial V} \delta V. \tag{3.48}$$

Let us now differentiate both sides of (3.47) logarithmically:

$$\delta(-\beta F) = \frac{\partial \log Z(\beta)}{\partial \beta} \delta\beta + \frac{\partial \log Z(\beta)}{\partial V} \delta V$$

$$= -\frac{1}{Z(\beta)} \sum_i E_i \exp(-\beta E_i)\delta\beta - \frac{\beta}{Z(\beta)} \sum_i \frac{\partial E_i}{\partial V} \exp(-\beta E_i)\delta V$$

$$= -\overline{E}\delta\beta + \beta p\delta V$$

$$= -\delta(\beta\overline{E}) + \beta\delta\overline{E} + \beta p\delta V, \tag{3.49}$$

where a statistical interpretation of the hydrostatic pressure is introduced:

$$p = -\left(\frac{\partial \overline{E}}{\partial V}\right)_S$$

$$= -\frac{1}{Z(\beta)} \sum_i \frac{\partial E_i}{\partial V} \exp(-\beta E_i). \tag{3.50}$$

Transferring the term $\delta(\beta E)$ to the left hand side of (3.49), we have

$$\delta(-\beta F + \beta E) = \beta(\delta E + p\delta V). \tag{3.51}$$

According to the thermodynamic relation (1.119), the quantity $\delta E + p\delta V$ is a small amount of heat, $\delta'Q$, transferred to the system during a quasistatic process, and (3.51) indicates that β is the integrating factor $(1/kT)$ which defines the total differential of the entropy, $\delta(-\beta F + \beta E)$, on the left hand side of the equation. The following identifications are now in order:

$$\beta = 1/kT, \quad -\beta F + \beta E = S/k, \tag{3.52}$$

where the Boltzmann constant k, which has the unit of entropy, is introduced to make the equations dimensionless. The quantity F is then identified as the *Helmholtz potential*,

$$F = E - TS. \tag{3.53}$$

We obtain an important relation:

$$F = -kT \log Z(\beta) = -kT \log \left[\sum_i \exp(-\beta E_i)\right]. \tag{3.54}$$

This is one of the most fundamental relations in the canonical ensemble formulation of statistical mechanics.

To summarize, the internal energy and the probability distribution function are given by (3.46) and (3.47),

$$\overline{E} = -\partial \log Z(\beta)/\partial\beta, \quad P(E_i) = \frac{1}{Z(\beta)} \exp(-\beta E_i), \tag{3.55}$$

and the entropy is given by [Exercise 3.3]

$$S = -k \sum_i P(E_i) \log P(E_i). \qquad (3.56)$$

3.9 Open systems

In experimental situations, the actual number of particles contained in a thermodynamic system is never measured. It is simply assumed that the number of particles, as a matter of principle, is fixed. So, we only have an approximate number of particles in the system. We consider a large amount of gas, and look at a small portion of it. This small amount of gas has volume V and is bounded by geometrically defined imaginary walls. Since in this situation there are no physical walls between the volume V and the rest of the gas, the gas particles can move in and out of volume V. The number of gas particles in this portion of the gas is, therefore, not fixed. This number can change over a wide range of values; however, we assume that there is a well defined average value of the number of gas particles contained in the volume V.

The situation is similar to a biological observation of bacteria under a microscope. The number of bacteria in the field of view may change over time; however, there should be a well defined average number of bacteria in the field of view which depends upon the concentration of the biological sample. The boundaries of the field of view are not the physical walls, but are only geometrically defined.

It would, therefore, be more convenient to generalize the ensemble formulation to the cases in which the number of particles in a physical system is also a statistical variable. Furthermore, we can apply statistical mechanics to systems which contain nonmaterial particles, such as photons, lattice phonons, magnons, and plasmons, in which the number of particles is indeed a nonconserving statistical variable.

Even in the simplest case of the quantum statistical mechanics of an ideal gas, we are concerned with the numbers of particles in different one-particle quantum states. Because of the indistinguishability of identical particles we must abandon the idea of defining a microstate by specifying which particle is occupying which of the one-particle states. Instead, a microstate is defined by specifying the number of particles occupying each of the one-particle quantum states. In this case, the number of particles in each one-particle quantum state is a variable, zero or unity in the case of Fermi–Dirac particles and from zero to infinity in the case of Bose–Einstein particles, even though the total number of particles in the entire system is fixed.

3.10 The grand canonical distribution

We can extend the canonical ensemble formulation of Sec. 3.7 to the case in which each member of the ensemble is a physical system characterized by a set of state variables $(N, V, E_i(N, V))$, where N is the number of particles in the system and is different from one system to another, V is the volume of the container, and $E_i(N, V)$ is the energy eigenvalue of the system. Fixing the volume V is essential because it will define the boundary conditions, and, in turn, the boundary conditions and the number of particles together will define all the possible eigenvalues $E_i(N, V)$ of the system Hamiltonian.

The grand canonical ensemble is now constructed. Let us suppose that there is a physical system which is made up of a huge number of particles, contained in a huge box, and isolated from the rest of the universe. Let this super-system be divided into a huge number of geometrically identical boxes, where V is the volume of each box. Since there are no physical boundaries between these boxes, the particles can move and the energy can be transferred from one box to another. Thus the entire system has been divided into many subsystems each contained in a box of volume V. The microstate of the subsystem in one of the boxes will be characterized by a set of variables (N, V, E_i). If we assume one of the systems in the super-gas to be the real system in the laboratory the remainder can be regarded as the heat-and-particle reservoir.

It should be noted that the only difference between the grand ensemble formulation and the canonical ensemble formulation is the way in which the partitioning walls define the volume V of each system in the super-gas. The diathermic walls in the canonical ensemble formulation are now replaced by semipermeable membranes which not only conduct heat but also allow particles through the walls from one box to another.

The super-gas is isolated from the universe, and its energy, E_0, is held at a constant value. The total number, N_0, of particles in the super-gas is also fixed to a constant value. Of course, the number of boxes or the number of subsystems, M_0, is fixed. With these three constraints there are many different quantum states of the super-gas. Again using the assumption of equal *a priori* probabilities, we can construct an ensemble in which all of the eigenstates of the super-gas are equally represented, i.e., by considering the identical replicas of the super-gas, one replica to one eigenstate of the super-gas.

3.11 The grand partition function

Applying the same mathematical formulation as in the case of the canonical ensemble, we can find the probability distribution. The probability of finding a physical system in the state $(N, V, E_i(N, V))$ will be given by

$$P(N, E_i(N, V)) = \exp(-\beta E_i(N, V) - \gamma N)/\Xi(V, \beta, \gamma), \qquad (3.57)$$

where β and γ are the Lagrange multipliers corresponding to the two constraints on the values of the total energy, E_0, and the total number of particles in the super-gas, N_0. The $\Xi(V, \beta, \gamma)$ term appearing in the denominator is the normalization factor; it is called the *grand partition function*,

$$\Xi(V, \beta, \gamma) = \sum_N \sum_i \exp(-\beta E_i(N, V) - \gamma N). \tag{3.58}$$

If we take the logarithmic differentials of both sides of the above equation, we find

$$\frac{1}{\Xi}\delta\Xi(V, \beta, \gamma) = -\frac{1}{\Xi}\sum_N \sum_i E_i(N, V)\exp(-\beta E_i(N, V) - \gamma N)\delta\beta$$

$$-\frac{1}{\Xi}\sum_N \sum_i \frac{\partial E_i(N, V)}{\partial V}\exp(-\beta E_i(N, V) - \gamma N)\delta V$$

$$-\frac{1}{\Xi}\sum_N \sum_i N\exp(-\beta E_i(N, V) - \gamma N)\delta\gamma. \tag{3.59}$$

Then we obtain the following thermodynamic relations:

$$\bar{E} = -\frac{\partial \log \Xi(V, \beta, \gamma)}{\partial\beta}, \quad p = \frac{\partial \log \Xi(V, \beta, \gamma)}{\partial V}, \quad \bar{N} = -\frac{\partial \log \Xi(V, \beta, \gamma)}{\partial\gamma}. \tag{3.60}$$

In the above relations $\bar{E}$ and $\bar{N}$ are the average energy and the average number of particles. Hereinafter, however, we may drop the bars for these quantities.

If we set

$$\Xi = \exp(-\beta\Omega), \tag{3.61}$$

then the logarithmic differentials of both sides of the equation give

$$-\delta(\beta\Omega) = -E\delta\beta + p\beta\delta V - N\delta\gamma$$
$$= -\delta(\beta E) + \beta\delta E - \beta p\delta V - \delta(\gamma N) + \gamma\delta N. \tag{3.62}$$

Transferring two complete differentials on the right hand side to the left hand side of the equation, we obtain

$$\delta(\beta E + \gamma N - \beta\Omega) = \beta(\delta E + p\delta V + (\gamma/\beta)\delta N). \tag{3.63}$$

If we look at this equation as a form of the second law of thermodynamics (see either (1.101), (1.121) or (1.144)),

$$\delta S = \frac{1}{T}\delta'Q = \frac{1}{T}(\delta E + p\delta V - \mu\delta N), \tag{3.64}$$

the absolute temperature is seen to be an integrating denominator which makes the small heat quantity a complete differential, and causes the chemical potential to be the strength parameter controlling the change of the number of particles. Then the following identifications result:

$$\beta = 1/kT, \quad \gamma/\beta = -\mu \tag{3.65}$$

and

$$\beta(E - \mu N - \Omega) = S/k, \tag{3.66}$$

and finally

$$\Omega = -pV. \tag{3.67}$$

The quantity Ω is called the grand potential discussed in Sec. 2.6.

3.12 The ideal quantum gases

Although the formulations in the preceding sections are applicable to any system made up of interacting particles, let us discuss the case of an ideal Fermi gas. The Hamiltonian for this system is given by (Sec. 6.2)

$$H = \sum_i n_i \epsilon_i, \quad N = \sum_i n_i, \tag{3.68}$$

where ϵ_i is the ith one-particle energy eigenvalue defined in a cube of volume V and n_i is the occupation number of the same state. n_i can take only two possible eigenvalues, 0 and 1. The summations in (3.68) are taken over all the accessible one-particle eigenstates. If, for example, there are some internal degrees of freedom they must be included in the definition of the eigenstate. Because of the indistinguishability of the particles the microstate of the entire system is characterized by a set of occupation numbers of different one-particle eigenstates. The microstate cannot be specified by knowing which of the one-particle eigenstates is occupied by which of the particles in the system.

Due to the summation property of the energy eigenvalues over individual one-particle states, the grand partition function becomes a product of the factors, each of which is the summation over the occupation numbers. Hence

$$\Xi = \exp(-\beta\Omega) = \sum_N \sum_{\sum n_i=N} \exp\left[-\sum_i(\beta\epsilon_i + \gamma)n_i\right]$$
$$= \prod_i \sum_{n_i} \exp[-(\beta\epsilon_i + \gamma)n_i]$$
$$= \prod_i [1 \pm \exp(-(\beta\epsilon_i + \gamma))]^{\pm}. \tag{3.69}$$

The transformation from the first line to the second line may need some explanation. The inner summation over the one-particle energy eigenstates must be stipulated in such a way that the sum of the particle numbers is equal to the total number N. The outer summation, however, runs over all values of N from zero to infinity. This means that the first stipulation can be essentially ignored. This process may be seen more explicitly by means of Cauchy's integral formula:

$$
\begin{aligned}
\Xi &= \sum_N \lambda^N \prod_i \frac{1}{2\pi i} \oint x^{-(N+1)} \sum_{n_i} [\exp(-\beta \epsilon_i) x]^{n_i} dx \\
&= \prod_i \frac{1}{2\pi i} \oint (x - \lambda)^{-1} \sum_{n_i} [\exp(-\beta \epsilon_i) x]^{n_i} dx \\
&= \prod_i \sum_{n_i} [\lambda \exp(-\beta \epsilon_i)]^{n_i}, \qquad \lambda = \exp(-\gamma).
\end{aligned}
\tag{3.70}
$$

The total energy of the system, the average number of particles, and the grand potential are found, respectively, as

$$
E = -\frac{\partial \log \Xi}{\partial \beta} = \sum_i \frac{\epsilon_i}{\frac{1}{\lambda} \exp(\beta \epsilon_i) \pm 1},
$$

$$
N = \lambda \frac{\partial \log \Xi}{\partial \lambda} = \sum_i \frac{1}{\frac{1}{\lambda} \exp(\beta \epsilon_i) \pm 1} = \sum_i n_i,
$$

$$
\Omega = -pV = -kT \log \Xi,
$$

$$
pV = \pm kT \sum_i \log[1 \pm \lambda \exp(-\beta \epsilon_i)].
\tag{3.71}
$$

Exercises

3.1 Derive the expression given by (3.22).

[**Hint**]

The plane-wave energy is given by

$$
\epsilon = \frac{\hbar^2 k^2}{2m} = \frac{\hbar^2}{2m} (k_x^2 + k_y^2 + k_z^2) = \frac{\hbar^2}{2m} \left(\frac{2\pi}{L} \right) (n_x^2 + n_y^2 + n_z^2),
\tag{E3.1}
$$

and hence the number of plane-wave states contained in a narrow range of energy given by n_x and $n_x + \Delta n_x, n_y$ ... etc. is given by

$$
\begin{aligned}
g_i &= \Delta n_x \Delta n_y \Delta n_z = \left(\frac{L}{2\pi} \right)^3 dk_x dk_y dk_z \\
&= \frac{V}{(2\pi)^3} 4\pi k^2 dk = \frac{V}{(2\pi)^3} 4\pi \left(\frac{2m}{\hbar^2} \right)^{\frac{3}{2}} \frac{1}{2} \epsilon^{\frac{1}{2}} d\epsilon.
\end{aligned}
\tag{E3.2}
$$

This can be expressed as

$$g_i = g(\epsilon)d\epsilon, \quad g(\epsilon) = 2\pi V \left(\frac{2m}{h^2}\right)^{\frac{3}{2}} \epsilon^{\frac{1}{2}}. \tag{E3.3}$$

$g(\epsilon)$ is called the plane-wave density of states.

The molecular partition function is now expressed as an integral:

$$q(\beta) = \sum_i g_i \exp(-\beta\epsilon_i)$$

$$= 2\pi V (2m/h^2)^{\frac{3}{2}} \int_0^\infty \exp(-\beta\epsilon)\epsilon^{\frac{1}{2}} \, d\epsilon$$

$$= 2\pi V (2m/h^2)^{\frac{3}{2}} (kT)^{\frac{3}{2}} \int_0^\infty \exp(-t)t^{\frac{3}{2}-1} \, dt$$

$$= \pi V (2m/h^2)^{\frac{3}{2}} (kT)^{\frac{3}{2}} \Gamma(\tfrac{1}{2})$$

$$= V (2\pi m/h^2)^{\frac{3}{2}} (kT)^{\frac{3}{2}}. \tag{E3.4}$$

3.2 Show that the variation of (3.29) leads to the Fermi–Dirac distribution function.

3.3 Derive the expression for the entropy (3.56) from the Helmholtz potential (3.54).

[**Hint**]

Since

$$P(E_i) = \exp(-\beta E_i)/Z(\beta), \quad F = -kT \log Z(\beta), \quad E = \sum_i E_i P(E_i), \tag{E3.5}$$

$$S = \frac{1}{T}(E - F)$$

$$= -k \sum_i P(E_i)[-\beta E_i - \log Z(\beta)]$$

$$= -k \sum_i P(E_i) \log P(E_i). \tag{E3.6}$$

4

System Hamiltonians

4.1 Representations of the state vectors

No attempt will be made to develop fully the formulation of quantum mechanics. Only some key definitions and thorems will be reviewed.

4.1.1 Coordinate representation

In quantum mechanics, a complex quantity called the wave function $\psi_a(x)$ is introduced in order to describe a dynamical state of a given system. The subscript a of the function indicates a set of values of physical quantities, or the corresponding quantum numbers, which define the state. The notation x represents a set of values of all variables upon which the wave function depends.

The description of the state by a function of the coordinates is called the *coordinate representation*, although this definition will be given a more rigorous meaning later in this section.

Since the wave function is a complex quantity, only the absolute square, $\psi_a(x)^*\psi_a(x)$, has a direct physical interpretation, representing the statistical probability of finding the system, which is in the state a, in an elementary volume dx at position x in the coordinate space. Because of this probabilistic description of the physical state the absolute square of the wave function is normalized to unity:

$$\int \psi_a(x)^*\psi_a(x)dx = 1. \tag{4.1}$$

This condition is called the *quadratic integrability condition*. Because of this, the wave function $\psi_a(x)$ is given another name, i.e., the *probability amplitude*. The probability amplitude is undetermined by an arbitrary *phase factor*, $\exp(i\phi)$, since both $\psi_a(x)$ and $\psi_a(x)\exp(i\phi)$ give rise to the same probability.

The concept of the average value or the *expectation value* of a physically observable (measurable) quantity, F, in a given state $\psi_a(x)$ also plays an important role. F, which is henceforth called the *observable* is defined, in general, as a function of the cartesian coordinates and their conjugate momenta, and is represented by a linear operator $\hat{F}$. In order to find the linear operator for the observable F, the cartesian coordinates and the conjugate momenta upon which F is defined are replaced by appropriate operators according to the rules called the *quantization*. In the coordinate representation, the rules are, for each degree of freedom,

$$\hat{x} = x, \quad \hat{p} = -i\hbar\frac{\partial}{\partial x}, \tag{4.2}$$

where $\hbar = h/2\pi$, and h is the *quantum of action* called *Planck's constant*.

Once the coordinate and the conjugate momentum are replaced by $\hat{x}$ and $\hat{p}$ they become noncommuting operators satisfying the commutation relation

$$[\hat{x}, \hat{p}] = x\hat{p} - \hat{p}x = i\hbar, \tag{4.3}$$

whereby the square bracket notation is called the *commutator*.

For the sake of notational simplicity, a system of only one degree of freedom is treated in the following formulation; however, the formulae should be translated into many-dimensional coordinate space whenever possible.

The *average value* or the *expectation value* of $\hat{F}$ in a given state, say $\psi_a(x)$, is defined by

$$\langle\hat{F}\rangle = \int \psi_a(x)^* F(\hat{x}, \hat{p})\psi_a(x)\mathrm{d}x. \tag{4.4}$$

For a given operator $\hat{F}$ one can define its *adjoint* (alternatively called the *Hermitian conjugate*) *operator* $\hat{F}^*$ by

$$\int \psi(x)^* \hat{F}^*\phi(x)\mathrm{d}x = \left\{\int \phi(x)^* \hat{F}\psi(x)\mathrm{d}x\right\}^*, \tag{4.5}$$

where ψ and ϕ are arbitrary functions.

Any operator $\hat{F}$ which satisfies the condition

$$\hat{F}^* = \hat{F} \tag{4.6}$$

is called the *self-adjoint* operator or the *Hermitian* operator. It can be shown that the Hermitian operator has a real average value. (The proof is left as an exercise.)

Using the definition of the average value, we can also evaluate the average value of the root mean square deviation from the average value in a given state $\psi_a(x)$. For an Hermitian operator $\hat{F}$ the deviation

$$\hat{\Delta F} = \hat{F} - \langle\hat{F}\rangle \tag{4.7}$$

is also an Hermitian operator.

The average value of $(\hat{\Delta F})^2$ is found to be

$$\langle (\hat{\Delta F})^2 \rangle = \int \psi_a(x)^*(\hat{\Delta F})(\hat{\Delta F})\psi_a(x)\mathrm{d}x. \qquad (4.8)$$

Since $\hat{\Delta F}$ is Hermitian, we can write

$$\langle (\hat{\Delta F})^2 \rangle = \int |(\hat{\Delta F})\psi_a(x)|^2\mathrm{d}x. \qquad (4.9)$$

If the above definition is extended, we can find a dynamical state $u(x)$ for which the average value of the mean square deviation vanishes, i.e., the state for which the physical quantity F has a well defined value. For such a state the above equation reduces to

$$0 = \int |(\hat{\Delta F})u(x)|^2\mathrm{d}x. \qquad (4.10)$$

Since the integrand of this equation is positive definite, the condition is satisfied only if

$$\hat{F}u(x) = Fu(x), \qquad (4.11)$$

where the F on the right hand side is a real number. In order for the solution $u(x)$ of this equation to represent a physical state, the function $u(x)$ must be interpreted as the probability amplitude, i.e., it should satisfy the *quadratic integrability condition*:

$$\int u(x)^*u(x)\mathrm{d}x = 1. \qquad (4.12)$$

In general, when the quadratic integrability condition is imposed, the solution of (4.11) exists only for a set of certain well defined values of the parameter $F = F_n$, and $u_n(x)$, $n = 1, 2, \ldots$ Such a value, F_n, and the corresponding solution, $u_n(x)$,

$$\hat{F}u_n(x) = F_nu_n(x), \qquad (4.13)$$

are called the *eigenvalue* and *eigenfunction*, respectively, of the operator $\hat{F}$. The eigenvalues can form either a discrete denumerably infinite set or a continuous non-denumerably infinite set, or a combination of both, depending upon the situation.

The eigenfunctions of an observable which are orthonormalized to unity by the quadratic integrability condition (4.12) can be used as a set of basis vectors in order to define an abstract Hilbert space. The Hilbert space thus defined is, as a matter of course, conceptually different from the many-dimensional *x*-coordinate space.

Let us, in the following, examine what happens when the coordinate operator $\hat{x}$ is chosen as the above-mentioned observable. Since the coordinate operator $\hat{x}$ itself is an Hermitian operator, we can set

$$\hat{x}X(x, x_0) = x_0X(x, x_0), \qquad (4.14)$$

where x_0 and $X(x, x_0)$ are the eigenvalue and eigenfunction of the operator $\hat{x}$.

According to (4.4),

$$\langle x \rangle = \int X(x, x_0)^* \hat{x} X(x, x_0) \, dx$$

$$= x_0 \int X(x, x_0)^* X(x, x_0) \, dx = x_0. \tag{4.15}$$

Both (4.14) and (4.15) are satisfied by the *Dirac delta function*

$$X(x, x_0) = \delta(x - x_0). \tag{4.16}$$

The orthogonalization between $\delta(x - x_1)$ and $\delta(x - x_0)$ is expressed by

$$\int \delta(x - x_1)\delta(x - x_0)dx = \delta(x_1 - x_0). \tag{4.17}$$

The right hand side of the equation is zero as long as x_0 and x_1 are different, no matter how close together they are. The condition is called the *delta function orthonormalization*.

As we have seen in the above, all the eigenfunctions, $X(x, x_0)$, of the coordinate operator are orthonormal to one another in the Hilbert space, and therefore they can be looked at as a set of basis vectors.

In the case of functions defined in many-dimensional *x*-space the delta function must be generalized to the multidimensional delta function

$$\delta(\mathbf{x} - \mathbf{x}_0) = \delta(x_1 - x_{10}) \cdots \delta(x_n - x_{n0}). \tag{4.18}$$

According to Dirac (1958), any state a of the system can be represented by an entity called the *state vector* or the *ket* vector denoted by $|a\rangle$ or $|\psi_a\rangle$ defined in the abstract Hilbert space. To each *ket* vector one can assign a *dual bra* state vector, denoted by the symbol $\langle a|$ and connected with the *ket* vector through the relation $\langle a| = |a\rangle^*$. Any state of a dynamical system can then be described either by a *ket* vector or by a *bra* vector. The *ket* and *bra* vectors have, therefore, a different nature and cannot be added together.

The scalar product of two *ket* vectors $|a\rangle$ and $|b\rangle$ is denoted by the bracket $\langle b|a\rangle$. The scalar product is formed by multiplying the *ket* vector $|a\rangle$ by the *bra* vector which is the dual of the *ket* vector $|b\rangle$ and usually integrated over the entire coordinate space *x*. The scalar product $\langle b|a\rangle$ is a complex number and satisfies $\langle b|a\rangle = \langle a|b\rangle^*$.

Let us now, for the sake of demonstration, find the scalar product of a state vector $|\psi_a\rangle$ with one of the eigenfunctions of the coordinate operator, say $|x_0\rangle$, which is also a vector in the Hilbert space:

$$\langle x_0|\psi_a\rangle = \int \delta(x - x_0)\psi_a(x) \, dx = \psi_a(x_0). \tag{4.19}$$

This equation has a rather interesting implication, i.e., the wave function $\psi_a(x)$ can be regarded as a component of the *ket* vector $|\psi_a\rangle$ in the direction of x in the abstract *Hilbert space*. If the wave function is given such an interpretation, then the *ket* vector $|\psi_a\rangle$ has nondenumerably infinite number of components, which are denoted by the *brackets*:

$$\langle x|\psi_a\rangle = \psi_a(x), \tag{4.20}$$

where the coordinate x takes on continuous values. The coordinate representation, $\psi_a(x)$, of the *ket* vector $|\psi_a\rangle$ must, therefore, be regarded as a collection of infinitely many components (wave functions) $\langle x|\psi_a\rangle$ of the state vector.

There is a rather simple yet interesting application of the eigenfunctions of the coordinate operator $\hat{x}$. For this purpose, let us expand the coordinate eigenfunction $\langle x|x_1\rangle$ in terms of the eigenfunctions $u_n(x)$ of the operator $\hat{F}$:

$$\langle x|x_1\rangle = \sum_n \langle x|u_n\rangle\langle u_n|x_1\rangle. \tag{4.21}$$

Construct the scalar product of this vector with another eigenfunction of the coordinate $|x_2\rangle$,

$$\langle x_2|x_1\rangle = \int \langle x_2|x\rangle\langle x|x_1\rangle dx = \sum_n \langle x_2|u_n\rangle\langle u_n|x_1\rangle. \tag{4.22}$$

Because of the orthonormality of the eigenfunctions of the coordinate $\hat{x}$, the left hand side of the equation is equal to the delta function, and hence we find

$$\delta(x_2 - x_1) = \sum_n \langle x_2|u_n\rangle\langle u_n|x_1\rangle, \tag{4.23}$$

the condition known as the completeness condition for any set of orthonormal eigenfunctions $\{|u_n\rangle\}$.

4.1.2 The momentum representation

The basis functions in the momentum representation are the eigenfunctions of the momentum operator,

$$\hat{p}|p\rangle = -i\hbar\frac{\partial}{\partial x}|p\rangle = p|p\rangle. \tag{4.24}$$

This defines the *ket* in the coordinate representation:

$$\langle x|p\rangle = (2\pi\hbar)^{-\frac{1}{2}}\exp(ipx/\hbar), \tag{4.25}$$

which satisfies the delta function orthonormalization,

$$\int \langle p'|x\rangle\langle x|p\rangle \mathrm{d}x = \langle p'|p\rangle = \delta(p'-p). \tag{4.26}$$

Expanding the wave function of the state $|a\rangle$ in terms of the complete set of the momentum eigenfunctions (basis functions), we find

$$\langle x|a\rangle = \int \langle x|p\rangle\langle p|a\rangle \mathrm{d}p, \tag{4.27}$$

where $\langle p|a\rangle = \psi_a(p)$ is the coefficient of expansion and determines the state vector $|a\rangle$ in the momentum representation. The absolute square of these functions is the probability density in the momentum representation,

$$\rho(p) = |\langle p|a\rangle|^2 = |\psi_a(p)|^2. \tag{4.28}$$

The transformation which is the inverse of (4.24) is found by

$$\langle p|a\rangle = \int \mathrm{d}x\,\langle p|x\rangle\langle x|a\rangle. \tag{4.29}$$

When the explicit form (4.25) is inserted into (4.27) and (4.29), those equations take the form of the Fourier transforms:

$$\langle x|a\rangle = (2\pi\hbar)^{-\frac{1}{2}}\int \exp(\mathrm{i}px/\hbar)\langle p|a\rangle \mathrm{d}p,$$

$$\langle p|a\rangle = (2\pi\hbar)^{-\frac{1}{2}}\int \exp(-\mathrm{i}px/\hbar)\langle x|a\rangle \mathrm{d}x. \tag{4.30}$$

The representation in which the orthonormal set of functions (4.25) is chosen as the basis kets is called the *momentum representation*.

4.1.3 The eigenrepresentation

Neither the coordinate representation nor the momentum representation is the most convenient one because of the delta function nature of the basis functions. In many cases the set of eigenfunctions of the system Hamiltonian or any similar Hermitian operator which has a set of discrete eigenvalues is employed as the basis of representation. Such a representation is called the *eigenrepresentation* or more specifically the *energy representation*, if the system Hamiltonian is employed.

Let the eigenfunctions be denoted by

$$u_n(x) = \langle x|E_n\rangle. \tag{4.31}$$

The basis function itself is given in the coordinate representation at this stage. We use the notation

$$u_n^*(x) = \langle E_n | x \rangle = \langle x | E_n \rangle^*.$$ (4.32)

The orthonormality of the eigenfunctions may, then, be written in the form

$$\int u_m^*(x) u_n(x)\, dx = \int \langle E_m | x \rangle \langle x | E_n \rangle\, dx = \langle E_m | E_n \rangle = \delta_{mn}.$$ (4.33)

In order to change the representation of the state vector $|a\rangle$ from the coordinate representation to the eigenrepresentation, the function is expanded in terms of the basis functions $u_n(x)$,

$$\psi_a(x) = \sum_n u_n(x) \phi_a(E_n),$$ (4.34)

where $\phi_a(E_n)$ is the coefficient of expansion given by

$$\phi_a(E_n) = \int u_n(x)^* \psi_a(x)\, dx = \langle E_n | a \rangle.$$ (4.35)

In terms of the bracket notation, the expansion (4.34) is seen as

$$\langle x | a \rangle = \sum_n \langle x | E_n \rangle \langle E_n | a \rangle.$$ (4.36)

Since the state vector in the coordinate representation is normalized, the same is true for the function in the new representation. This is clear from the following:

$$1 = \int dx\, \langle a | x \rangle \langle x | a \rangle$$

$$= \sum_m \sum_n \int \langle a | E_m \rangle \langle E_m | x \rangle \langle x | E_n \rangle \langle E_n | a \rangle\, dx.$$ (4.37)

If the orthonormality condition (4.33) is employed,

$$\sum_n \langle a | E_n \rangle \langle E_n | a \rangle = \sum_n |\phi_a(E_n)|^2 = 1.$$ (4.38)

The collection of the entire set of complex numbers,

$$\langle E_1 | a \rangle, \langle E_2 | a \rangle, \langle E_3 | a \rangle, \ldots,$$ (4.39)

is called the energy representation of the state vector $|a\rangle$, and each member $\langle E_i | a \rangle$ may be called the wave function in the energy representation. Equation (4.38) indicates the fact that the state vector $|a\rangle$ is normalized also in the eigenrepresentation. The state vector normalization is independent of the representation.

4.2 The unitary transformation

Since there are many choices of the basic Hermitian operators other than the system Hamiltonian, there are many different representations possible. To find the most convenient representation for a given state vector $|a\rangle$, we often try to transform from one representation to another. Suppose we have selected two Hermitian operators E and F, with sets of eigenfunctions $u_n(x)$ and $v_\alpha(x)$, respectively,

$$\hat{E}\langle x|u_n\rangle = E_n\langle x|u_n\rangle, \quad n, m, \ldots,$$
$$\hat{F}\langle x|v_\alpha\rangle = F_\alpha\langle x|v_\alpha\rangle, \quad \alpha, \beta, \ldots, \tag{4.40}$$

where both basis functions are orthonormalized,

$$\int \langle u_m|x\rangle\langle x|u_n\rangle \, \mathrm{d}x = \delta_{mn}, \quad \int \langle v_\alpha|x\rangle\langle x|v_\beta\rangle \mathrm{d}x = \delta_{\alpha\beta}. \tag{4.41}$$

We now attempt a transformation from one set of basis functions to the other. The transformation between two sets of orthonormal basis functions, where the functions are complex numbers, is called the *unitary transformation* in the Hilbert space.

Let us expand the u functions in terms of the v functions:

$$\langle x|n\rangle = \sum_\alpha \langle x|\alpha\rangle\langle\alpha|n\rangle, \quad |n\rangle \equiv |u_n\rangle, \quad |\alpha\rangle \equiv |v_\alpha\rangle, \tag{4.42}$$

where $\langle\alpha|n\rangle$ is the coefficient of expansion called the $\alpha - n$ element of transformation matrix given by

$$\langle\alpha|n\rangle = \int \langle\alpha|x\rangle\langle x|n\rangle\mathrm{d}x. \tag{4.43}$$

Similarly the inverse transform of (4.42) is

$$\langle x|\alpha\rangle = \sum_n \langle x|n\rangle\langle n|\alpha\rangle, \tag{4.44}$$

which gives the coefficient of expansion as

$$\langle n|\alpha\rangle = \int \langle n|x\rangle\langle x|\alpha\rangle\mathrm{d}x$$
$$= \left(\int \langle\alpha|x\rangle\langle x|n\rangle\mathrm{d}x\right)^*$$
$$= \langle\alpha|n\rangle^*. \tag{4.45}$$

Because of the orthonormality of the basis functions, the following conditions for the elements of the transformation matrix hold:

$$\delta_{mn} = \int \langle m|x\rangle\langle x|n\rangle dx$$

$$= \sum_\alpha \sum_\beta \int \langle m|\alpha\rangle\langle \alpha|x\rangle\langle x|\beta\rangle\langle \beta|n\rangle dx$$

$$= \sum_\alpha \sum_\beta \langle m|\alpha\rangle\delta_{\alpha\beta}\langle \beta|n\rangle$$

$$= \sum_\alpha \langle m|\alpha\rangle\langle \alpha|n\rangle,$$

$$\delta_{\alpha\beta} = \int \langle \alpha|x\rangle\langle x|\beta\rangle dx$$

$$= \sum_m \sum_n \int \langle \alpha|m\rangle\langle m|x\rangle\langle x|n\rangle\langle n|\beta\rangle dx$$

$$= \sum_m \sum_n \langle \alpha|m\rangle\delta_{mn}\langle n|\beta\rangle$$

$$= \sum_n \langle \alpha|n\rangle\langle n|\beta\rangle. \tag{4.46}$$

The matrix for which the elements satisfy the above properties is called the *unitary matrix*.

4.3 Representations of operators

In Sec. 4.1, we introduced the coordinate representation of linear operators. We will now examine more general representations of the linear operators.

Let us assume that there is a set of orthonormal basis functions $|n\rangle$, $|m\rangle$,..... These $|n\rangle$'s are the eigenfunctions of a certain Hermitian operator whose nature does not have to be identified for the present situation.

Now consider an Hermitian operator or an observable $\hat{F} = F(\hat{x}, \hat{p})$. Since $|n\rangle$'s are not the eigenfunctions of $\hat{F}$, both the average values (diagonal elements) F_{nn} and the off-diagonal elements F_{mn} can be found:

$$F_{nn} = \int \langle n|x\rangle F(\hat{x}, \hat{p})\langle x|n\rangle dx, \tag{4.47}$$

and

$$F_{mn} = \int \langle m|x\rangle F(\hat{x}, \hat{p})\langle x|n\rangle dx. \tag{4.48}$$

It can be easily seen that the above two definitions are equivalent with the following representation of the operator $\hat{F}$:

$$\hat{F}(x, p) = F(\hat{x}, \hat{p}) = \sum_m \sum_n |m\rangle F_{mn} \langle n|. \tag{4.49}$$

This is called the $|n\rangle$-*representation* of the operator $\hat{F}$. An important feature of this representation is that it is valid with any basis set of eigenfunctions, $|n\rangle$.

In particular we can define a unit operator or an identity operator by

$$\hat{1} = \sum_n |n\rangle\langle n| = \int |x\rangle\langle x| dx. \tag{4.50}$$

The coordinate representation of this relation is simply the completeness condition (4.23), i.e.,

$$\delta(x_1 - x_2) = \sum_n \langle x_1|n\rangle\langle n|x_2\rangle = \int \langle x_1|x\rangle\langle x|x_2\rangle dx. \tag{4.51}$$

The identity operator has an interesting and most convenient feature:

$$\langle u|u\rangle = \int \langle u|x\rangle\langle x|u\rangle dx = 1, \tag{4.52}$$

which is equivalent to (4.12);

$$\langle x|a\rangle = \sum_n \langle x|n\rangle\langle n|a\rangle, \tag{4.53}$$

which is equivalent to (4.34);

$$\langle a|a\rangle = \sum_n \langle a|E_n\rangle\langle E_n|a\rangle = |\phi_a(E_n)|^2 = 1, \tag{4.54}$$

which is equivalent to (4.38);

$$\langle x|n\rangle = \langle x|\alpha\rangle\langle \alpha|n\rangle, \tag{4.55}$$

which is equivalent to (4.42).

4.4 Number representation for the harmonic oscillator

The classical Hamiltonian for the linear harmonic oscillator is given by

$$H_{\text{osc}} = \frac{p^2}{2m} + \frac{K}{2}x^2, \tag{4.56}$$

where p and x are the canonically conjugate momentum and coordinate, and m and K are the mass of the oscillator and the constant of the restoring force, respectively.

In quantum mechanics, x and p are replaced by the operators in the coordinate representation, i.e.,

$$\hat{x} = x, \quad \hat{p} = -i\hbar \frac{\partial}{\partial x}. \tag{4.57}$$

The $\hat{x}$ and $\hat{p}$ satisfy the well known commutation relation:

$$[\hat{x}, \hat{p}] = i\hbar. \tag{4.58}$$

Then, the quantized Hamiltonian is given by

$$\hat{H} = -\frac{\hbar^2}{2m}\frac{d^2}{dx^2} + \frac{K}{2}\hat{x}^2. \tag{4.59}$$

For the sake of mathematical convenience, a dimensionless coordinate ξ is introduced;

$$\hat{\xi} = \hat{x}\left(\frac{m\omega}{\hbar}\right)^{\frac{1}{2}}, \tag{4.60}$$

where $\omega = (K/m)^{\frac{1}{2}}$, and the Hamiltonian is cast into the following form:

$$\hat{H} = \frac{1}{2}\hbar\omega\left(\hat{\xi}^2 - \frac{d^2}{d\xi^2}\right). \tag{4.61}$$

Instead of the coordinate and the momentum operators, $\hat{\xi}$ and $\hat{p}_\xi = -i\partial/\partial\xi$, we can introduce a new set of operators defined by

$$\hat{a} = \frac{1}{\sqrt{2}}(\hat{\xi} + i\hat{p}_\xi) = \frac{1}{\sqrt{2}}\left(\hat{\xi} + \frac{\partial}{\partial\xi}\right), \tag{4.62}$$

$$\hat{a}^* = \frac{1}{\sqrt{2}}(\hat{\xi} - i\hat{p}_\xi) = \frac{1}{\sqrt{2}}\left(\hat{\xi} - \frac{\partial}{\partial\xi}\right). \tag{4.63}$$

These operators satisfy the following commutation relation:

$$[\hat{a}, \hat{a}^*] = \hat{a}\hat{a}^* - \hat{a}^*\hat{a} = 1. \tag{4.64}$$

In terms of these new operators the Hamiltonian is cast into the following form:

$$\hat{H} = \tfrac{1}{2}\hbar\omega(\hat{a}\hat{a}^* + \hat{a}^*\hat{a}) = \hbar\omega(\hat{a}^*\hat{a} + \tfrac{1}{2}). \tag{4.65}$$

Let us now introduce another operator $\hat{n}$ defined by

$$\hat{n} = \hat{a}^*\hat{a}, \quad \text{and} \quad \hat{H} = \hbar\omega(\hat{n} + \tfrac{1}{2}). \tag{4.66}$$

The eigenvalues of this new operator may be found by the following rather heuristic approach. Let us assume that one of the eigenfunctions, $|n'\rangle$, and the corresponding eigenvalue, n', of $\hat{n}$ are known, i.e.,

$$\hat{n}|n'\rangle = n'|n'\rangle. \tag{4.67}$$

Firstly, it can be seen that the eigenvalue n' is a non-negative number. We have

$$\langle n'|\hat{n}|n'\rangle = n'\langle n'|n'\rangle, \tag{4.68}$$

and

$$\langle n'|\hat{n}|n'\rangle = \langle n'|\hat{a}^*\hat{a}|n'\rangle = |(\hat{a}|n'\rangle)|^2\rangle 0. \tag{4.69}$$

It should be noted that $\langle n'|n'\rangle$ is positive, and from (4.69) we find that $\langle n'|\hat{n}|n'\rangle$ is non-negative. On combining the two pieces of information we conclude that

$$n' \geq 0. \tag{4.70}$$

$n' = 0$ only if the *ket* $|\Xi\rangle = \hat{a}|n'\rangle = 0$.

Secondly, by means of the commutation relation (4.64) the following identities will be established [Exercise 4.1]:

$$[\hat{n}, \hat{a}] = -\hat{a}, \quad \text{and} \quad [\hat{n}, \hat{a}^*] = \hat{a}^*. \tag{4.71}$$

If these operator identities operate on the eigenket $|n'\rangle$, we find

$$[\hat{n}, \hat{a}]|n'\rangle = -\hat{a}|n'\rangle, \quad \Rightarrow \quad \hat{n}\hat{a}|n'\rangle = (n' - 1)\hat{a}|n'\rangle, \tag{4.72}$$

$$[\hat{n}, \hat{a}^*]|n'\rangle = \hat{a}^*|n'\rangle, \quad \Rightarrow \quad \hat{n}\hat{a}^*|n'\rangle = (n' + 1)\hat{a}^*|n'\rangle, \tag{4.73}$$

and (4.72) states that if $|n'\rangle$ is an eigenket of $\hat{n}$ belonging to the eigenvalue n', then $\hat{a}|n'\rangle$ is another eigenket of $\hat{n}$ for which the eigenvalue is $n' - 1$. In this way, a hierarchy of eigenkets whose eigenvalues are successively smaller, each time by unity, will be generated by successive application of the operator $\hat{a}$ from the left onto the $|n'\rangle$. A similar situation is seen in (4.73) except that the eigenvalues are successively increased, each time by unity, when the operator $\hat{a}^*$ is applied. The operators $\hat{a}$ and $\hat{a}^*$ are called the *annihilation* and *creation* operators, respectively.

Since we are dealing with the harmonic oscillator, the eigenvalue, and hence the excitation energy, will not be bound from above; however, the excitation energy cannot be decreased indefinitely, and hence there must be a state $|0\rangle$ below which there is no state. This implies that

$$|\Xi\rangle = \hat{a}|0\rangle = 0. \tag{4.74}$$

Thus, we can conclude that the ground state (the lowest energy state) $|0\rangle$ is the state for which the eigenvalue of $\hat{n}$ is zero (see (4.70)).

$$\hat{n}|0\rangle = 0|0\rangle, \tag{4.75}$$

and all the higher eigenvalues of $\hat{n}$ are the successive positive integers.

For the sake of illustration, let us find the ground state wave function, $|0\rangle$. This will be the solution of the differential equation

$$\hat{a}|\phi_0\rangle = 0, \quad |\phi_0\rangle = |0\rangle. \tag{4.76}$$

From (4.62),

$$\xi\phi_0 + \frac{d}{d\xi}\phi_0 = 0 \quad \Rightarrow \quad \phi_0 = \frac{1}{\sqrt{[4]\pi}}\exp\left(-\tfrac{1}{2}\xi^2\right). \tag{4.77}$$

From (4.72) and (4.73) [Exercise 4.2],

$$\hat{a}|n\rangle = n^{\frac{1}{2}}|n-1\rangle, \quad \hat{a}^*|n\rangle = (n+1)^{\frac{1}{2}}|(n+1)\rangle. \tag{4.78}$$

We can also show [Exercise 4.3]

$$|n\rangle = \frac{1}{(n!)^{\frac{1}{2}}}(\hat{a}^*)^n|0\rangle. \tag{4.79}$$

Going back to the Hamiltonian (4.66) the energy eigenvalues are

$$E_n = \hbar\omega\left(n + \tfrac{1}{2}\right) \quad n = 0, 1, 2, \ldots. \tag{4.80}$$

In this formulation, the nth eigenstate is characterized by its energy eigenvalue having n units of excitation of magnitude $\hbar\omega$. This is interpreted as the nth eigenstate having n identical energy quanta, each of magnitude $\hbar\omega$. These energy quanta are called the *phonons* of the harmonic oscillator.

Those energy quanta or particles which, in a generalized sense, can be excited any number of times in the same energy state are called *bosons* or particles obeying *Bose–Einstein statistics*. The characteristics of the boson-type excitations are one of the consequences of the commutation relation (4.64).

The entire set of the eigenkets

$$|0\rangle, |1\rangle, |2\rangle, |3\rangle, \ldots, \tag{4.81}$$

is used as the set of orthonormal basis kets of representation. This is the occupation number representation or simply the number representation. The operators $\hat{n}$, $\hat{a}$, and $\hat{a}^*$ are defined in the number representation.

It would be an interesting exercise to calculate, for instance, $\langle 0|\hat{x}^4|0\rangle$.

4.5 Coupled oscillators: the linear chain

Suppose that not just one simple harmonic oscillator but a whole assembly of vibrating atoms interact with each other through nearest neighbors in a one-dimensional chain with an equal spacing of length a.

We set the coordinate axis along the chain and denote the displacement of the nth mass from its equilibrium position by x_n. We find the classical Hamiltonian of the vibrating masses in the harmonic approximation to be given by

$$H = \frac{1}{2m} \sum_n p_n^2 + \frac{1}{2} K \sum_n (x_n - x_{n-1})^2, \tag{4.82}$$

where p_n is the canonical conjugate momentum of the coordinate x_n, g is the constant of the restoring force of the linear chain, and μ is the mass of the vibrating atoms making up the chain. To simplify the calculations we introduce periodic boundary conditions

$$x_n = x_{n+N}. \tag{4.83}$$

The classical theory of such a system is well known. Let us work it out in quantum mechanics, starting with the quantization by replacing the coordinates and canonical momenta by the corresponding operators satisfying the commutation relations,

$$[\hat{x}_m, \hat{p}_n] = i\hbar \delta_{mn}. \tag{4.84}$$

The commutation relation introduced in (4.58) is now extended to the system of many degrees of freedom, and an assertion is made that different degrees of freedom do not interfere with one another.

To diagonalize the Hamiltonian, we introduce a Fourier transformation to find the momentum representation of the operators

$$\hat{u}_k = \frac{1}{\sqrt{N}} \sum_n \exp(ikna)\hat{x}_n \quad \hat{p}_k = (N)^{-\frac{1}{2}} \sum_n \exp(-ikna)\hat{p}_n, \tag{4.85}$$

where k is the wave number given, by virtue of the periodic boundary conditions (4.83), by

$$k = \frac{2\pi}{Na}n, \quad -\tfrac{1}{2}N \le n \le \tfrac{1}{2}N. \tag{4.86}$$

When N happens to be an even integer, one of the boundary values in (4.86) should be excluded.

These new operators in the momentum representation are found to satisfy the following commutation relations:

$$[\hat{u}_k, \hat{p}_{k'}] = \frac{1}{N} \sum_{n,n'} \exp i(kn - k'n')a[\hat{x}_n, \hat{p}_{n'}]$$

$$= \frac{1}{N} \sum_{n,n'} \exp i(kn - k'n')a\, i\hbar\delta_{nn'}$$

$$= i\hbar \frac{1}{N} \sum_n \exp i(k - k')na$$

$$= i\hbar\delta_{k,k'}. \tag{4.87}$$

Thus, the new coordinate and momentum operators are canonically conjugate.

The transformed coordinates and momenta turn out not to be Hermitian, and they satisfy somewhat more complicated conjugation relations,

$$\hat{u}_k^* = (N)^{-\frac{1}{2}} \sum_n \exp(-ikna)\hat{x}_n = \hat{u}_{-k},$$

$$\hat{p}_k^* = \hat{p}_{-k}. \tag{4.88}$$

This does not present a problem, because both the products $\hat{p}_k^* \hat{p}_k$ and $\hat{u}_k^* \hat{u}_k$ are Hermitian operators.

The transformations are now introduced into the Hamiltonian (4.82), and we find

$$\hat{H} = \frac{1}{2} \sum_k \left[\frac{1}{m} \hat{p}_k^* \hat{p}_k + m\omega_k^2 \hat{u}_k^* \hat{u}_k \right], \tag{4.89}$$

where

$$\omega_k = \left[\frac{2K(1 - \cos ka)}{m} \right]^{\frac{1}{2}} = 2(K/m)^{\frac{1}{2}} \sin \frac{1}{2}|ka|. \tag{4.90}$$

As in the case of the single harmonic oscillator in Sec. 4.4, the Hamiltonian is now diagonalized by the following transformations:

$$\hat{a}_k = (2\hbar\omega_k\mu)^{-\frac{1}{2}}(\hat{p}_k - i\mu\omega_k\hat{u}_k^*)$$

$$\hat{a}_k^* = (2\hbar\omega_k\mu)^{-\frac{1}{2}}(\hat{p}_k^* + i\mu\omega_k\hat{u}_k), \tag{4.91}$$

which can be seen to satisfy the commutation relations

$$[\hat{a}_k, \hat{a}_{k'}^*] = \delta_{k,k'}, \quad [\hat{a}_k, \hat{a}_{k'}] = 0, \quad [\hat{a}_k^*, \hat{a}_{k'}^*] = 0. \tag{4.92}$$

After some manipulations, the Hamiltonian can be given in the form:

$$\hat{H} = \frac{1}{2} \sum_k \hbar\omega(\hat{a}_k^*\hat{a}_k + \hat{a}_k\hat{a}_k^*) = \sum_k \hbar\omega(\hat{a}_k^*\hat{a}_k + \frac{1}{2}). \tag{4.93}$$

From this point on, the argument is more or less the same as with the single harmonic oscillator. There are more phonons distributed over the spectrum given by the dispersion relation (4.90).

If we introduce the phonon numbers $v_k = 0, 1, 2, \ldots$, i.e., the occupation numbers of the quantum states of different oscillators, the wave functions of the various vibrational states in the number representation will be given by

$$| \cdots v_j \cdots v_k \cdots \rangle, \tag{4.94}$$

or, if one defines the vacuum state or the ground state of the system by

$$|0\rangle = | \cdots 0 \cdots 0 \cdots \rangle, \tag{4.95}$$

then

$$| \cdots v_j \cdots v_k \cdots \rangle = (\cdots v_j! \cdots v_k! \cdots)^{-\frac{1}{2}}(\cdots (\hat{a}_j^*)^{v_j} \cdots (\hat{a}_k^*)^{v_k} \cdots)|0\rangle. \tag{4.96}$$

The phonon excitation of the vibrating linear chain is another example of the system of boson particles, although they are not material particles.

4.6 The second quantization for bosons

There are many systems in nature made up of identical bosons (identical particles obeying Bose–Einstein statistics). One of the most notable examples of such a system is liquid helium-4. Others include nonmaterial particles such as phonons in crystals, photons in a radiation field, the spin-waves in insulating ferromagnetic crystals, and so on. As long as the high energy phenomena are excluded, the bosons are assumed to be made up of an even number combination of electrons, protons, and neutrons. The electron, proton, and neutron are themselves fermions if they exist individually. The hydrogen atom, hydrogen molecule, and the helium-4 atom are typical examples of bosons, and the deuteron atom and the helium-3 atom are examples of fermions. A system made up of identical fermions behaves differently from that comprising bosons and will be treated later. The reason that the system of hydrogen molecules does not show any peculiar phenomena, unlike liquid helium-4, is because the intermolecular interaction of hydrogen molecules is rather strong and the system becomes a solid at about 14 K, before the so-called Bose–Einstein condensation can set in.

In this section we shall try to find the Hamiltonian for the system of bosons in the occupation number representation, so that the dynamical state of the system can be characterized by the wave functions in the number representation

$$\Psi = | \ldots, n_i, \ldots, n_j, \ldots, n_k, \ldots \rangle, \tag{4.97}$$

where the state Ψ means that the one-particle state $|i\rangle$ is occupied by n_i particles, $|j\rangle$ is occupied by n_j particles, and so on. One should not conclude from this that the system is made up only of noninteracting particles; rather, a complete set of those functions can form a basis set in the Hilbert space so that the dynamical state of the system of interacting particles can be represented by a certain linear combination of those basis functions. Furthermore, we are trying to describe these different states without writing down the state functions explicitly.

For this purpose, let us introduce a set of one-particle orthonormalized eigenfunctions defined in a cube of side length L with periodic boundary conditions. These functions are the *plane-wave eigenfunctions*;

$$\langle r|k\rangle = V^{-\frac{1}{2}} \exp(ik \cdot r), \tag{4.98}$$

where $V = L^3$ is the volume of the cube and

$$k = (k_x, k_y, k_y) = \frac{2\pi}{L}(n_1, n_2, n_3), \quad r = (x, y, x), \tag{4.99}$$

where n_1, n_2, n_3 take the integer values $0, 1, 2, \ldots, \infty$.

We assume that the one-particle plane-wave states can be occupied by any number of bosons just as in the case of the phonons in the vibrating chain. For this purpose, as before, the creation, $\hat{a}_k^*$, and annihilation, $\hat{a}_k$, operators, which satisfy the following commutation relations, are introduced:

$$[\hat{a}_k, \hat{a}_{k'}^*] = \delta_{kk'}, \quad [\hat{a}_k^*, \hat{a}_{k'}^*] = 0, \quad [\hat{a}_k, \hat{a}_{k'}] = 0, \tag{4.100}$$

except for generalization to many single-particle states.

The properties of the operators for each one-particle state have been established as

$$\begin{aligned} &\hat{a}_k|n\rangle = n^{\frac{1}{2}}|(n-1)\rangle, \hat{a}_k^*|n\rangle = (n+1)^{\frac{1}{2}}|(n+1)\rangle, \\ &\hat{a}_k^*\hat{a}_k|n\rangle = n^{\frac{1}{2}}\hat{a}_k^*|n-1\rangle = n|n\rangle, \\ &\hat{a}_k\hat{a}_k^*|n\rangle = (n+1)^{\frac{1}{2}}\hat{a}_k|(n+1)\rangle = (n+1)|n\rangle, \\ &\hat{n}|n\rangle = n|n\rangle, \end{aligned} \tag{4.101}$$

and hence $\hat{n}_k = \hat{a}_k^*\hat{a}_k$ is the occupation number operator in the plane-wave state $\hat{a}_k^*|0\rangle = \langle r|k\rangle$ as before.

In this number representation, the basis function Ψ takes the form

$$\Psi = \cdots \hat{a}_{k_i}^* \cdots \hat{a}_{k_j}^* \cdots |0\rangle, \tag{4.102}$$

where $|0\rangle$ is the vacuum state and the operators $\hat{a}_{k_i}^*$ appear any number of times. Due to the commutation relations (4.100) in the above, the creation operators appearing next to each other in the basis function can be transposed (exchanged) and hence any two creation operators, no matter how far apart in the product, can be interchanged without causing any change in the wave function. Such a wave function is called the *symmetrized wave function*.

For the system of noninteracting particles, therefore, the Hamiltonian operator can be expressed in terms of the occupation number operators,

$$\hat{H}_0 = \sum_k \epsilon_k \hat{n}_k, \quad \epsilon_k = \frac{\hbar^2 k^2}{2m}, \tag{4.103}$$

where $k^2 = k_x^2 + k_y^2 + k_z^2$ and m is the mass of the particle.

We also introduce the so-called *field operators* and the density operator, defined in terms of the creation and annihilation operators, by

$$\hat{\psi}(r)^* = \sum_k \hat{a}_k^* \langle k | r \rangle,$$

$$\hat{\psi}(r) = \sum_k \langle r | k \rangle \hat{a}_k,$$

$$\hat{\rho}(r) = \hat{\psi}(r)^* \hat{\psi}(r). \tag{4.104}$$

By virtue of the commutation relations for the $\hat{a}$-operators, the following commutation relation for the field operators is obtained:

$$\hat{\psi}(r)\hat{\psi}(r')^* - \hat{\psi}(r')^*\hat{\psi}(r)$$

$$= \frac{1}{V} \sum_{kk'} (\hat{a}_k \hat{a}_{k'}^* - \hat{a}_{k'}^* \hat{a}_k) \exp(ik \cdot r - ik' \cdot r')$$

$$= \frac{1}{V} \sum_k \exp(ik \cdot (r - r'))$$

$$= \delta(r - r'). \tag{4.105}$$

In order to see the physical interpretation of the field operators, we may integrate the density operator over the entire coordinate space:

$$\int \hat{\rho}(r) dr$$

$$= \int \hat{\psi}(r)^* \hat{\psi}(r) dr$$

$$= \sum_{k'k} \int \hat{a}_{k'}^* \hat{a}_k \exp(i(k - k') \cdot r) dr$$

$$= \sum_k \hat{a}_k^* \hat{a}_k. \tag{4.106}$$

This means that $\hat{\rho}(r)$ is the operator for the particle number density. For any physical system in which the total number of particles is conserved, the particle number density operator $\hat{\rho}(r)$ can be normalized to the total number of particles N. With this interpretation of the field operators we may try to find the kinetic energy of the entire system using the following expression:

$$\hat{H}_0 = \frac{1}{2m} \int \hat{\psi}(r)^*(-i\hbar\nabla)^2\hat{\psi}(r)dr = \sum_k \epsilon_k\hat{a}_k^*\hat{a}_k. \qquad (4.107)$$

This equation looks very much like the average value of the single-particle kinetic energy in the Schrödinger mechanics, where $\hat{\psi}$ is the wave function. In the current formulation, however, $\hat{\psi}$ is the field operator, and it looks as if the wave function is once more quantized. For this reason the present formulation is called the *second quantization*. The integrand in (4.107) is called the kinetic energy part of the *Hamiltonian density*. Let us go back to (4.104), where $\hat{\psi}$ is expanded in terms of the momentum eigenfunctions $\langle r|k\rangle$ and the $\hat{a}_k$ are the coefficients of expansion. In the Schrödinger quantum mechanics, $\hat{\psi}$ is the wave function and the a_k are the classical quantities (called the *c-number*), while in the second quantization, $\hat{\psi}$ is the field operator and the $\hat{a}_k$ are the operators (called the *q-number*).

Let us look at the interaction part of the Hamiltonian density in the second quantization formulation. Suppose that the bosons in the system interact with one another via a two-particle interaction potential given by $v(r_i - r_j)$. Judging from the kinetic energy part of the Hamiltonian in the second quantization formulation the interaction energy part of the Hamiltonian would be given by

$$\hat{H}_{int} = \frac{1}{2} \int v(r_1 - r_2)\hat{\rho}(r_1)\hat{\rho}(r_2)dr_1dr_2. \qquad (4.108)$$

This is exactly the expression for the interaction energy in the second quantization. The coefficient $\frac{1}{2}$ is to correct an over-counting of the free energy, because for any relative configuration of r_1 and r_2 in space there will be another configuration in which r_1 and r_2 are interchanged.

When manipulating the operators it is more convenient to arrange different operators in the so-called normal order, in which all the creation operators appear to the left and all the annihilation operators appear to the right in any product expression. In rearranging any pair of operators one has to comply carefully with the commutation relations (4.100). In this way the interaction Hamiltonian will be brought into the following form:

$$H_{int} = \frac{1}{2} \int \hat{\psi}(r_1)^*v(r_1 - r_2)\delta(r_1 - r_2)\hat{\psi}(r_2)dr_1dr_2$$
$$+ \frac{1}{2} \int \hat{\psi}(r_1)^*\hat{\psi}(r_2)^*v(r_1 - r_2)\hat{\psi}(r_2)\hat{\psi}(r_1)dr_1dr_2. \qquad (4.109)$$

The first integral will be reduced further to

$$\frac{1}{2}v(0)\int \hat{\psi}(r_1)^* \hat{\psi}(r_1) dr_1 = \frac{1}{2}v(0)N, \tag{4.110}$$

which is just a constant and does not play any important role in physics. When the interaction free energy has a hard core, $v(0)$ becomes infinity, and some justification must be required in order to drop such a term.

The interaction Hamiltonian will now be examined in more detail:

$$\hat{H}_{\text{int}} = \frac{1}{2}\int \hat{\psi}(r_1)^* \hat{\psi}(r_2)^* v(r_1 - r_2)\hat{\psi}(r_2)\hat{\psi}(r_1) dr_1 dr_2. \tag{4.111}$$

Since the concept of annihilation and creation of particles has been introduced in the number representation, the effect of the interaction between particles can be looked at as a process in which two particles are eliminated from some states and created in some other states. In order to see this more explicitly it is necessary to transform the field operators into the plane-wave representation,

$$H_{\text{int}} = \frac{1}{2}\sum_{k_1 k_2 k_3 k_4} v(k_3 - k_2)\hat{a}_{k_4}^* \hat{a}_{k_3}^* \hat{a}_{k_2}\hat{a}_{k_1}\delta(k_1 + k_2 - k_3 - k_4), \tag{4.112}$$

where $v(k)$ is the Fourier transform of the interaction potential,

$$v(r) = \frac{1}{V}\sum_k v(k)\exp(ik \cdot r), \quad v(k) = \int v(r)\exp(-ik \cdot r)dr. \tag{4.113}$$

The momentum conservation is a result of two facts: the first is the interaction $v(r)$ being the function only of the distance between two particles; the second is, of course, due to the orthogonality of the plane-wave eigenfunctions; both are the consequence of the translational invariance of the physical space. Because of this momentum conservation there are only three independent wave vectors in the interaction Hamiltonian,

$$\hat{H}_{\text{int}} = \frac{1}{2}\sum_{qk_1 k_2} v(q)\hat{a}_{k_1+q}^* \hat{a}_{k_2-q}^* \hat{a}_{k_2}\hat{a}_{k_1}. \tag{4.114}$$

4.7 The system of interacting fermions

To construct an occupation number representation of the Hamiltonian for the system of identical particles which obey Fermi–Dirac statistics is superficially similar to the case for bosons. There are, however, somewhat more complicated conceptual problems related to a nonintegral value of the spin quantum number. In some of the formulations for the fermion system, the spin state of particles should be explicitly taken into account, but otherwise the spin variables may be suppressed implicitly.

The first step towards the establishment of the wave function describing a physical state for such a system is to set up the orthonormal basis functions of representation. An obvious and the simplest choice would naturally be to use the product functions of single-particle plane-wave functions. The most stringent condition imposed upon the wave function for the many-fermion system is, however, the antisymmetricity of the function, i.e., the function must change its sign when any pair of coordinates are interchanged, known as the *Pauli exclusion principle*:

$$\Psi(r_1, \ldots, r_i, \ldots, r_j, \ldots, r_N) = -\Psi(r_1, \ldots, r_j, \ldots, r_k, \ldots, r_N). \quad (4.115)$$

Functions of this type have been long known in introductory analysis; for example, the determinant comprising the single-particle plane-wave functions (4.98):

$$\Psi(r_1, \ldots, r_i, \ldots, r_N)$$
$$= (N!)^{-\frac{1}{2}} \det |\langle r_1|k_1\rangle, \ldots, \langle r_i|k_i\rangle, \ldots, \langle r_N|k_N\rangle|, \quad (4.116)$$

where

$$\langle r|k\rangle = V^{-\frac{1}{2}} \exp(\mathrm{i}k \cdot r). \quad (4.117)$$

Because of the properties of the determinant, an interchange of any two columns amounts to the interchange of the two associated coordinates, and hence causes a change of the sign. One sees, therefore, that the choice of the determinants comprising the single-particle plane waves is quite appropriate. One difficulty, which is rather serious, is the bulkiness of the mathematical expression of the determinant. Any substantial mathematical manipulations that are required are likely to be beyond the scope of human resources; and therefore second quantization brings about gratifying simplification.

As for fermions, the particle annihilation and creation operators are introduced by means of the following commutation relations:

$$\hat{a}_k \hat{a}_{k'}^* + \hat{a}_{k'}^* \hat{a}_k = \delta(kk'),$$
$$\hat{a}_k^* \hat{a}_{k'}^* + \hat{a}_{k'}^* \hat{a}_k^* = 0,$$
$$\hat{a}_k \hat{a}_{k'} + \hat{a}_{k'} \hat{a}_k = 0. \quad (4.118)$$

In order to avoid any confusion with the case of the commutation relations for bosons, the above relations are often called the *anticommutation relations*, specifically referring to the fermion operators. The anticommutation relation may be conveniently represented by

$$\{\hat{A}, \hat{B}\} = \hat{A}\hat{B} + \hat{B}\hat{A}. \quad (4.119)$$

One immediate consequence of the commutation relations is that the operator $\hat{n}_k = \hat{a}_k^* \hat{a}_k$ satisfies the following identities:

$$\hat{n}_k^2 = \hat{n}_k, \quad \hat{n}_k(1 - \hat{n}_k) = 0. \tag{4.120}$$

A very important property is that the operator n_k only has eigenvalues of 0 and 1 for any one-particle state $|k\rangle$, i.e., the Pauli exclusion principle.

n_k is called the *fermion occupation number operator*. Furthermore, any properly normalized eigenket represented by

$$|\Psi\rangle = |\cdots \hat{a}_k^* \cdots \hat{a}_{k'}^* \cdots |0\rangle$$
$$= -|\cdots \hat{a}_{k'}^* \cdots \hat{a}_k^* \cdots |0\rangle \tag{4.121}$$

will represent the determinantal or antisymmetrized eigenket of the occupation number operators. This fact can easily be observed by noting that any transposition (exchange) of two neighboring daggered operators brings in one sign change (the second relation of (4.100)), and there will be $2n + 1$ times of the sign change when the $\hat{a}_k^*$ and $\hat{a}_{k'}^*$ are interchanged in the above ket, where n is the number of asterisked operators in between $\hat{a}_k^*$ and $\hat{a}_{k'}^*$.

The remainder of the formulation is more or less the same as for bosons. The field operators are defined by

$$\hat{\psi}(r)^* = \sum_k \hat{a}_k^* \langle k|r\rangle,$$
$$\hat{\psi}(r) = \sum_k \langle r|k\rangle \hat{a}_k, \tag{4.122}$$

which satisfy the anticommutation relation

$$\hat{\psi}(r)\hat{\psi}(r'^*) + \hat{\psi}(r')\hat{\psi}(r) = \delta(r - r'). \tag{4.123}$$

The total Hamiltonian for the interacting fermions will then be the sum of the kinetic energy and the interaction potential, expressed by

$$\hat{H}_0 = \sum_k \epsilon_k \hat{n}_k, \quad \epsilon_k = \frac{\hbar^2 k^2}{2m}, \tag{4.124}$$

where $k^2 = k_x^2 + k_y^2 + k_z^2$ and m is the mass of the particle, and

$$\hat{H}_{int} = \frac{1}{2} \sum_{qk_1k_2} v(q)\hat{a}_{k_1+q}^* \hat{a}_{k_2-q}^* \hat{a}_{k_2} \hat{a}_{k_1}, \tag{4.125}$$

where $v(q)$ is the Fourier transform of the pair potential $v(r)$.

4.8 Some examples exhibiting the effect of Fermi–Dirac statistics

4.8.1 The Fermi hole

Let us consider a system made up of noninteracting fermions contained in a cube of volume $V = L^3$. One further assumes that the system is in its lowest energy state. Because of Fermi statistics the particles occupy the one-particle plane-wave states, one particle per one state, from the lowest energy state up to the states which are characterized by the Fermi momentum k_F. This is the ground state and will be denoted by $|0\rangle$. Since the one-particle state is given by

$$\langle r|k\rangle = V^{-\frac{1}{2}} \exp(ik \cdot r), \tag{4.126}$$

where

$$k = (k_x, k_y, k_y) = \frac{2\pi}{L}(n_1, n_2, n_3), \tag{4.127}$$

the number of one-particle states contained in a volume element $\Delta k_x \Delta k_y \Delta k_z$ in the wave number k-space is given by

$$\Delta n_1 \Delta n_2 \Delta n_3 = \left(\frac{L}{2\pi}\right)^3 \Delta k_x \Delta k_y \Delta k_z. \tag{4.128}$$

If the summation over k is replaced by an integration in the k-space, the Fermi momentum k_F is determined by the condition

$$\frac{V}{(2\pi)^3}\frac{4\pi}{3}k_F^3 = N \quad \text{or} \quad k_F^3 = 6\pi^2 \frac{N}{V}. \tag{4.129}$$

In the ground state, the probabilty of finding two particles, one at 0 and the other at r, is

$$P(r) = \left(\frac{V}{N}\right)^2 \langle 0|\hat{\rho}(0)\hat{\rho}(r)|0\rangle$$

$$= \left(\frac{V}{N}\right)^2 \langle 0|\hat{\psi}(0)^*\hat{\psi}(0)\hat{\psi}(r)^*\hat{\psi}(r)|0\rangle$$

$$= -\left(\frac{V}{N}\right)^2 \langle 0|\hat{\psi}(0)^*\hat{\psi}(r)^*\hat{\psi}(0)\hat{\psi}(r)|0\rangle$$

$$= -\frac{1}{N^2} \sum_{k_1 k_2 k_3 k_4} \langle 0|\hat{a}_{k_1}^*\hat{a}_{k_2}^*\hat{a}_{k_3}\hat{a}_{k_4}|0\rangle \exp(i(k_4 - k_2) \cdot r), \tag{4.130}$$

where the plane-wave expansion (4.104) has been employed.

In the above four-fold summation there are two group of terms which yield nonzero contributions, i.e.,

$$k_1 = k_3, \ k_2 = k_4 \quad \text{and} \quad k_1 = k_4, \ k_2 = k_3. \qquad (4.131)$$

These two groups yield the following relation:

$$P(r) = 1 - \left[\frac{1}{N} \sum_{|k| \leq k_{\mathrm{F}}} \exp(i k \cdot r) \right]^2$$

$$= 1 - \left[\frac{V}{N\pi^2} \cdot \frac{1}{r^3} (\sin k_{\mathrm{F}} r - k_{\mathrm{F}} r \cos k_{\mathrm{F}} r) \right]^2. \qquad (4.132)$$

So, $P(r) \Rightarrow 0$ as $r \Rightarrow 0$, i.e., one fermion repels other neighboring fermions, creating a hole called the *Fermi hole*. Since

$$P(r) = 1 - \mathrm{O}\left(\frac{1}{r^6} \right), \quad P(r) \Rightarrow 1 \quad \text{as} \quad r \Rightarrow \infty. \qquad (4.133)$$

4.8.2 The hydrogen molecule

As an example of the electron–electron interaction in which the electronic spin plays an essential role, let us examine the binding mechanism of a hydrogen molecule. The first attempt, by Heitler & London (1927) was fairly successful in explaining the stable formation of the hydrogen molecule based on the first-order perturbation theory. Heitler & London assumed the following two wave functions. The first is a symmetrical wave function in the coordinate space,

$$\Psi_s(r_1, r_2) = 2^{-\frac{1}{2}}[1 + |\langle a|b\rangle|^2]^{-\frac{1}{2}}[\psi_a(r_1)\psi_b(r_2) + \psi_a(r_2)\psi_b(r_1)], \quad (4.134)$$

which is multiplied by an antisymmetrized spin function,

$$2^{-\frac{1}{2}}[\alpha(1)\beta(2) - \alpha(2)\beta(1)]. \qquad (4.135)$$

The entire function is called the singlet state.

The second wave function is antisymmetric in the coordinate space,

$$\Psi_a(r_1, r_2) = 2^{-\frac{1}{2}}[1 - |\langle a|b\rangle|]^{-\frac{1}{2}}[\psi_a(r_1)\psi_b(r_2) - \psi_a(r_2)\psi_b(r_1)], \quad (4.136)$$

which is multiplied by one of the three symmetric spin functions,

$$\alpha(1)\alpha(2), \quad 2^{-\frac{1}{2}}[\alpha(1)\beta(2) + \alpha(2)\beta(1)], \quad \text{or} \quad \beta(1)\beta(2). \qquad (4.137)$$

The entire function is called the triplet state. In the above expressions ψ_a is an isolated free atom of hydrogen with its proton at the position R_a, and ψ_b is an isolated free atom of hydrogen with its proton at the position R_b. $\langle a|b\rangle$ is called the *overlap integral* defined by

$$\langle a|b \rangle = \int \psi_a(\mathbf{r})^* \psi_b(\mathbf{r}) d\mathbf{r}. \tag{4.138}$$

Except when the two protons are separated by a large distance the two hydrogen wave functions $|a\rangle$ and $|b\rangle$ are not orthogonal, a fact which created many difficulties in the theories describing the formation of the hydrogen molecule.

In spite of such formulational difficulties which were subsequently resolved by a rather careful construction of the approximate wave functions, let us in the following explain the Heitler–London formulation which led to an initial success in the interpretation of the mechanism of the molecular hydrogen formation.

In terms of the above wave functions the energy of the system made up of two protons and two electrons is evaluated for different values of the proton–proton distance. The total Hamiltonian is separated into two parts,

$$H = H_0 + \Delta H,$$
$$H_0 = -\tfrac{1}{2}\nabla_1^2 - \tfrac{1}{2}\nabla_1^2 + \frac{e^2}{r_{a1}} + \frac{e^2}{r_{b2}},$$
$$\Delta H = \frac{e^2}{r_{12}} + \frac{e^2}{r_{ab}} - \frac{e^2}{r_{b1}} - \frac{e^2}{r_{a2}}, \tag{4.139}$$

where H_0 is the Hamiltonian for two independent hydrogen atoms and will give rise to twice the energy of the ground state hydrogen atom.

Since the total Hamiltonian does not contain any spin variable, the spin part of the wave function simply gives rise to the spin degeneracy factors of 1 (singlet) and 3 (triplet).

We now evaluate the energy difference between the singlet and triplet states:

$$\langle H \rangle_{\text{sing}} - \langle H \rangle_{\text{trip}} = 2\langle ab|\Delta H|ba \rangle. \tag{4.140}$$

This integral is called the *exchange integral* because the roles of atomic orbitals $|a\rangle$ and $|b\rangle$ are exchanged between the initial state and the final state. Heitler–London's crude calculation showed that this difference is negative for all the reasonable values of the inter-proton distance R. Thus two hydrogen atoms can form a molecule when they come together in the singlet spin state. The equilibrium distance, R_0, between the protons must correspond to the minimum of the energy $\langle ab|\Delta H|ba \rangle$ as a function of R. Heitler–London's initial calculation gives the value $1.51a = 0.80$ Å for R_0. Experimentally, $R_0 = 0.7395$ Å. The agreement between the theoretical and experimental values is rather poor. Since the Heitler–London formulation of the binding energy of the hydrogen molecule, numerous papers have been published, mostly based on variational calculations with a substantial number of variational parameters. The numerical value of the binding energy has been greatly improved. The significance of the Heitler–London formulation, however, is its conceptual novelty, i.e., the molecular binding is favored when the two electrons are bound in the singlet state, and the energy lowering is due to the negative sign of the exchange integral.

4.9 The Heisenberg exchange Hamiltonian

It was later pointed out by Heisenberg (1928) that the electron spin alignment in the ferromagnetic insulators containing incomplete d-shell transition metal ions or f-shell rare-earth ions is due to the positive sign of the exchange integral and by Dirac (1958) that the exchange interaction energy can be expressed in terms of the spin operators in the form

$$\text{exchange energy} = -\tfrac{1}{2}J(1 + 4\hat{s}_1 \cdot \hat{s}_2), \tag{4.141}$$

where J is the exchange integral, and $\hat{s}_1$ and $\hat{s}_2$ are the spins of the electrons on the neighboring ions. As is easily seen by the formula

$$2\hat{s}_1 \cdot \hat{s}_2 = (\hat{s}_1 + \hat{s}_2)^2 - 2s(s+1), \tag{4.142}$$

the exchange energy is expressed as

$$\text{exchange energy} = J[1 - s(s+1)] = \pm J, s = 0, \text{ or } s = 1,$$
$$\hat{s}_1 + \hat{s}_2 = \hat{S}, \quad \hat{S}^2 = S(S+1), \tag{4.143}$$

where S is the magnitude of the composite spin $\hat{S}$.

When there are many electrons in each incomplete magnetic ion, the above exchange interaction must be summed over all possible pairs of electrons on the neighboring ions, and an effective interaction between two neighboring ions will be given by

$$\hat{H} = -2\sum_{i,j} J_{ab}^{ij}(\hat{s}_i^a \cdot \hat{s}_j^b). \tag{4.144}$$

According to the well studied atomic theories of structure of magnetic ions, there are several mechanisms, such as the Russel–Saunders coupling, the Hund rules, and the intra-ionic exchange interaction, according to which there is a tendency for the electron spins within a single ion to couple parallel to one another and so yield a maximum possible total spin S. We are led to the effective interaction given by

$$\hat{H} = -2\sum_{\langle i,j \rangle} J_{ij}\hat{S}_i \cdot \hat{S}_j, \tag{4.145}$$

where $\hat{S}_i$, $\hat{S}_j$ are the total spin of the ions i and j, respectively, and $J_{i,j}$ is the effective exchange integral between the neighboring ions i and j.

This form of interaction is called the *Heisenberg exchange interaction* after the idea suggested by Heisenberg, but the spin operator representation is due to Dirac, and its standard use in the theories of magnetic phenomena has been strongly promoted after extensive and successful studies by Van Vleck (1932).

The effectiveness and applicability of the exchange interaction have been also examined rather extensively, and it has been found that the magnetic insulators seem to be the materials for which the interaction is most suitable. In particular, insulator ferromagnets EuO and EuS are the ideal systems for which the Heisenberg exchange interaction represents most accurately the nature of the interaction. Experimentally, the numerical value of the exchange integral J has been determined for those systems by Charap and Boyd (1964).

The standard form of the system Hamiltonian for insulator ferromagnets is given by

$$\hat{H} = -g\mu_{\rm B}B \sum_i \hat{S}_i^z - 2\sum_{\langle i,j\rangle} J_{ij}\hat{\mathbf{S}}_i \cdot \hat{\mathbf{S}}_j, \tag{4.146}$$

where g is the Landé g-factor,[†] $\mu_{\rm B} = e\hbar/2mc$ is the Bohr magneton, and J_{ij} is the exchange integral.

There is a celebrated model system called the *Ising model* (Ising, 1925). The system Hamiltonian for the Ising model is given by

$$\hat{H} = -g\mu_{\rm B}B \sum_i \hat{S}_i^z - 2\sum_{\langle i,j\rangle} J_{i,j}\hat{S}_i^z\hat{S}_j^z. \tag{4.147}$$

An important difference between the Heisenberg model and the Ising model is that, in the latter, the spin operator does not have transverse components and hence behaves as a classical scalar operator. Ising himself treated only a system of linear chain of scalar spins, and hence was not able to find any phase transition. As will be seen in the remainder of the book, the Ising model has been studied extensively and has played an important role in the development of statistical mechanics.

4.10 The electron–phonon interaction in a metal

When an electron is placed in a purely periodic electric field, such as in the three-dimensional crystalline field given by

$$U(\mathbf{r}) = \sum_{\mathbf{m}=m_1,m_2,m_3} w(\mathbf{r} - m_1\mathbf{a}_1 - m_2\mathbf{a}_2 - m_3\mathbf{a}_3), \tag{4.148}$$

then

$$U(\mathbf{r} + n_1\mathbf{a}_1 + n_2\mathbf{a}_2 + n_3\mathbf{a}_3) = U(\mathbf{r}), \tag{4.149}$$

where $\mathbf{a}_1, \mathbf{a}_2, \mathbf{a}_3$ are the primitive translation vectors of the lattice and n_1, n_2, n_3 are integers the eigenstate wave functions will be given by the so-called *Bloch function*,

$$\phi_{k\sigma} = V^{-\frac{1}{2}} \exp(i\mathbf{k} \cdot \mathbf{r})u_k(\mathbf{r})\chi_\sigma, \tag{4.150}$$

where

$$u_k(\mathbf{r} + \mathbf{a}_1) = u_k(\mathbf{r} + \mathbf{a}_2) = u_k(\mathbf{r} + \mathbf{a}_3) = u(\mathbf{r}), \tag{4.151}$$

and χ_σ is the electronic spin eigenfunction.

[†] $g = 2$ for electrons in the $s = \frac{1}{2}$ state.

If ϵ_k is the energy eigenvalue of the Bloch state, the Hamiltonian of the system of electrons, in the absence of other interactions, is given by

$$\hat{H}_0 = \sum_{k\sigma} \epsilon_k \hat{a}_{k,\sigma}^* \hat{a}_{k,\sigma}, \tag{4.152}$$

where $\hat{a}_k^*$ and $\hat{a}_k$ are the creation and annihilation operators of the one-particle eigenstate $\phi_{k\sigma}$.

If, however, one of the positive ions moves out of the regular lattice site n by a small displacement $\hat{\xi}$, the periodicity of the lattice is destroyed and hence the electrons cannot stay in those Bloch states. The small vibrations of the positive ions create some interaction with the system of electrons. This is called the *electron–phonon interaction* and acts as the source of the electrical resistivity.

Let us, in the following, find this interaction in more detail. Due to the displacement $\hat{\xi}_n$ of the positive ion at the lattice site n the periodic potential of the lattice will change to

$$\sum_n [w(r - na) + \xi_n \cdot \nabla_n w(r - na)]. \tag{4.153}$$

Then the electron–phonon interaction energy is given by

$$\hat{H}_{int} = \int \hat{\psi}^*(r) \sum_n \hat{\xi}_n \cdot \nabla_n w(r - na) \hat{\psi}(r) dr$$

$$= -\int \hat{\psi}^*(r) \sum_n w(r - na)(\nabla_r \cdot \hat{\xi}_n) \hat{\psi}(r) dr, \tag{4.154}$$

where $\hat{\psi}^*(r)$ and $\hat{\psi}(r)$ are the field operators for the Bloch electrons given by

$$\hat{\psi}(r) = V^{-\frac{1}{2}} \sum_{k,\sigma} \hat{a}_{k,\sigma} \exp(ik \cdot r) u_k(r) \chi_\sigma. \tag{4.155}$$

When the ionic displacement operator $\hat{\xi}_n$ is expanded in terms of the momentum space operators, we find that

$$(\nabla_n \cdot \hat{\xi}_n) = \sum_q \left(\frac{\hbar q}{2MNs}\right)^{\frac{1}{2}} (\hat{b}_q^* \exp(iq \cdot n) + \hat{b}_q \exp(-iq \cdot n)), \tag{4.156}$$

where $\hat{b}_q^*$ and $\hat{b}_q$ are the creation and annihilation operators for the phonons, s is the velocity of a longitudinal sound wave of frequency ω, and q is the magnitude of a wave vector q. They are related by $\omega = sq$.

Substituting this into (4.154):

$$\hat{H}_{int} = \sum_q \hat{H}_q, \quad \hat{H}_q = D(q)\hat{b}_q \hat{\rho}_q + \text{H.c}, \tag{4.157}$$

where H.c. refers to the Hermitian conjugate operator, and

$$\hat{\rho}_q^* = \sum_{k,\sigma} \hat{a}_{k+q,\sigma}^* \hat{a}_{k,\sigma}, \tag{4.158}$$

and

$$D(q) = \left(\frac{\hbar q}{2MVs}\right)^{\frac{1}{2}} \int u_{k+q}^* w u_k \mathrm{d}r, \tag{4.159}$$

where the integral is over the unit cell of the crystalline lattice because u_k is normalized to unity in the primitive cell.

For the sake of simplicity we now assume that only the longitudinal phonons interact with electrons and drop the k dependence of $D(q)$ in the remainder of the formulation.

The energy of the electrons and the phonons change because of the interaction between them; however, we are interested in the behavior of the electrons only. The change on the phonon spectrum may be taken into account indirectly through the experimental value of the sound velocity s.

The entire energy is made up of the sum of energies of electrons and phonons, and is given by

$$\hat{H} = \hat{H}_0 + \hat{H}_{\mathrm{int}}, \quad \hat{H}_0 = \sum_{k,\sigma} \epsilon_k \hat{a}_{k,\sigma}^* \hat{a}_{k,\sigma} + \sum_q \hbar\omega_q \left(\tfrac{1}{2} + \hat{b}_q^* \hat{b}_q\right), \tag{4.160}$$

and

$$\hat{H}_{\mathrm{int}} = \sum_{k,\sigma,q} D(q) \hat{a}_{k+q,\sigma}^* \hat{a}_{k,\sigma} \hat{b}_q + \mathrm{H.c.} \tag{4.161}$$

In order to see the influence of the electron–phonon interaction upon the state of electrons, we can estimate the interaction in the second-order perturbation. This will be accomplished most conveniently by the method of a transformation suggested by Fröhlich (1950).[†] The transformed Hamiltonian has the form

$$\hat{H}' = e^{-\hat{S}} \hat{H} e^{\hat{S}} = \hat{H} + [\hat{H}, \hat{S}] + \tfrac{1}{2}[[\hat{H}, \hat{S}], \hat{S}] + \cdots,$$

$$= \hat{H}_0 + \hat{H}_{\mathrm{int}} + [\hat{H}_0, \hat{S}] + [\hat{H}_{\mathrm{int}}, \hat{S}] + \tfrac{1}{2}[[\hat{H}_0, \hat{S}], \hat{S}] + \cdots. \tag{4.163}$$

We may choose the transformation operator $\hat{S}$ in such a way that

$$\hat{H}_{\mathrm{int}} + [\hat{H}_0, \hat{S}] = 0, \tag{4.164}$$

and then $\hat{H}'$ is given, up to the second order in the perturbation, by

$$\hat{H}' = \hat{H}_0 + \tfrac{1}{2}[\hat{H}_{\mathrm{int}}, \hat{S}]. \tag{4.165}$$

[†] The eigenvalues of the Hamiltonian are invariant under this transformation because of the following relation:

$$\mathrm{Det}[\hat{H} - \lambda I] = \mathrm{Det}[e^{-\hat{S}}(\hat{H}' - \lambda I)e^{\hat{S}}]$$
$$= \mathrm{Det}[e^{-\hat{S}}]\mathrm{Det}[\hat{H}' - \lambda I]\mathrm{Det}[e^{\hat{S}}]$$
$$= \mathrm{Det}[\hat{H}' - \lambda I]. \tag{4.162}$$

We can now set the transformation operator $\hat{S}$ in the following form:

$$\hat{S} = \hat{S}^* = \sum_q \hat{S}_q, \quad \hat{S}_q = \hat{\gamma}_q \hat{b}_q + \hat{\gamma}_q^* \hat{b}_q^* \tag{4.166}$$

with

$$\hat{\gamma}_q = \sum_{k,\sigma} \Phi(k, q) \hat{a}_{k+q,\sigma}^* \hat{a}_{k,\sigma}, \tag{4.167}$$

where $\Phi(k, q)$ will be determined presently. Equation (4.164) can be solved rather easily. It is found that

$$[\hat{H}_0, \hat{S}_q] + \hat{H}_q$$
$$= \sum_{k,\sigma} [(\epsilon_k - \epsilon_{k+q} + \hbar\omega_q)\Phi(k, q) + D(q)]\hat{a}_{k+q,\sigma}^* \hat{a}_{k,\sigma} \hat{b}_q, +\text{H.c.}$$

$$\tag{4.168}$$

$\Phi(k, q)$ is then determined as

$$\Phi(k, q) = -\frac{D(q)}{\epsilon_k - \epsilon_{k+q} + \hbar\omega_q}. \tag{4.169}$$

Substituting definitions for the $\hat{S}_q$ operators, we find that

$$[\hat{H}_{\text{int}}, \hat{S}] = \sum_q [\hat{H}_q, \hat{S}_q]$$
$$= \sum_q \sum_{k,\sigma} D(q)(\hat{a}_{k+q,\sigma}^* \hat{a}_{k,\sigma} \hat{b}_q \hat{\gamma}_q^* \hat{b}_q^* - \hat{\gamma}_q^* \hat{b}_q^* \hat{a}_{k+q,\sigma}^* \hat{a}_{k,\sigma} \hat{b}_q) + \text{H.c.}$$

$$\tag{4.170}$$

This second-order energy is now evaluated in the ground state of the phonon Hamiltonian. Since

$$\langle 0_{\text{ph}} | \hat{b}_q^* \hat{b}_q | 0_{\text{ph}} \rangle = 0, \quad \text{and} \quad \langle 0_{\text{ph}} | \hat{b}_q \hat{b}_q^* | 0_{\text{ph}} \rangle = 1, \tag{4.171}$$

the second-order perturbation $\hat{H}_2$ will be given by

$$\hat{H}_2 = \frac{1}{2} \sum_{q,k,\sigma} D(q) \hat{a}_{k+q,\sigma}^* \hat{a}_k \hat{\gamma}_q^* + \text{H.c.}, \tag{4.172}$$

and if $\hat{\gamma}_q^*$ is substituted from (4.166) and (4.168), finally the Hamiltonian for the electron–phonon interaction is given in the form:

$$\hat{H} = \sum_{k,\sigma} \epsilon_k \hat{a}_{k,\sigma}^* \hat{a}_{k,\sigma}$$
$$- \frac{1}{2} \sum_q \sum_{k,\sigma} \sum_{k',\sigma'} \frac{|D(q)|^2}{\epsilon_{k'} - \epsilon_{k'+q} + \hbar\omega_q} \hat{a}_{k+q,\sigma}^* \hat{a}_{k',\sigma'}^* \hat{a}_{k'+q,\sigma'} \hat{a}_{k,\sigma}. \tag{4.173}$$

From this it can be seen that some electrons excited near the Fermi surface have an attractive interaction, i.e., if both $\epsilon_{k'}$ and $\epsilon_{k'+q}$ lie near the Fermi surface

$$\epsilon_{k'} - \epsilon_{k'+q} \ll \hbar\omega_q, \tag{4.174}$$

and if we look at the term of $q = -k - k'$ and $\sigma' = -\sigma$, i.e.,

$$\hat{H}_{\text{int}} = -\frac{1}{2}\sum_{k,k'\sigma}\frac{|D(k'-k)|^2}{\hbar\omega_{k'-k}}\hat{a}_{k',\sigma}^*\hat{a}_{-k',-\sigma}^*\hat{a}_{-k,-\sigma}\hat{a}_{k,\sigma}. \tag{4.175}$$

As will be discussed in Chapter 9, this attractive interaction leads to the stability of a superconducting state at low temperatures. The combined operators

$$\hat{a}_{k,\sigma}^*\hat{a}_{-k,-\sigma}^* \quad \text{and} \quad \hat{a}_{-k,-\sigma}\hat{a}_{k,\sigma} \tag{4.176}$$

are called the *Cooper pair creation* and *annihilation operators*.

4.11 The dilute Bose gas

One of the characteristic properties of the system of bosons is the fact that any one-particle state can be occupied by any number of particles. If there is no physical interaction whatsoever between the Bose particles, in the ideal Bose gas the ground state of the system will be the state in which all the particles will occupy the lowest energy one-particle state. This phenomenon is called the *Bose–Einstein condensation* in the ideal boson gas. Fritz London (1938) was the first to attribute the salient superfluidity observed in liquid helium-4 below 2.19 K to the manifestation of the Bose–Einstein condensation in a real physical system. In reality, however, helium-4 exhibits peculiar behavior in the liquid phase below the gas–liquid critical temperature of about 5.2 K, and, therefore, it is not appropriate to compare various properties exhibited by liquid helium-4 with those of the Bose–Einstein condensed ideal gas.

Since the work of London, many authors started working on the application of Bose–Einstein statistics to the liquid helium-4 system. In the theory of liquid helium, however, we immediately encounter a serious difficulty, i.e, how to treat the hard-core repulsive interaction between the helium atoms at shorter distances, though the attractive interaction at larger distances is rather weak. In the general formulation of second quantization of the boson Hamiltonian we saw a need for $v(q)$ (see (4.114)). The Fourier transform of the hard-core repulsive interaction, however, is infinity, and hence $v(q)$ cannot be defined. Most theories of liquid helium-4, for this reason, are based on the somewhat unrealistic model of soft-core bosons, and even the fact that the system is in the liquid phase is altogether ignored.[†] One of the more successful theories of liquid helium is presented by Bogoliubov (1947).

[†] There is one exception to this: there is a lattice gas theory of hard-core bosons proposed by Matsubara and Matsuda (1956). They assumed that the helium atoms satisfy Fermi–Dirac statistics on the same lattice site (the statistical repulsion hard-core) and obey the Bose–Einstein statistics on different lattice sites.

Consider, for simplicity, a system of N bosons described by the Hamiltonian

$$H = \sum_k \frac{\hbar^2 k^2}{2m} \hat{a}_k^* \hat{a}_k + \frac{1}{2} g \sum_{k+k'=k''+k'''} \hat{a}_{k'''}^* \hat{a}_{k''}^* \hat{a}_{k'} \hat{a}_k. \qquad (4.177)$$

The annihilation and creation operators satisfy the commutation relations

$$\hat{a}_k \hat{a}_{k'}^* - \hat{a}_{k'}^* \hat{a}_k = \delta_{kk'},$$

$$\hat{a}_k^* \hat{a}_{k'}^* - \hat{a}_{k'}^* \hat{a}_k^* = 0,$$

$$\hat{a}_k \hat{a}_{k'} - \hat{a}_{k'} \hat{a}_k = 0. \qquad (4.178)$$

At this moment, it is useful to give some thought to the operators for the zero-momentum (hence zero-energy) state even though they satisfy the same relations; i.e.,

$$\hat{a}_0 \hat{a}_0^* = \hat{a}_0^* \hat{a}_0 + 1. \qquad (4.179)$$

Since the average number of zero-momentum particles can be as large as the total number of particles, once the Bose–Einstein condensation sets in,

$$\hat{a}_0^* \hat{a}_0 = N_0 \approx N. \qquad (4.180)$$

The 1 on the right hand side of (4.178) is so negligibly small compared with N_0, that we can set

$$\hat{a}_0 \hat{a}_0^* = \hat{a}_0^* \hat{a}_0, \qquad (4.181)$$

i.e., the operators $\hat{a}_0$ and $\hat{a}_0^*$ are commuting (classical) quantities which are of the order of $N_0^{\frac{1}{2}}$.

Now, since a large number of particles are in the zero-momentum state, the number of excited particles is much smaller than N_0, and hence their interaction energy among themselves may be ignored in the lowest order approximation. Based on this argument, the most important terms can be extracted from the interaction Hamiltonian as follows:

$$\hat{H}_{\text{int}} \approx \frac{1}{2} g \left\{ \hat{a}_0^* \hat{a}_0^* \hat{a}_0 \hat{a}_0 + \sum_{k \neq 0} [2 \hat{a}_k^* \hat{a}_0^* \hat{a}_k \hat{a}_0 + 2 \hat{a}_{-k}^* \hat{a}_0^* \hat{a}_{-k} \hat{a}_0 \right.$$

$$\left. + \hat{a}_k^* \hat{a}_{-k}^* a_0 a_0 + \hat{a}_0^* \hat{a}_0^* \hat{a}_k \hat{a}_{-k}] \right\}. \qquad (4.182)$$

Since the operators $\hat{a}_0$ and $\hat{a}_0^*$ behave like classical quantities, they may be replaced by the c-number, $N_0^{\frac{1}{2}}$, and the interaction Hamiltonian is further simplified to

$$\hat{H}_{\text{int}} \approx \frac{1}{2} g \left[N_0^2 + 2 N_0 \sum_{k \neq 0} (\hat{a}_k^* \hat{a}_k + \hat{a}_{-k}^* \hat{a}_{-k}) \right.$$

$$\left. + N_0 \sum_{k \neq 0} (\hat{a}_k^* \hat{a}_{-k}^* + \hat{a}_k \hat{a}_{-k}) \right]. \qquad (4.183)$$

In the above there is still one unknown parameter N_0, and the quantity must be determined before we can proceed. In order to accomplish this, the following condition is used:

$$N = N_0 + \sum_{k \neq 0} (\hat{a}_k^* \hat{a}_k + \hat{a}_{-k}^* \hat{a}_{-k}), \qquad (4.184)$$

i.e., the total number of particles is a quantity which should be conserved in any representation. When this condition is substituted for N_0, the total Hamiltonian is given by

$$
\begin{aligned}
\hat{H} \approx {} & \frac{1}{2} g N^2 + \frac{1}{2} \sum_{k \neq 0} \left(\frac{\hbar^2 k^2}{2m} + g N \right) (\hat{a}_k^* \hat{a}_k + \hat{a}_{-k}^* \hat{a}_{-k}) \\
& + g N \sum_{k \neq 0} (\hat{a}_k^* \hat{a}_{-k}^* + \hat{a}_k \hat{a}_{-k}) \\
& - \frac{3}{8} g \sum_{kk'} (\hat{a}_k^* \hat{a}_k + \hat{a}_{-k}^* \hat{a}_{-k})(\hat{a}_{k'}^* \hat{a}_{k'} + \hat{a}_{-k'}^* \hat{a}_{-k'}) \\
& - \frac{1}{4} g \sum_{kk'} (\hat{a}_k^* \hat{a}_k + \hat{a}_{-k}^* \hat{a}_{-k})(\hat{a}_{k'}^* \hat{a}_{-k'}^* + \hat{a}_{k'} \hat{a}_{-k'}). \qquad (4.185)
\end{aligned}
$$

This Hamiltonian looks rather complicated at first glance; however, it has a very special structure, i.e., there is no triple summation, in contrast to the original Hamiltonian.

4.12 The spin-wave Hamiltonian

In order to study the low temperature behavior of the Heisenberg ferromagnet it is necessary to find the low-lying energy eigenvalues exactly. The topic was a subject of intensive investigation for about 35 years (1930–65), and the correct behavior of the Heisenberg ferromagnet has been, at least at low temperatures, fully understood since then. This particular system seems to offer some pedagogically significant example in the application of statistical mechanics.

Firstly, one should realize the fact that the ground state, $|0\rangle$, is a well defined state, i.e., the state in which all spins are lined up parallel in the same direction. In the absence of an external magnetic field, however, the direction of quantization is not determined uniquely. To avoid this ambiguity an infinitesimally small external magnetic field, B, is applied along the z-direction. The Heisenberg exchange Hamiltonian is usually expressed as

$$
\begin{aligned}
\hat{H} &= -2 \sum_{f < g} J_{fg} \hat{\mathbf{S}}_f \cdot \hat{\mathbf{S}}_g - B \sum_f \hat{S}_f^z \\
&= -\sum_{f < g} J_{fg} \left(\hat{S}_f^+ \hat{S}_g^- + \hat{S}_f^- \hat{S}_g^+ + 2 \hat{S}_f^z \hat{S}_g^z \right) - B \sum_f \hat{S}_f^z, \qquad (4.186)
\end{aligned}
$$

where J_{fg} is equal to the exchange integral J when f and g are the nearest neighbor sites and zero otherwise. $\hat{S}_f^+$ and $\hat{S}_f^-$ are the step-up and step-down operators and they, together with $\hat{S}_f^z$, satisfy the commutation relations:

$$[\hat{S}_f^z, \hat{S}_g^+] = \hat{S}_f^+ \delta_{fg}, \quad [\hat{S}_f^z, \hat{S}_g^-] = -\hat{S}_f^- \delta_{fg}, \quad [\hat{S}_f^+, \hat{S}_f^-] = 2\hat{S}_f^z \delta_{fg}. \quad (4.187)$$

The Hamiltonian can be given a more convenient form:

$$\hat{H} = -\sum_f \sum_g J_{fg} \hat{S}_f^+ \hat{S}_g^- - \sum_f \sum_g J_{fg}(\hat{S}_f^z \hat{S}_g^z - S^2) - B \sum_f (\hat{S}_f^z - S),$$

$$(4.188)$$

where two constant terms are added so that the eigenvalue is zero for the ground state $|0\rangle$, and $\hat{S}_f^+ \hat{S}_g^- |0\rangle$ is zero for all sites f. The double summation is defined to cover both (f, g) and (g, f) terms. The low temperature behavior of the Heisenberg ferromagnet is described well by the theory of spin waves first introduced by Bloch (1930).

The Fourier transforms of the exchange integral and the operator $\hat{S}_f^-$ are defined by

$$J(k) = \sum_f J_{fg} \exp[ik \cdot (r_f - r_g)], \quad \hat{S}_k^- = N^{-\frac{1}{2}} \sum_f \hat{S}_f^- \exp(ik \cdot r_{fg}),$$

$$(4.189)$$

where N is the total number of lattice sites in the system.

The first step is to make an important observation, i.e.,

$$[\hat{H}, \hat{S}_0^-] = B\hat{S}_0^-, \quad (4.190)$$

where $\hat{S}_0^-$ is the $k = 0$ value of the operator S_k. Equation (4.190) implies that if $|\rangle$ is an eigenfunction of $\hat{H}$ with the eigenvalue E, then $S_0^- |\rangle$ is also an eigenfunction with the eigenvalue $E + B$. In particular,

$$\hat{S}_0^- |\rangle, \quad \hat{S}_0^- \hat{S}_0^- |\rangle, \quad \cdots, \quad \hat{S}_0^{-n} |\rangle, \quad (4.191)$$

are exact eigenfunctions of H with the eigenvalues $B, 2B, \ldots, nB$, respectively.

The commutation relation between the Hamiltonian $\hat{H}$ and $\hat{S}_k^-$ is given by [Exercise 4.4]

$$[\hat{H}, \hat{S}_k^-]$$

$$= B\hat{S}_k^- - 2N^{-\frac{1}{2}} \sum_f \sum_g J_{fi} \exp(ik \cdot r_g)[\exp(ik \cdot (r_f - r_g)) - 1]\hat{S}_f^z \hat{S}_g^-$$

$$= B\hat{S}_k^- + O(k). \quad (4.192)$$

If $k_1, k_2, \ldots, k_n$ are very small,

$$\hat{S}_{k_1}^- |0\rangle, \quad \hat{S}_{k_1}^- \hat{S}_{k_2}^- |0\rangle, \quad \cdots, \quad \hat{S}_{k_1}^- \hat{S}_{k_2}^- \cdots \hat{S}_{k_n}^- |0\rangle \quad (4.193)$$

are the eigenfunctions of H corresponding to the eigenvalues $B, 2B, \ldots, nB$ with errors, both of $O(k_1)$, $O(k_1, k_2)$, $\ldots$, $O(k_1, \ldots, k_n)$, respectively.[†] Denoting the errors by $\phi(k_1, \ldots, k_n)$ and $\epsilon(k_1, \ldots, k_n)$, the eigenfunctions and eigenvalues are expressed as

$$\Psi(k_1) = \hat{S}_{k_1}^- |0\rangle + \phi(k_1),$$
$$\Psi(k_1, k_2) = \hat{S}_{k_1}^- \hat{S}_{k_2}^- |0\rangle + \hat{S}_{k_1}^- \phi(k_2) + \hat{S}_{k_2}^- \phi(k_1) + \phi(k_1, k_2),$$
$$\Psi(k_1, k_2, k_3) = \hat{S}_{k_1}^- \hat{S}_{k_2}^- \hat{S}_{k_3}^- |0\rangle$$
$$+ \hat{S}_{k_1}^- \hat{S}_{k_2}^- \phi(k_3) + \hat{S}_{k_2}^- \hat{S}_{k_3}^- \phi(k_1) + \hat{S}_{k_3}^- \hat{S}_{k_1}^- \phi(k_2)$$
$$+ \hat{S}_{k_1}^- \phi(k_2, k_3) + \hat{S}_{k_2}^- \phi(k_3, k_1) + \hat{S}_{k_3}^- \phi(k_1, k_2)$$
$$+ \phi(k_1, k_2, k_3), \tag{4.194}$$

$$E(k_1) = B + \epsilon(k_1),$$
$$E(k_1, k_2) = 2B + \epsilon(k_1) + \epsilon(k_2) + \epsilon(k_1, k_2),$$
$$E(k_1, k_2, k_3) = 3B + \epsilon(k_1) + \epsilon(k_2) + \epsilon(k_3)$$
$$+ \epsilon(k_1, k_2) + \epsilon(k_2, k_3) + \epsilon(k_3, k_1) + \epsilon(k_1, k_2, k_3). \tag{4.195}$$

The quantities $\phi(k_1, \ldots, k_n)$ and $\epsilon(k_1, \ldots, k_n)$ vanish if one of the arguments is zero, because a state in which one of the k's is zero represents an exact eigenstate with one less argument and hence there is no need for correction. Note that if $k_1 = 0$ in $\Psi(k_1$ and $E(k_1)$, then

$$\Psi(0) = \hat{S}_0^- |0\rangle + \phi(0), \quad E(0) + B + \epsilon(0), \tag{4.196}$$

where $\hat{S}_0^- |0\rangle$ is an exact eigenfunction which belongs to the eigenvalue B; and hence $\phi(0) = 0$ and $\epsilon(0) = 0$.

Now if $k_2 = 0$ in $\Psi(k_1, k_2)$ and $E(k_1, k_2)$, then

$$\Psi(k_1, 0) = \hat{S}_0^- [\hat{S}_{k_1}^- |0\rangle + \phi(k_1)] + \phi(k_1, 0) = \hat{S}_0^- \Psi(k_1) + \phi(k_1, 0),$$
$$E(k_1, 0) = 2B + \epsilon(k_1, 0) = B + E(k_1) + \epsilon(k_1, 0), \tag{4.197}$$

where $\hat{S}_0^- \Psi(k_1)$ is also an exact eigenfunction whose eigenvalue is $B + E(k_1)$; hence $\phi(k_1, 0) = 0$ and $\epsilon(k_1, 0) = 0$. Then, by mathematical induction, the statement is true. Thus the orders of magnitude of these corrections are estimated to be

$$\phi(k_1, \ldots, k_n) = O(k_1) \cdots O(k_n),$$
$$\epsilon(k_1, \ldots, k_n) = O(k_1) \cdots O(k_n). \tag{4.198}$$

The order of magnitude of $\epsilon(k_1, \ldots, k_n)$ as a function of the size of the system is evaluated as of $O(N^{-n+1})$ under the assumption that there exist no low-lying

---

[†] $O(k_1, k_2, \cdots, k_n) = O(k_1) + O(k_2) + \cdots + O(k_n) + O(k_1)O(k_2) + \cdots + O(k_1)O(k_2)O(k_3)$
$+ \cdots + O(k_1)O(k_2) \cdots O(k_n).$

bound states. The order of magnitude of the ϵ's is estimated as follows. As far as the bound state does not exist, the shift of eigenvalue of the Hamiltonian appears when two spin waves come to the same site or nearest-neighbor lattice sites of each other. This situation changes the effective number of lattice sites available for two noninteracting spin waves from N^2 by a number of order N. This means that $\epsilon(k, k')$ is one order of magnitude smaller than $\epsilon(k)$, i.e., $\epsilon(k, k') = O(1/N)$, and so $\sum_k \sum_{k'} \epsilon(k, k')$ becomes a quantity of the order N. In the three-spin-wave problem, $\epsilon(k, k', k'')$ is due to the change of the effective number of lattice sites, caused by the cases when all three come to the same or nearest neighbors of one another; i.e., a number of $O(N)$ in N^3. This results in the correction of $O(1/N^2)$ to the energy eigenvalue, i.e., $\epsilon(k, k', k'') = O(1/N^2)$. Similarly, $\epsilon(k_1, \ldots, k_n)$ is estimated to be of $O(1/N^{-n+1})$.

For cubic lattices,

$$\phi(k) = 0, \ \epsilon(k) = 2S[J(0) - J(k)] = O(k^2),$$
$$\phi(k_1, k_2) = O(k_1, k_2) + O\left(k_1^2, k_2^2\right),$$
$$\epsilon(k_1, k_2) = O(k_1, k_2) + O(k_1, k_2)^2 + O\left(k_1^2, k_2^2\right) + O(k_1, k_2)^3. \quad (4.199)$$

The eigenfunctions of the Hamiltonian for the Heisenberg ferromagnet can be specified by a set $\{n_k\}$ and are expressed as

$$\Psi(\{n_k\}) = \prod_k (\hat{S}_k^-)^{n_k} |0\rangle + \cdots. \quad (4.200)$$

The corresponding energies are expressed as

$$E(\{n_k\}) = \sum_k (B + \epsilon(k)) + \sum_{k \langle k'} n_k n_{k'} \epsilon(k, k')$$
$$+ \sum_{k \langle k' \langle k''} \epsilon(k, k', k'') n_k n_{k'} n_{k''} + \cdots. \quad (4.201)$$

This expression for the excitation energy of the Heisenberg ferromagnet has rather special features:[†]

- the spin-wave excitations are bosons, i.e., any number of excitations of the same wave number can be excited;
- the excitation energy is completely specified by the set of spin-wave excitation numbers, i.e., the excitation energy is diagonal with respect to the spin-wave number operators $\{n_k\}$;
- as an important consequence of the above statements it will be established that the spin-wave excitations are exactly the quasiparticles. This implies that the statistical distribution

---

[†] For a detailed analysis of $\phi(k_1, k_2)$ and $\epsilon(k_1, k_2)$, see Morita & Tanaka (1965).

function for the number of spin waves is exactly given by the ideal Bose–Einstein distribution function in spite of the presence of two-body, three-body, and many-body interaction energies;

• the most convincing and clearest proof of the above statement is given by Wentzel (1960), which will be fully explained in Chapter 8.

Exercises

4.1 Show that the overlap integral is given by

$$\langle a|b \rangle = \left(1 + \rho + \tfrac{1}{3}\rho^2\right), \tag{E4.1}$$

where $\rho = R/a$.

4.2 Calculate the average value of the fourth power of the displacement, x^4, in the ground state of the linear harmonic oscillator.

4.3 Derive the commutation relation between the Heisenberg Hamiltonian and the spin-wave operator $\hat{S}_k^-$ given by (4.192).

[Hint]

Let us look at the commutation relation $[\hat{H}, \hat{S}_j^-]$, where $\hat{S}_j^-$ is the operator on the jth site. From (4.188),

$$[\hat{H}, \hat{S}_j^-] = -\sum_g J_{jg}([\hat{S}_j^+, \hat{S}_j^-]\hat{S}_g^- + 2[\hat{S}_j^z, \hat{S}_j^-]\hat{S}_g^z - B[\hat{S}_j^z, \hat{S}_j^-], \tag{E4.2}$$

where the factor 2 in front of the second term is due to the fact that the site j could be either f or g. Because of the commutation relations (4.186),

$$[\hat{H}, \hat{S}_j^-] = -\sum_g 2J_{jg}\left[\left(\hat{S}_j^z\hat{S}_g^- - \hat{S}_j^-\hat{S}_g^z\right)\right] + B\hat{S}_j^-,$$

$$\hat{H}S_j^-|0\rangle = -\sum_g 2J_{jg}S[\hat{S}_g^- - \hat{S}_j^-]|0\rangle + B\hat{S}_j^-|0\rangle. \tag{E4.3}$$

This result indicates that a single site excitation $S_j^-|0\rangle$ cannot be an eigenstate of the Hamiltonian. If, however, both sides of the equation are multiplied by $N^{-\frac{1}{2}}\exp(i\mathbf{k}\cdot\mathbf{r}_j)$ and summed over j, we find

$$H\hat{S}_k^-|0\rangle = E(k)\hat{S}_k^-|0\rangle, \quad E(k) = B + \epsilon(k),$$

$$\epsilon(\mathbf{k}) = 2JS\sum_\rho (1 - \exp(i\mathbf{k}\cdot\rho)), \tag{E4.4}$$

where ρ is a vector joining nearest neighbors in the lattice.

5

The density matrix

5.1 The canonical partition function

In the canonical ensemble theory (see Chapter 3) it is assumed that all the eigen-states and eigenvalues of a given system are known. This is seldom the case, and instead one has to be satisfied with some kind of approximation. This is the situation in quantum mechanics, in which many different perturbation techniques have been developed. Using perturbation theory, therefore, we are able to find an approximate eigenfunction and eigenvalue starting from an unperturbed state of the system. In statistical mechanics applications, however, thermodynamical quantities such as the Helmholtz potential, entropy, specific heat, etc. are the quantities of interest. We are not really interested in changes in the particular eigenvalue or eigenfunction due to a perturbation. This means that the techniques of perturbation theory must be blended naturally into the calculation of the partition function. This is most effectively ac-complished by direct application of the perturbation techniques to the *density matrix*.

Let us first develop a rigorous formulation and introduce approximations later. For this purpose, consider a physical system containing a macroscopic number, N, of interacting particles. The system Hamiltonian[†] is of the type described in Chapter 4, and all the eigenfunctions and eigenvalues are assumed to be known. Thus, one has

$$H \Psi_i = E_i \Psi_i. \tag{5.1}$$

It should be noted here that a capital letter E_i is used for the eigenvalue to indicate that the system is macroscopic. Each of the eigenstates is a microstate, and one can construct the canonical ensemble as described in Chapter 3.

The probability of finding the system under consideration in one of the microstates having energy eigenvalue E_i is given by the canonical distribution

$$P(E_i) = \frac{1}{Z(\beta)} \exp(-\beta E_i), \quad Z(\beta) = \sum_i \exp(-\beta E_i). \tag{5.2}$$

[†] It should be noted that a hat symbol above an operator will be suppressed in this and subsequent chapters.

Utilizing the normalization condition for the eigenfunctions $\{\Psi_i\}$, we see that

$$Z(\beta) = \sum_i \int \Psi_i^* \Psi_i \exp(-\beta E_i)\, d\tau$$

$$= \sum_i \int \Psi_i^* \exp(-\beta H)\Psi_i\, d\tau$$

$$= \sum_i \langle \Psi_i | \exp(-\beta H)|\Psi_i \rangle, \qquad (5.3)$$

where the exponential operator is defined in terms of its Taylor series, and then the transformation from the first line to the second line in (5.3) becomes obvious, i.e.,

$$\exp(-\beta H)\Psi_i = \left[1 - (\beta H) + \frac{1}{2!}(\beta H)^2 - \frac{1}{3!}(\beta H)^3 + \cdots \right]\Psi_i$$

$$= \left[1 - (\beta E_i) + \frac{1}{2!}(\beta E_i)^2 - \frac{1}{3!}(\beta E_i)^3 + \cdots \right]\Psi_i$$

$$= \exp(-\beta E_i)\Psi_i. \qquad (5.4)$$

The canonical partition function is then expressed as the trace (diagonal sum) of the exponential operator:

$$Z(\beta) = \sum_i \langle \Psi_i | \exp(-\beta H)|\Psi_i \rangle. \qquad (5.5)$$

The exponential operator

$$\rho(\beta) = \exp(-\beta H) \qquad (5.6)$$

is called the *statistical operator* or the *probability density matrix*, or simply the *density matrix*.

5.2 The trace invariance

In practice, it is almost impossible to find exact eigenfunctions of the total Hamiltonian. In order to be able to introduce approximations, we can use some simpler and well defined basis functions such as the plane-wave functions. Let us define a set of such orthonormal functions $\{\Phi_a\}$ which are defined in the same function space as the set $\{\Psi_i\}$, and expand Ψ_i in terms of the set $\{\Phi_a\}$;

$$\Psi_i = \sum_a c_{ia} \Phi_a. \qquad (5.7)$$

Because both $\{\Psi_i\}$ and $\{\Phi_a\}$ are the normalized sets, the expansion coefficients c

satisfy the unitary conditions:

$$\sum_i c^*_{ia} c_{ib} = \delta_{ab}, \quad \sum_a c^*_{ia} c_{ja} = \delta_{ij}. \tag{5.8}$$

When the expansion (5.7) is introduced into (5.5) and the above unitary conditions are used, we find that

$$Z(\beta) = \sum_i \langle \Psi_i | \exp(-\beta H) | \Psi_i \rangle$$

$$= \sum_a \sum_b \sum_i c^*_{ia} c_{ib} \langle \Phi_a | \exp(-\beta H) | \Phi_b \rangle$$

$$= \sum_a \langle \Phi_a | \exp(-\beta H) | \Phi_a \rangle. \tag{5.9}$$

This is the invariance of the trace of the density matrix under unitary transformations. The proof is given for the statistical operator herein; however, the same is true with any operator.

Similarly it can be shown that the trace of the product of three or more operators is invariant with respect to cyclic permutations of the operators. The proof for the product of three operators is as follows:

$$\text{tr}[ABC] = \sum_i \sum_j \sum_k \langle i|A|j \rangle \langle j|B|k \rangle \langle k|C|i \rangle$$

$$= \sum_i \sum_j \sum_k \langle k|C|i \rangle \langle i|A|j \rangle \langle j|B|k \rangle$$

$$= \text{tr}[CAB]. \tag{5.10}$$

If the statistical operator is differentiated with respect to β, what results is the so-called *Bloch equation*,

$$\frac{\partial \rho}{\partial \beta} = -H\rho. \tag{5.11}$$

Let us now define the probability density matrix, which is normalized, by

$$\rho^*(\beta) = \exp(-\beta H)/Z(\beta), \tag{5.12}$$

and then differentiate it with respect to β. We then find the modified Bloch equation given by

$$\frac{\partial \rho^*}{\partial \beta} = -(H - E)\rho^*. \tag{5.13}$$

5.3 The perturbation expansion

Even though the exact partition function can be calculated as the trace of the statistical operator in (5.9), in which very simple basis functions can be used, it is still very difficult to calculate the diagonal matrix elements of $\exp(-\beta H)$. For this purpose, a perturbation expansion for the density operator is desirable.

Let us assume that the Hamiltonian can be decomposed as

$$H = H_0 + H',\qquad(5.14)$$

hence

$$\rho(\beta) = \exp(-\beta(H_0 + H')).\qquad(5.15)$$

Now define $U(\beta)$ by

$$\exp(-\beta(H_0 + H')) = \exp(-\beta H_0)U(\beta),\qquad(5.16)$$

solve for $U(\beta)$,

$$U(\beta) = \exp(+\beta H_0)\exp(-\beta(H_0 + H')),\qquad(5.17)$$

differentiate both sides of the equation with respect to β, and then

$$\frac{\partial U(\beta)}{\partial \rho} = -H'(\beta)U(\beta),\qquad(5.18)$$

where

$$H'(\beta) = \exp(+\beta H_0)H'\exp(-\beta H_0).\qquad(5.19)$$

Integrating (5.18) formally yields

$$U(\beta) = 1 - \int_0^\beta H'(u)U(u)\,du.\qquad(5.20)$$

This is the solution of (5.18) which satisfies the initial condition $U(0) = 1$.

Now an iterative solution of (5.20) will be found:

$$U(\beta) = 1 + \sum_{n=1}^\infty (-1)^n$$

$$\cdot \int_0^\beta \int_0^{u_1} \cdots \int_0^{u_{n-1}} H'(u_1)H'(u_2)\cdots H'(u_n)\,du_1 du_2 \cdots du_n.$$

$$(5.21)$$

This expression may be given a more useful form:

$$U(\beta) = 1 + \sum_{n=1}^\infty \frac{(-1)^n}{n!}$$

$$\cdot \int_0^\beta \int_0^\beta \cdots \int_0^\beta P[H'(u_1)H'(u_2)\cdots H'(u_n)]\,du_1 du_2 \cdots du_n,\quad(5.22)$$

where P is called the *chronological ordering operator*. Since in the above multiple integration, $u_1, u_2, \ldots, u_n$ are all in the range $0 < u_1, u_2, \ldots, u_n < \beta$, their values can be in any order. The operator P rearranges the $H'(u_j)$ operators in the product in such a way that the $H'(u_j)$ which has the largest u_j is to the extreme left of the product, the $H'(u_j)$ which has the next largest u_j is in the second position from the left, and so on. The transformation from (5.21) to (5.22) may be obtained by the method of mathematical induction starting with the case of $n = 2$; however, a direct proof that (5.22) is indeed the solution of the differential equation (5.18) is more instructive.

If (5.22) is differentiated with respect to β, there appear n terms, since there are n integrations, each of which is differentiated with respect to the upper limit. If we recall the very definition of differentiation, the differentiation of the jth integration means

$$\lim_{\Delta\beta\to 0} \left[\int_0^{\beta+\Delta\beta} du_j - \int_0^{\beta} du_j \right] \Big/ \Delta\beta. \tag{5.23}$$

This shows that the value u_j in the factor $H'(u_j)$ is larger than β, larger than any other u's, and hence the $H'(u_j = \beta_+)$ should be taken to the extreme left. Such a contribution appears n times, and hence one obtains the differential equation (5.18).

5.4 Reduced density matrices

In Sec. 4.9 the Heisenberg exchange Hamiltonian was presented as appropriate for ferromagnetic insulators containing incomplete d-shell transition metal ions or f-shell rare-earth ions. In such an insulating ferromagnet, magnetic dipoles (called spins for simplicity hereafter) are permanently attached to their respective lattice sites in the crystal, and the dynamical states of the entire system can be characterized by the spin states of individual lattice sites.

We must be cautious about the statement in the above 'can be characterized by the spin states of individual lattice sites', because a wave function characterized by z-components of all the spin operators in the system is not necessarily an eigenfunction of the Heisenberg exchange Hamiltonian except in one case, i.e., the state in which all the z-components are up (the ground state) is an exact eigenfunction. Nevertheless, it is most convenient to choose a set of distinct functions, each of which is characterized by the z-components of all the spin operators in the system, and there are 2^N of them for spin $\frac{1}{2}$, as the bases of representation [Exercise 5.1].

The total density matrix for the system is, by definition, given by

$$\rho^{(N)} = \frac{\exp(-\beta H)}{Z_N(\beta)}, \tag{5.24}$$

where H is the Heisenberg Hamiltonian

$$H = -\sum_{f<g} J_{fg}\left(S_f^+ S_g^- + S_f^- S_g^+ + 2S_f^z S_g^z\right) - B\sum_f S_f^z; \tag{5.25}$$

B is an infinitesimal external magnetic field in the z-direction given in appropriate units, and $Z_N(\beta)$ is the partition function given by

$$Z_N(\beta) = \text{tr} \exp(-\beta H). \tag{5.26}$$

Now the reduced density matrices which satisfy the following conditions are introduced:

$$\text{tr}_n \rho^{(n)}(S_1, S_2, \ldots, S_n) = \rho^{(n-1)}(S_1, S_2, \ldots, S_{n-1}),$$

$$n = N, N-1, \ldots, 2,$$

$$\text{tr}_1 \rho^{(1)}(S_1) = 1. \tag{5.27}$$

It is not possible to perform successive reductions of the density matrix to smaller reduced density matrices starting directly from $\rho^{(N)}$; however, the general structure of the density matrix of any size, which satisfies all the required reducibility and normalization conditions, can be found rather easily in terms of some unknown parameters.

The required conditions for the density matrix are:

- it must satisfy the symmetries required by the system Hamiltonian;
- it must satisfy the crystallographic symmetries of the background lattice;
- it must satisfy the reducibility to smaller reduced density matrices;
- it should be normalized to unity.

In the following sections some smaller reduced density matrices will be constructed for the Heisenberg ferromagnet.

5.5 One-site and two-site density matrices

For the sake of simplicity let us limit our formulation to the system of $S = \frac{1}{2}$ spins; also simplified operator notations will be used.

The most general form of the one-site density matrix is now found.

5.5.1 One-site density matrix

The one-site density matrix for the ith site is a 2×2 matrix, because there are only two states: either up or down of each spin. Here, it is tacitly assumed that the representation in which the total magnetization is diagonal is the most convenient one.[†] Let this be

$$\rho^{(1)}(i) = \begin{bmatrix} a & 0 \\ 0 & 1-a \end{bmatrix}, \tag{5.28}$$

[†] The pair exchange interaction $S_i^+ S_j^- + S_i^- S_j^+ + 2S_i^z S_j^z$ conserves the total magnetization.

where a is the probability that the spin is up and $1 - a$ is the probability that the spin is down. The trace of the matrix is unity, i.e., it is normalized. The two off-diagonal elements represent the probability that the spin is changed from down to up or vice versa. Since the total magnetization should be conserved by the nature of the total Hamiltonian, the off-diagonal elements should be set equal to zero.

Let us introduce an operator which has the eigenvalue unity in the state in which the spin is up and the eigenvalue zero in the state in which the spin is down. Such an operator is called the *projection operator* in the state in which the spin is equal to $\frac{1}{2}$, and is given by

$$P(+) = \tfrac{1}{2}\left(1 + 2S_i^z\right). \tag{5.29}$$

Let us next calculate the statistical average of $P(+)$ with the density matrix (5.28), i.e.,

$$
\begin{aligned}
\langle P(+)\rangle &= \mathrm{tr}\left[P(+)\rho^{(1)}(i)\right]\\
&= \left\langle + \left|\tfrac{1}{2}\left(1 + 2S_i^z\right)\right| + \right\rangle \cdot a + \left\langle - \left|\tfrac{1}{2}\left(1 + 2S_i^z\right)\right| - \right\rangle \cdot (1 - a)\\
&= 1 \cdot a + 0 \cdot (1 - a)\\
&= a.
\end{aligned}
\tag{5.30}
$$

Similarly,

$$
\begin{aligned}
\langle P(-)\rangle &= \mathrm{tr}[P(-)\rho^{(1)}(i)]\\
&= \left\langle + \left|\tfrac{1}{2}\left(1 - 2S_i^z\right)\right| + \right\rangle \cdot a + \left\langle - \left|\tfrac{1}{2}\left(1 - 2S_i^z\right)\right| - \right\rangle \cdot (1 - a)\\
&= 0 \cdot a + 1 \cdot (1 - a)\\
&= 1 - a.
\end{aligned}
\tag{5.31}
$$

It has been established, in this way, that the probability that the spin is up is equal to the thermal average of the projection operator for that state. The one-site density matrix is now expressed as

$$\rho^{(1)}(i) = \frac{1}{2}\begin{bmatrix}\left(1 + 2\langle S_i^z\rangle\right) & 0\\ 0 & \left(1 - 2\langle S_i^z\rangle\right)\end{bmatrix}, \tag{5.32}$$

where $\langle S_i^z\rangle$ is the statistical average of the operator S_i^z, and the value is independent of the site number i because of the lattice translational invariance. The value of $\langle S_i^z\rangle$ is not known at this point. According to the Heisenberg exchange Hamiltonian, $\rho^{(1)}(i)$ is the exact form of the one-site reduced density matrix which will be used in the cluster variation method in Chapter 6, regardless of the degree of approximation (one-site, two-site, or even eight-site). The two elements of the one-site density matrix are, therefore, given as

$$R_1(1) = \tfrac{1}{2}(1 + x_1),$$
$$R_1(2) = \tfrac{1}{2}(1 - x_1),$$
$$x_1 = 2\langle S_i^z \rangle. \tag{5.33}$$

Note the change in notation from $\rho^{(1)}(i)$ to $R_1(1)$ and $R_1(2)$; however, there is no essential difference between the two notations at this point.

5.5.2 Two-site density matrix

It is a straightforward matter to show that the two-site reduced density matrix is a 4×4 matrix. There are 16 elements of which only four diagonal and two off-diagonal elements are nonzero and conserve the total magnetization. These elements are the statistical averages of the following projection operators:

$$\tfrac{1}{4}\left(1 \pm 2S_i^z\right)\left(1 \pm 2S_j^z\right): \qquad \text{diagonal}$$
$$S_i^+ S_j^-, \quad S_i^- S_j^+ : \qquad \text{off-diagonal} \tag{5.34}$$

where the two off-diagonal operators have the following matrix elements:

$$\langle + - |S_i^+ S_j^-| - + \rangle = \langle - + |S_i^- S_j^+| + - \rangle = 1. \tag{5.35}$$

The two-site density matrix is then given by

$$\rho^{(2)}(i, j) = \begin{bmatrix} A & 0 & 0 & 0 \\ 0 & B & E & 0 \\ 0 & F & C & 0 \\ 0 & 0 & 0 & D \end{bmatrix}, \tag{5.36}$$

where

$$A = N_2\langle\left(1 + 2S_i^z\right)\left(1 + 2S_j^z\right)\rangle,$$
$$B = N_2\langle\left(1 + 2S_i^z\right)\left(1 - 2S_j^z\right)\rangle,$$
$$C = N_2\langle\left(1 - 2S_i^z\right)\left(1 + 2S_j^z\right)\rangle,$$
$$D = N_2\langle\left(1 - 2S_i^z\right)\left(1 - 2S_j^z\right)\rangle,$$
$$E = \langle S_i^+ S_j^- \rangle,$$
$$F = \langle S_i^- S_j^+ \rangle,$$
$$N_2 = \tfrac{1}{4}. \tag{5.37}$$

Because of the translational and rotational symmetry of the underlying lattice, the conditions

$$B = C \quad \text{and} \quad E = F \tag{5.38}$$

must be met. It is important to confirm that the reducibility and normalization conditions are satisfied [Exercise 5.2]:

$$\text{tr}_i \rho^{(2)}(i, j) = \rho^{(1)}(j), \quad \text{tr}_j \rho^{(2)}(i, j) = \rho^{(1)}(i), \quad \text{and} \quad \text{tr}_{i,j} \rho^{(2)}(i, j) = 1. \tag{5.39}$$

It will be shown in Chapter 6 that the diagonalization of reduced density matrices is required in order to perform any statistical mechanics calculation. The 2×2 center block matrix of (5.36) is easily diagonalized, and four eigenvalues of the two-site density matrix are given by

$$\begin{aligned}
R_2(1) &= N_2(1 + 2x_1 + x_2), \\
R_2(2) &= N_2(1 - x_2 + 4y_1), \\
R_2(3) &= N_2(1 - x_2 - 4y_1), \\
R_2(4) &= N_2(1 - 2x_1 + x_2),
\end{aligned} \tag{5.40}$$

where $x_1 = 2\langle S_i^z \rangle = 2\langle S_j^z \rangle$, $x_2 = 4\langle S_i^z S_j^z \rangle$, and $y_1 = \langle S_i^- S_j^+ \rangle$.

Here again the notation has been changed from $\rho^{(2)}$ to R_2; however, the change is essential because the R_2's are for the diagonalized elements of the two-site density matrix.

This is again the exact form of the two-site reduced density matrix for any distance between the two sites (i, j), for any crystallographic structure of the underlying lattice, and in any degree of approximation.

5.6 The four-site reduced density matrix

$\rho^{(4)}(S_1, S_2, S_3, S_4)$ is a 16×16 matrix, in which the diagonal elements are given by the statistical averages of 16 different mutually orthogonal projection operators:[†]

$$\tfrac{1}{16}\langle (1 \pm 2S_1^z)(1 \pm 2S_2^z)(1 \pm 2S_3^z)(1 \pm 2S_4^z) \rangle. \tag{5.41}$$

As far as the diagonal elements are concerned, it is easily seen that the above $\rho^{(4)}(S_1, S_2, S_3, S_4)$ matrix is reduced to a smaller $\rho^{(3)}(S_1, S_2, S_3)$ matrix by taking a partial trace with respect to the fourth spin variable S_4, i.e.,

$$\begin{aligned}
&\tfrac{1}{16}\langle (1 \pm 2S_1^z)(1 \pm 2S_2^z)(1 \pm 2S_3^z)(1 + 2S_4^z) \rangle \\
&+ \tfrac{1}{16}\langle (1 \pm 2S_1^z)(1 \pm 2S_2^z)(1 \pm 2S_3^z)(1 - 2S_4^z) \rangle \\
&= \tfrac{1}{8}\langle (1 \pm 2S_1^z)(1 \pm 2S_2^z)(1 \pm 2S_3^z) \rangle,
\end{aligned} \tag{5.42}$$

which are the eight diagonal elements of $\rho^{(3)}(S_1, S_2, S_3)$ [Exercise 5.3].

[†] The formulation in this section was developed by Tanaka & Libelo (1975).

Many off-diagonal elements are zero because of nonconservation of the total magnetization, and those elements which are not equal to zero are:

$$\tfrac{1}{4}\langle S_1^- S_2^+ (1 \pm 2S_3^z)(1 \pm 2S_4^z)\rangle,$$
$$\tfrac{1}{4}\langle S_1^- (1 \pm 2S_2^z) S_3^+ (1 \pm 2S_4^z)\rangle,$$
$$\tfrac{1}{4}\langle (1 \pm 2S_1^z) S_2^- S_3^+ (1 \pm 2S_4^z)\rangle,$$
$$\tfrac{1}{4}\langle S_1^- (1 \pm 2S_2^z)(1 \pm 2S_3^z) S_4^+\rangle,$$
$$\tfrac{1}{4}\langle (1 \pm 2S_1^z) S_2^- (1 \pm 2S_3^z) S_4^+\rangle,$$
$$\tfrac{1}{4}\langle (1 \pm 2S_1^z)(1 \pm 2S_2^z) S_3^- S_4^+\rangle,$$
$$\langle S_1^- S_2^- S_3^+ S_4^+\rangle,$$
$$\langle S_1^- S_2^+ S_3^- S_4^+\rangle,$$
$$\langle S_1^- S_2^+ S_3^+ S_4^-\rangle, \tag{5.43}$$

and their complex conjugates.

By taking the partial trace with respect to the fourth spin variable S_4 in the top three expressions and ignoring the remaining six expressions of (5.43), it can be confirmed that the reducibility of $\rho^{(4)}$ to $\rho^{(3)}$ is satisfied with respect to the off-diagonal elements as well [Exercise 5.3].

The total 16×16 matrix will be broken down to

* one 1×1 matrix in the total spin $S = 2$ subspace;
* one 4×4 matrix in the total spin $S = 1$ subspace;
* one 6×6 matrix in the total spin $S = 0$ subspace;
* one 4×4 matrix in the total spin $S = -1$ subspace;
* one 1×1 matrix in the total spin $S = -2$ subspace.

These submatrices will be constructed, step by step, in the following:

1×1 matrix in the total $S^z = 2$ subspace

The diagonal element is

$$\tfrac{1}{16}\langle (1 + 2S_1^z)(1 + 2S_2^z)(1 + 2S_3^z)(1 + 2S_4^z)\rangle. \tag{5.44}$$

4×4 matrix in the total $S^z = 1$ subspace

$$\begin{bmatrix} A_{11} & A_{12} & A_{13} & A_{14} \\ A_{21} & A_{22} & A_{23} & A_{24} \\ A_{31} & A_{32} & A_{33} & A_{34} \\ A_{41} & A_{42} & A_{43} & A_{44} \end{bmatrix}, \tag{5.45}$$

where

$$A_{11} = \tfrac{1}{16}\langle(1 - 2S_1^z)(1 + 2S_2^z)(1 + 2S_3^z)(1 + 2S_4^z)\rangle,$$
$$A_{22} = \tfrac{1}{16}\langle(1 + 2S_1^z)(1 - 2S_2^z)(1 + 2S_3^z)(1 + 2S_4^z)\rangle,$$
$$A_{33} = \tfrac{1}{16}\langle(1 + 2S_1^z)(1 + 2S_2^z)(1 - 2S_3^z)(1 + 2S_4^z)\rangle,$$
$$A_{44} = \tfrac{1}{16}\langle(1 + 2S_1^z)(1 + 2S_2^z)(1 + 2S_3^z)(1 - 2S_4^z)\rangle,$$
$$A_{12} = \tfrac{1}{4}\langle S_1^- S_2^+ (1 + 2S_3^z)(1 + 2S_4^z)\rangle = A_{21},$$
$$A_{13} = \tfrac{1}{4}\langle S_1^- (1 + 2S_2^z) S_3^+ (1 + 2S_4^z)\rangle = A_{31},$$
$$A_{14} = \tfrac{1}{4}\langle S_1^- (1 + 2S_2^z)(1 + 2S_3^z) S_4^+ \rangle = A_{41},$$
$$A_{23} = \tfrac{1}{4}\langle(1 + 2S_1^z) S_2^- S_3^+ (1 + 2S_4^z)\rangle = A_{32},$$
$$A_{24} = \tfrac{1}{4}\langle(1 + 2S_1^z) S_2^- (1 + 2S_3^z) S_4^+ \rangle = A_{42},$$
$$A_{34} = \tfrac{1}{4}\langle(1 + 2S_1^z)(1 + 2S_2^z) S_3^- S_4^+ \rangle = A_{43}. \tag{5.46}$$

6×6 matrix in the total $S^z = 0$ subspace

$$\begin{bmatrix} B_{11} & B_{12} & B_{13} & B_{14} & B_{15} & B_{16} \\ B_{21} & B_{22} & B_{23} & B_{24} & B_{25} & B_{26} \\ B_{31} & B_{32} & B_{33} & B_{34} & B_{35} & B_{36} \\ B_{41} & B_{42} & B_{43} & B_{44} & B_{45} & B_{46} \\ B_{51} & B_{52} & B_{53} & B_{54} & B_{55} & B_{56} \\ B_{61} & B_{62} & B_{63} & B_{64} & B_{65} & B_{66} \end{bmatrix}, \tag{5.47}$$

where

$$B_{11} = \tfrac{1}{16}\langle(1 + 2S_1^z)(1 + 2S_2^z)(1 - 2S_3^z)(1 - 2S_4^z)\rangle,$$
$$B_{22} = \tfrac{1}{16}\langle(1 + 2S_1^z)(1 - 2S_2^z)(1 + 2S_3^z)(1 - 2S_4^z)\rangle,$$
$$B_{33} = \tfrac{1}{16}\langle(1 - 2S_1^z)(1 + 2S_2^z)(1 + 2S_3^z)(1 - 2S_4^z)\rangle,$$
$$B_{44} = \tfrac{1}{16}\langle(1 + 2S_1^z)(1 - 2S_2^z)(1 - 2S_3^z)(1 + 2S_4^z)\rangle,$$
$$\text{etc.},$$
$$B_{12} = \tfrac{1}{4}\langle(1 + 2S_1^z) S_2^- S_3^+ (1 - 2S_4^z)\rangle,$$
$$\text{etc.},$$
$$B_{34} = \langle(S_1^- S_2^+ S_3^+ S_4^-)\rangle,$$
$$\text{etc.} \tag{5.48}$$

4×4 matrix in the total $S^z = -1$ subspace

$$\begin{bmatrix} C_{11} & C_{12} & C_{13} & C_{14} \\ C_{21} & C_{22} & C_{23} & C_{24} \\ C_{31} & C_{32} & C_{33} & C_{34} \\ C_{41} & C_{42} & C_{43} & C_{44} \end{bmatrix}, \tag{5.49}$$

where

$$C_{11} = \tfrac{1}{16}\langle(1 + 2S_1^z)(1 - 2S_2^z)(1 - 2S_3^z)(1 - 2S_4^z)\rangle,$$
$$C_{22} = \tfrac{1}{16}\langle(1 - 2S_1^z)(1 + 2S_2^z)(1 - 2S_3^z)(1 - 2S_4^z)\rangle,$$
$$C_{33} = \tfrac{1}{16}\langle(1 - 2S_1^z)(1 - 2S_2^z)(1 + 2S_3^z)(1 - 2S_4^z)\rangle,$$
$$C_{44} = \tfrac{1}{16}\langle(1 - 2S_1^z)(1 - 2S_2^z)(1 - 2S_3^z)(1 + 2S_4^z)\rangle,$$
$$C_{12} = \tfrac{1}{4}\langle S_1^- S_2^+ (1 - 2S_3^z)(1 - 2S_4^z)\rangle = C_{21},$$
$$C_{13} = \tfrac{1}{4}\langle S_1^- (1 - 2S_2^z)S_3^+ (1 - 2S_4^z)\rangle = C_{31},$$
$$C_{14} = \tfrac{1}{4}\langle S_1^- (1 - 2S_2^z)(1 - 2S_3^z)S_4^+\rangle = C_{41},$$
$$C_{23} = \tfrac{1}{4}\langle(1 - 2S_1^z)S_2^- S_3^+ (1 - 2S_4^z)\rangle = C_{32},$$
$$C_{24} = \tfrac{1}{4}\langle(1 - 2S_1^z)S_2^- (1 - 2S_3^z)S_4^+\rangle = C_{42},$$
$$C_{34} = \tfrac{1}{4}\langle(1 - 2S_1^z)(1 - 2S_2^z)S_3^- S_4^+\rangle = C_{43}. \tag{5.50}$$

1×1 matrix in the total $S^z = -2$ subspace

The diagonal element is

$$\tfrac{1}{16}\langle(1 - 2S_1^z)(1 - 2S_2^z)(1 - 2S_3^z)(1 - 2S_4^z)\rangle. \tag{5.51}$$

The structure, reducibility conditions, and normalization condition of the density matrix developed in the above are valid for any geometrical locations of the four sites and are independent of the structure of the underlying lattice.

5.6.1 The reduced density matrix for a square cluster

If the shape of the four-site figure is square, the structure of the four-site density matrix will be drastically simplified because of the geometrical symmetry of the square.

The statistical averages, $\langle S_1^z\rangle$, $\langle S_1^z S_2^z\rangle$, $\langle S_1^+ S_2^-\rangle$, $\langle S_1^z S_2^z S_3^z\rangle$ and similar quantities which appear throughout the book, are, henceforth, called the one-site, two-site, three-site, etc. *correlation functions*. Strictly speaking, the correlation function in statistical mechanics is defined in a more rigorous way. The two-site correlation

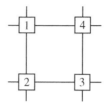

Fig. 5.1. Square cluster.

function, for example, is defined by $\langle S_1^z S_2^z \rangle - \langle S_1^z \rangle \langle S_2^z \rangle$, which approaches zero in the limit as the distance between two sites becomes infinite.

For the sake of notational convenience, let us define the x- and y-correlation functions. The y-correlation functions representing the off-diagonal quantities are zero for the Ising model.

$$
\begin{aligned}
&x_1 = 2\langle S_1^z \rangle, \quad x_2 = 4\langle S_1^z S_2^z \rangle, \quad x_3 = 8\langle S_1^z S_2^z S_3^z \rangle, \\
&x_4 = 16\langle S_1^z S_2^z S_3^z S_4^z \rangle, \quad x_5 = 4\langle S_1^z S_3^z \rangle, \\
&y_1 = \langle S_1^+ S_2^- \rangle, \quad y_2 = 2\langle S_1^z S_2^+ S_3^- \rangle, \\
&y_3 = 4\langle S_1^z S_2^z S_3^+ S_4^- \rangle, \quad y_4 = \langle S_1^+ S_2^- S_3^+ S_4^- \rangle, \\
&y_5 = \langle S_1^+ S_3^- \rangle, \quad y_6 = 2\langle S_1^z S_2^+ S_4^- \rangle, \\
&y_7 = 4\langle S_1^z S_3^z S_2^+ S_4^- \rangle, \quad y_8 = \langle S_1^+ S_2^+ S_3^- S_4^- \rangle.
\end{aligned}
\tag{5.52}
$$

The numerical factors in these definitions are introduced so that both the x- and y-functions are found between zero and unity.

The density matrices constructed in this section are also applicable to the Ising model by setting all the off-diagonal correlation functions, y_1 through y_8, equal to zero.

When the additional square symmetry is imposed, the density matrices take simpler forms.

1×1 matrix in the total $S^z = 2$ subspace

The diagonal element is

$$
\begin{aligned}
&\tfrac{1}{16}\langle (1 + 2S_1^z)(1 + 2S_2^z)(1 + 2S_3^z)(1 + 2S_4^z) \rangle \\
&= \tfrac{1}{16}(1 + 4x_1 + 4x_2 + 4x_3 + x_4 + 2x_5).
\end{aligned}
\tag{5.53}
$$

4×4 matrix in the total S^z = 1 subspace

$$
\begin{bmatrix}
a_1 & a_2 & a_3 & a_2 \\
a_2 & a_1 & a_2 & a_3 \\
a_3 & a_2 & a_1 & a_2 \\
a_2 & a_3 & a_2 & a_1
\end{bmatrix},
\tag{5.54}
$$

where

$$
\begin{aligned}
a_1 &= \tfrac{1}{16}\langle(1 - 2S_1^z)(1 + 2S_2^z)(1 + 2S_3^z)(1 + 2S_4^z)\rangle, \\
&= \tfrac{1}{16}(1 + 2x_1 - 2y_1 - x_4), \\
a_2 &= \tfrac{1}{4}\langle S_1^- S_2^+(1 + 2S_3^z)(1 + 2S_4^z)\rangle \\
&= \tfrac{1}{4}(y_1 + 2y_2 + y_3), \\
a_3 &= \tfrac{1}{4}\langle S_1^-(1 + 2S_2^z)S_3^+(1 + 2S_4^z)\rangle \\
&= \tfrac{1}{4}(y_5 + 2y_6 + y_7).
\end{aligned}
\tag{5.55}
$$

6×6 matrix in the total S^z = 0 subspace

$$
\begin{bmatrix}
b_1 & b_3 & b_4 & b_4 & b_3 & b_6 \\
b_3 & b_2 & b_3 & b_4 & b_5 & b_3 \\
b_4 & b_3 & b_1 & b_6 & b_3 & b_4 \\
b_4 & b_3 & b_6 & b_1 & b_3 & b_4 \\
b_3 & b_5 & b_3 & b_3 & b_2 & b_3 \\
b_6 & b_3 & b_4 & b_4 & b_3 & b_1
\end{bmatrix},
\tag{5.56}
$$

where

$$
\begin{aligned}
b_1 &= \tfrac{1}{16}\langle(1 - 2S_1^z)(1 - 2S_2^z)(1 + 2S_3^z)(1 + 2S_4^z)\rangle \\
&= \tfrac{1}{16}(1 + x_4 - 2x_5), \\
b_2 &= \tfrac{1}{16}\langle(1 - 2S_1^z)(1 + 2S_2^z)(1 - 2S_3^z)(1 + 2S_4^z)\rangle \\
&= \tfrac{1}{16}(1 - 4x_2 + x_4 + 2x_5), \\
b_3 &= \tfrac{1}{4}\langle(1 + 2S_1^z)S_2^- S_3^+(1 - 2S_4^z)\rangle = \tfrac{1}{4}(y_1 - y_3), \\
b_4 &= \tfrac{1}{4}\langle(1 + 2S_1^z)S_2^-(1 - 2S_3^z)S_4^+\rangle = \tfrac{1}{4}(y_5 - y_7), \\
b_5 &= \langle(S_1^- S_2^+ S_3^- S_4^+)\rangle = y_7, \quad b_6 = \langle(S_1^- S_2^- S_3^+ S_4^+)\rangle = y_8.
\end{aligned}
\tag{5.57}
$$

4×4 matrix in the total $S^z = -1$ subspace

$$
\begin{bmatrix}
c_1 & c_2 & c_3 & c_2 \\
c_2 & c_1 & c_2 & c_3 \\
c_3 & c_2 & c_1 & c_2 \\
c_2 & c_3 & c_2 & c_1
\end{bmatrix},
\tag{5.58}
$$

where

$$
c_1 = \tfrac{1}{16}(1 - 2x_1 + 2y_1 - x_4),
$$
$$
c_2 = \tfrac{1}{4}(y_1 - 2y_2 + y_3),
$$
$$
c_3 = \tfrac{1}{4}(y_5 - 2x_6 + x_7).
\tag{5.59}
$$

1×1 matrix in the total $S^z = -2$ subspace

The diagonal element is

$$
\tfrac{1}{16}\langle(1 - 2S_1^z)(1 - 2S_2^z)(1 - 2S_3^z)(1 - 2S_4^z)\rangle
$$
$$
= \tfrac{1}{16}(1 - 4x_1 + 4x_2 - 4x_3 + x_4 + 2x_5).
\tag{5.60}
$$

Diagonalization of the preceding matrices can be accomplished analytically, and the 16 eigenvalues are given as follows [Exercise 5.4]:

$$
\begin{aligned}
R_4(1) &= N_4(1 + 4x_1 + 4x_2 + 4x_3 + x_4 + 2x_5), \\
R_4(2) &= N_4(1 - 4x_1 + 4x_2 - 4x_3 + x_4 + 2x_5), \\
R_4(3) &= N_4(1 + 2x_1 - 2x_3 - x_4 + 8y_1 + 16y_2 + 8y_3 + 4y_5 + 8y_6 + 4y_7), \\
R_4(4) &= N_4(1 - 2x_1 + 2x_3 - x_4 + 8y_1 - 16y_2 + 8y_3 + 4y_5 - 8y_6 + 4y_7), \\
R_4(5) &= N_4(1 + 2x_1 - 2x_3 - x_4 - 8y_1 - 16y_2 - 8y_3 + 4y_5 + 8y_6 + 4y_7), \\
R_4(6) &= N_4(1 - 2x_1 + 2x_3 - x_4 - 8y_1 + 16y_2 - 8y_3 + 4y_5 - 8y_6 + 4y_7), \\
R_4(7) &= N_4(1 + 2x_1 - 2x_3 - x_4 - 4y_5 - 8y_6 - 4x_7), \quad g = 2, \\
R_4(8) &= N_4(1 - 2x_1 + 2x_3 - x_4 - 4y_5 + 8y_6 - 4x_7), \quad g = 2, \\
R_4(9) &= N_4(1 + x_4 - 2x_5 - 8y_5 + 8y_7 + 16y_8), \\
R_4(10) &= N_4(1 - 4x_2 + x_4 + 2x_5 - 16y_4), \\
R_4(11) &= N_4(1 + x_4 - 2x_5 - 16y_8), \quad g = 2, \\
R_4(12) &= N_4(1 - 2x_2 + x_4 + 8y_4 + 8y_8 + 4y_5 - 4y_7 + \Gamma), \\
R_4(13) &= N_4(1 - 2x_2 + x_4 + 8y_4 + 8y_8 + 4y_5 - 4y_7 - \Gamma),
\end{aligned}
\tag{5.61}
$$

where $N_4 = \tfrac{1}{16}$ and $g = 2$ means the multiplicity of the particular eigenvalue and all other eigenvalues without g are the singlets. The quantity Γ is

$$\Gamma = [(2x_2 - 2x_5 + 4y_5 - 4y_7 - 8y_4 + 8y_8)^2 + 128(y_1 - y_3)^2]^{\frac{1}{2}}. \quad (5.62)$$

5.6.2 The reduced density matrix for a regular tetrahedron cluster

The preceding formulation can be applied, with only a minor modification, to a face center cubic lattice, whereby the smallest and most compact four-site cluster is the regular tetrahedron. Because of the additional geometrical symmetry of the tetrahedron the following degeneracies can be imposed:

$$x_5 = 4\langle S_1^z S_3^z \rangle = 4\langle S_1^z S_2^z \rangle = x_2,$$
$$y_5 = \langle S_1^+ S_3^- \rangle = \langle S_1^+ S_2^- \rangle = y_1,$$
$$y_6 = 2\langle S_1^z S_2^+ S_4^- \rangle = 2\langle S_1^z S_2^+ S_3^- \rangle = y_2,$$
$$y_7 = 4\langle S_1^z S_3^z S_2^+ S_4^- \rangle = 4\langle S_1^z S_2^z S_3^+ S_4^- \rangle = y_3,$$
$$y_8 = 4\langle S_1^+ S_2^+ S_3^- S_4^- \rangle = 4\langle S_1^+ S_2^- S_3^+ S_4^- \rangle = y_4. \quad (5.63)$$

When these degeneracies are taken into account, the eigenvalues of the four-site density matrices are further simplified [Exercise 5.5]:

$$R_4(1) = N_4(1 + 4x_1 + 6x_2 + 4x_3 + x_4),$$
$$R_4(2) = N_4(1 - 4x_1 + 6x_2 - 4x_3 + x_4),$$
$$R_4(3) = N_4(1 + 2x_1 - 2x_3 - x_4 + 12y_1 + 24y_2 + 12y_3),$$
$$R_4(4) = N_4(1 - 2x_1 + 2x_3 - x_4 + 12y_1 - 24y_2 + 12y_3),$$
$$R_4(5) = N_4(1 + 2x_1 - 2x_3 - x_4 - 4y_1 - 8y_2 - 4y_3), \quad g = 3,$$
$$R_4(6) = N_4(1 - 2x_1 + 2x_3 - x_4 - 4y_1 + 8y_2 - 4y_3), \quad g = 3,$$
$$R_4(7) = N_4(1 - 2x_2 + x_4 + 16y_1 - 16y_3 + 16y_4),$$
$$R_4(8) = N_4(1 - 2x_2 + x_4 - 8y_1 + 8y_3 + 16y_4), \quad g = 2,$$
$$R_4(9) = N_4(1 - 2x_2 + x_4 - 16y_4), \quad g = 3. \quad (5.64)$$

5.7 The probability distribution functions for the Ising model

In the case of the classical Ising ferromagnet all the density matrices are diagonal, and hence we can drop the quantum mechanical terminologies altogether. The elements of the reduced density matrices are now called the *reduced distribution functions*.

The Hamiltonian for the Ising model ferromagnet is given as

$$H = -J \sum_{\langle i,j \rangle} \mu_i \mu_j, \quad \mu_i = \pm 1, \quad (5.65)$$

where $\langle i, j \rangle$ means that the summation covers all the pairs of the nearest-neighbor sites in the lattice.

The correlation functions for the Ising ferromagnet are defined in conjunction with the corresponding correlation functions for the Heisenberg ferromagnet:

$$x_1 = \langle \mu_i \mu_j \rangle = 4\langle S_i^z S_j^z \rangle,$$

$$y_1 = \langle \mu_i \rangle = 2\langle S_i^z \rangle, \quad \text{etc.} \tag{5.66}$$

In anticipation of the computer manipulation of the statistical mechanics calculations in later chapters, the correlation function of the product of an even number of variables is given an x name, and that of the product of an odd number of variables is given a y name. This is convenient because at high temperatures all the y-variables (odd correlations) vanish and one can solve only the x-problem, first at high temperatures. Only below a certain temperature (the phase transition temperature) do the y-variables begin to develop nonzero values, and hence the total number of unknown variables increases.

All the elements of the reduced distribution function for the Ising model may be found from the corresponding elements of the reduced density matrix for the Heisenberg model by dropping all the off-diagonal elements, i.e., the y-correlation functions, and then by replacing all the odd number correlation functions for the Heisenberg model with the y-correlation functions.

5.7.1 The one-site reduced distribution function

There are two elements of the one-site distribution function, the statistical averages of the corresponding projection operators. These are

$$R_1(1) = N_1 \langle (1 + \mu_i) \rangle = N_1(1 + y_1),$$
$$R_1(2) = N_1 \langle (1 - \mu_i) \rangle = N_1(1 - y_1), \tag{5.67}$$

where $N_1 = \frac{1}{2}$ is the normalization constant, and $y_1 = \langle \mu_i \rangle$.

5.7.2 The two-site distribution function

There are four elements of the two-site distribution function. These are found readily from (5.40) to be

$$R_2(1) = N_2 \langle (1 + \mu_1)(1 + \mu_2) \rangle = N_2(1 + x_1 + 2y_1),$$
$$R_2(2) = N_2 \langle (1 + \mu_1)(1 - \mu_2) \rangle = N_2(1 - x_1),$$
$$R_2(3) = N_2 \langle (1 - \mu_1)(1 + \mu_2) \rangle = N_2(1 - x_1),$$
$$R_2(4) = N_2 \langle (1 - \mu_1)(1 - \mu_2) \rangle = N_2(1 + x_1 - 2y_1), \tag{5.68}$$

where $N_2 = \frac{1}{4}$ is the normalization constant for the two-site distribution function, and $x_1 = \langle \mu_1 \mu_2 \rangle$.

5.7.3 The equilateral triangle distribution function

$$R_3(1) = N_3(1 + 3x_1 + 3y_1 + y_2),$$
$$R_3(2) = N_3(1 + 3x_1 - 3y_1 - y_2),$$
$$R_3(2) = N_3(1 - x_1 + y_1 - y_2), \quad g = 3,$$
$$R_3(2) = N_3(1 - x_1 - y_1 + y_2), \quad g = 3, \tag{5.69}$$

where $N_3 = \frac{1}{8}$ and the equilateral triangle correlation function is defined by

$$y_2 = \langle \mu_1 \mu_2 \mu_3 \rangle. \tag{5.70}$$

5.7.4 The four-site (square) distribution function

The elements of the distribution function can be found as the statistical average of appropriate projection operators. They are, in this case,

$$N_4 \langle (1 \pm \mu_1)(1 \pm \mu_2)(1 \pm \mu_3)(1 \pm \mu_4) \rangle, \tag{5.71}$$

where $N_4 = \frac{1}{16}$ is the normalization constant for the four-site distribution function. The square cluster (Fig. 5.2) is repeated here from p. 118 for the reader's convenience. In the case of the square these are expanded to give [Exercise 5.6]

$$R_4(1) = N_4(1 + 4x_1 + 2x_2 + x_4 + 4y_1 + 4y_3),$$
$$R_4(2) = N_4(1 + 4x_1 + 2x_2 + x_4 - 4y_1 - 4y_3),$$
$$R_4(3) = N_4(1 - x_4 + 2y_1 - 2y_3), \quad g = 4,$$
$$R_4(4) = N_4(1 - x_4 - 2y_1 + 2y_3), \quad g = 4,$$
$$R_4(5) = N_4(1 - 2x_2 + x_4), \quad g = 4,$$
$$R_4(6) = N_4(1 - 4x_1 + 2x_2 + x_4), \quad g = 2. \tag{5.72}$$

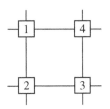

Fig. 5.2. Square cluster.

The new x- and y-correlation functions are defined by

$$x_2 = \langle \mu_1 \mu_3 \rangle,$$
$$x_4 = \langle \mu_1 \mu_2 \mu_3 \mu_4 \rangle,$$
$$y_3 = \langle \mu_1 \mu_2 \mu_3 \rangle. \tag{5.73}$$

5.7.5 The four-site (tetrahedron) distribution function

If the further symmetry condition holds that all the four sites are the nearest neighbors of one another, the preceding four-site distribution function will degenerate into the tetrahedron distribution. This means that $x_1 = x_2$, and hence

$$R_4(1) = N_4(1 + 6x_1 + x_4 + 4y_1 + 4y_2),$$
$$R_4(2) = N_4(1 + 6x_1 + x_4 - 4y_1 - 4y_2),$$
$$R_4(3) = N_4(1 - x_4 + 2y_1 - 2y_2), \quad g = 4,$$
$$R_4(4) = N_4(1 - x_4 - 2y_1 + 2y_2), \quad g = 4,$$
$$R_4(5) = N_4(1 - 2x_1 + x_4), \quad g = 6. \tag{5.74}$$

5.7.6 The six-site (regular octahedron) distribution function [Exercise 5.7]

$$R_6(1) = N_6(S_{12} + L_{12}),$$
$$R_6(2) = N_6(S_{12} - L_{12}),$$
$$R_6(3) = N_6(S_{34} + L_{34}), \quad g = 6,$$
$$R_6(4) = N_6(S_{34} - L_{34}), \quad g = 6,$$
$$R_6(5) = N_6(S_{56} + L_{56}), \quad g = 12,$$
$$R_6(6) = N_6(S_{56} - L_{56}), \quad g = 12,$$
$$R_6(7) = N_6(S_{78} + L_{78}), \quad g = 3,$$
$$R_6(8) = N_6(S_{78} - L_{78}), \quad g = 3,$$
$$R_6(9) = N_6(1 - 3x_2 + 3x_4 - x_6), \quad g = 8,$$
$$R_6(10) = N_6(1 - 4x_1 + x_2 - x_4 + 4x_5 - x_6), \quad g = 12, \tag{5.75}$$

where $N_6 = \frac{1}{64}$, and

$$S_{12} = 1 + 12x_1 + 3x_2 + 3x_4 + 12x_5 + x_6,$$
$$L_{12} = 6y_1 + 8y_2 + 12y_3 + 6y_4,$$
$$S_{34} = 1 + 4x_1 + x_2 - x_4 - 4x_5 - x_6,$$
$$L_{34} = 4y_1 - 4y_4,$$
$$S_{56} = 1 - x_2 - x_4 + x_6,$$
$$L_{56} = 2y_1 - 4y_3 + 2y_4,$$

$$S_{78} = 1 - 4x_1 + 3x_2 + 3x_4 - 4x_5 + x_6,$$
$$L_{78} = 2y_1 - 8y_2 + 4y_3 + 2y_4. \tag{5.76}$$

The newly introduced correlation functions are

$$\begin{aligned}
x_5 &= \langle \mu_1 \mu_2 \mu_3 \mu_6 \rangle, &\text{elongated tetrahedron,} \\
x_6 &= \langle \mu_1 \mu_2 \mu_3 \mu_5 \mu_6 \mu_7 \rangle, &\text{octahedron,} \\
y_4 &= \langle \mu_1 \mu_2 \mu_3 \mu_5 \mu_7 \rangle, &\text{pyramid.}
\end{aligned} \tag{5.77}$$

Exercises

5.1 Show that the state in which all the spins are parallel along the z-direction is the ground state of the Heisenberg Hamiltonian, and show that the eigenvalue of the Hamiltonian in the ground state is given by $E = -zNJS^2 - NBS$.

5.2 Confirm that the reducibility conditions and the normalization are satisfied;

$$\text{tr}_i \rho^{(2)}(i, j) = \rho^{(1)}(j), \quad \text{tr}_j \rho^{(2)}(i, j) = \rho^{(1)}(i),$$
$$\text{tr}_{i,j} \rho^{(2)}(i, j) = 1. \tag{E5.1}$$

5.3 By looking at (5.41), (5.42), and (5.43) confirm that the reducibility and normalization conditions are satisfied:

$$\begin{aligned}
&\text{tr}_1 \text{tr}_2 \text{tr}_3 \text{tr}_4 \rho^{(4)}(S_1, S_2, S_3, S_4), \\
&= \text{tr}_1 \text{tr}_2 \text{tr}_3 \rho^{(3)}(S_1, S_2, S_3), \\
&= \text{tr}_1 \text{tr}_2 \rho^{(2)}(S_1, S_2), \\
&= \text{tr}_1 \rho^{(1)}(S_1), \\
&= 1. \tag{E5.2}
\end{aligned}$$

[Hint]
The reduction steps from $\rho^{(4)}$ to $\rho^{(3)}$ were shown in the text. Similar steps may be followed to prove the above statement.

5.4 For the square lattice, find the 16 eigenvalues of the four-site density matrix.
[Hint]
As an example, let us find four eigenvalues of the total $S = 1$ submatrix given by (5.54):

$$\begin{bmatrix} a_1 & a_2 & a_3 & a_2 \\ a_2 & a_1 & a_2 & a_3 \\ a_3 & a_2 & a_1 & a_2 \\ a_2 & a_3 & a_2 & a_1 \end{bmatrix}. \tag{E5.3}$$

The secular determinant of this matrix is given by

$$\begin{vmatrix} a_1 - \lambda & a_2 & a_3 & a_2 \\ a_2 & a_1 - \lambda & a_2 & a_3 \\ a_3 & a_2 & a_1 - \lambda & a_2 \\ a_2 & a_3 & a_2 & a_1 - \lambda \end{vmatrix} = 0. \tag{E5.4}$$

If the elements of the first row are subtracted from the elements of the third row, and we follow the same procedure with the second and the fourth rows, we obtain

$$
\begin{vmatrix}
a_1 - \lambda & a_2 & a_3 & a_2 \\
a_2 & a_1 - \lambda & a_2 & a_3 \\
a_3 - a_1 + \lambda & 0 & a_1 - \lambda - a_3 & 0 \\
0 & a_3 - a_1 + \lambda & 0 & a_1 - \lambda - a_3
\end{vmatrix} = 0. \qquad (E5.5)
$$

This gives one doubly degenerate root

$$
\lambda = a_1 - a_3. \qquad (E5.6)
$$

If this doubly degenerate root is factored out, the remaining determinant looks like the following:

$$
\begin{vmatrix}
a_1 - \lambda & a_2 & a_3 & a_2 \\
a_2 & a_1 - \lambda & a_2 & a_3 \\
1 & 0 & -1 & 0 \\
0 & 1 & 0 & -1
\end{vmatrix} = 0. \qquad (E5.7)
$$

The next step is to add the elements of the third column to the elements of the first column, and to do the same with the fourth and second columns; we see that

$$
\begin{vmatrix}
a_1 + a_3 - \lambda & 2a_2 & a_3 & a_2 \\
2a_2 & a_1 + a_3 - \lambda & a_2 & a_3 \\
0 & 0 & -1 & 0 \\
0 & 0 & 0 & -1
\end{vmatrix} = 0. \qquad (E5.8)
$$

This is equivalent to the 2×2 determinant

$$
\begin{vmatrix}
a_1 + a_3 - \lambda & 2a_2 \\
2a_2 & a_1 + a_3 - \lambda
\end{vmatrix} = 0, \qquad (E5.9)
$$

which will yield the remaining two roots

$$
\lambda = a_1 + a_3 + 2a_2 \quad \text{and} \quad \lambda = a_1 + a_3 - 2a_2. \qquad (E5.10)
$$

Very similar row/column operations can yield the eigenvalues of the total $S = 0$, 6×6 submatrix and the total $S = -1$, 4×4 submatrix.

5.5 Find the nine eigenvalues (5.64) for the tetrahedron density matrix.

5.6 Find the six elements (5.72) of the four-site (square) distribution function.

5.7 Find the ten elements (5.75) of the regular octahedron distribution function.

6

The cluster variation method

6.1 The variational principle

Let us define a *variational potential F* as given by

$$F = \text{tr}[H\rho_t] + kT \ \text{tr}[\rho_t \log \rho_t], \tag{6.1}$$

where H, k, and T are the total Hamiltonian, the Boltzmann constant, and the absolute temperature, respectively, ρ_t is a trial density matrix, and tr is the abbreviation for the trace. As was explained in Sec. 5.2, the trace can be taken in any convenient representation because of the invariance of the trace under unitary transformations. The density matrix is assumed to be normalized to unity,

$$\text{tr} \ \rho_t = 1. \tag{6.2}$$

When the normalization is imposed as a subsidiary condition, with a Lagrange multiplier α, the variational potential is modified, i.e.,

$$F' = \text{tr}[H\rho_t] + kT \ \text{tr}[\rho_t \log \rho_t] - (\alpha + kT)(\text{tr} \ \rho_t - 1). \tag{6.3}$$

The variational principle is now stated as

The canonical density matrix is the one which minimizes the variational potential

$$dF' = 0. \tag{6.4}$$

If the variational potential, F', is minimized with respect to the trial density matrix ρ_t, one finds the canonical density matrix results:

$$\rho = \exp \beta(\alpha - H), \tag{6.5}$$

where $\beta = 1/kT$ and the subscript t has been suppressed since this is the canonical, rather than the trial, density matrix [Exercise 6.1].

It has been established, therefore, that the variational principle leads to the exact canonical density matrix.

When the normalization condition (6.2) is imposed, the Lagrange multiplier α will be fixed:

$$Z(\beta) = \exp(-\beta\alpha) = \text{tr}[\exp(-\beta H)], \tag{6.6}$$

where $Z(\beta)$ is the partition function for the system described by the total Hamiltonian H.

6.2 The cumulant expansion

It should be pointed out that the variational method, which will be developed in this chapter, is most suitable for *localized systems*. The localized system is a system which can be characterized by the dynamical variables attached to well defined crystalline lattice sites. It is assumed that the physical system is made up of N lattice sites which satisfy a certain translational symmetry.

Let us now define the *reduced density matrices* successively, each for a set of lattice sites, by the following relations:

$$\rho_t^{(n)}(1, 2, \ldots, n) = \text{tr}_{n+1}\rho_t^{(n+1)}(1, 2, \ldots, n, n+1),$$
$$n = 1, 2, \ldots, N - 1, \tag{6.7}$$

and

$$\text{tr}\,\rho_t^{(1)}(i) = 1, \quad n = 1, 2, \ldots, N. \tag{6.8}$$

Next, we define the *cluster functions*, $G^{(n)}$, and the *cumulant functions*, $g^{(n)}$, successively, by

$$G^{(1)}(i) = \text{tr}\left[\rho_t^{(1)}(i)\log\rho_t^{(1)}(i)\right] = g^{(1)}(i), \tag{6.9}$$

$$G^{(2)}(i, j) = \text{tr}\left[\rho_t^{(2)}(i, j)\log\rho_t^{(2)}(i, j)\right]$$
$$= g^{(1)}(i) + g^{(1)}(j) + g^{(2)}(i, j), \tag{6.10}$$

$$G^{(3)}(i, j, k) = \text{tr}\left[\rho_t^{(3)}(i, j, k)\log\rho_t^{(3)}(i, j, k)\right]$$
$$= g^{(1)}(i) + g^{(1)}(j) + g^{(1)}(k) + g^{(2)}(i, j)$$
$$+ g^{(2)}(j, k) + g^{(2)}(i, k) + g^{(3)}(i, j, k), \tag{6.11}$$

and, finally, the largest cluster function, $G^{(N)}$, which is the entropy term in the variational potential, is broken up as the sum of all the cumulant functions in the following fashion:

$$
\begin{aligned}
G^{(N)}(1, 2, \ldots, N) \\
= \mathrm{tr}\big[\rho_t^{(N)}(1, 2, \ldots, N) \log \rho_t^{(N)}(1, 2, \ldots, N)\big] \\
= \sum_i g^{(1)}(i) + \sum_{i<j} g^{(2)}(i, j) + \sum_{i<j<k} g^{(3)}(i, j, k) + \cdots + g^{(N)}(1, 2, \ldots, N).
\end{aligned}
$$

(6.12)

In this way the variational potential is expressed as

$$
\begin{aligned}
F = \mathrm{tr}\big[H\rho_t^{(N)}(1, 2, \ldots, N)\big] \\
+ kT \Bigg[\sum_i g^{(1)}(i) + \sum_{i<j} g^{(2)}(i, j) \\
+ \sum_{i<j<k} g^{(3)}(i, j, k) + \cdots + g^{(N)}(1, 2, \ldots, N) \Bigg].
\end{aligned}
$$

(6.13)

The expansion (6.12) of the largest cluster function $G^{(N)}$ in terms of the cumulant functions, $g^{(i)}$, $i = 1, 2, \ldots, N$, is called the *cumulant expansion*. The variational principle is now restated.

The canonical density matrix and the reduced equilibrium density matrices are the ones which minimize the variational potential (6.13), with N successive reducibility conditions (6.7), (6.8) imposed.

It looks as if the number of degrees of freedom has been increased in the new form of the variational principle, because we now have N reduced density matrices, which are the variational functions, compared with just one trial density matrix in the original variational function. This is not the case, however, because of the additional subsidiary conditions, i.e., there are N successive reducibility conditions imposed upon the reduced density matrices. The two forms of the variational principle are, therefore, equivalent since (6.12) is the definition of the $g^{(N)}$ in terms of the $G^{(N)}$ and is hence a tautology unless some approximations in the cumulant expansion are introduced. For the purposes of the actual applications of the variational principle, however, the second form is more convenient.

One important property of the cumulant functions should be pointed out. For instance, if the equation defining $g^{(2)}(i, j)$ is solved for itself, we obtain

$$
\begin{aligned}
g_2(i, j) = \mathrm{tr}\big[\rho_t^{(2)}(i, j) \log \rho_t^{(2)}(i, j)\big] - \mathrm{tr}\big[\rho_t^{(1)}(i) \log \rho_t^{(1)}(i)\big] \\
- \mathrm{tr}\big[\rho_t^{(1)}(j) \log \rho_t^{(1)}(j)\big].
\end{aligned}
$$

(6.14)

When the distance between the two sites, i and j, becomes large, $\rho_t^{(2)}(i, j)$ breaks down into the product of $\rho_t^{(1)}(i)$ and $\rho_t^{(1)}(j)$,

$$\rho_t^{(2)}(i, j) = \rho_t^{(1)}(i)\rho_t^{(1)}(j), \tag{6.15}$$

and hence

$$g^{(2)}(i, j) = 0. \tag{6.16}$$

This is why $g^{(2)}(i, j)$ is called the two-site cumulant function, which is expected to be a small quantity. $g^{(3)}(i, j, k)$ has a similar property, i.e., when one of the sites (i, j, k) is located at a large distance from the remaining two sites, $g^{(3)}(i, j, k)$ approaches zero. All the cumulant functions are successively smaller quantities.

6.3 The cluster variation method

The cumulant expansion is one of the most convenient ways of calculating the thermodynamical quantities in some desired degree of approximation by means of the variational method. The degree of approximation is decided upon by the number of cumulant functions retained in the variational potential. Approximation is due to this truncation of the cumulant expansion. In this method, therefore, the largest reduced density matrix which is included in the variational potential must first be chosen. This is called the *parent cluster* (density matrix) because the density matrix $\rho_t^{(n)}(1, 2, \ldots, n)$ is defined by a cluster of inclusive lattice sites.

Then we must consider all the reduced density matrices which are constructed by picking the lattice sites included in the parent cluster. The clusters representing these reduced density matrices are called the *subclusters* of the parent cluster. We can then express the variational potential in terms of the cumulant functions $g^{(i)}$, corresponding to the parent cluster and its subclusters.

The next step is to express all the cumulant functions which are retained in terms of the corresponding cluster functions $G^{(i)}$, $i = 1, 2, \ldots, n$. In this way the final form of the variational potential is determined.

In some approximations, more than one parent cluster may be chosen to be included in the variational potential. In such a case all the subcluster cumulant functions of the parent clusters must be included in the variational potential.

It usually happens that some of the cluster functions do not appear in the final potential in spite of the fact that the corresponding cumulant functions are included in the variational potential to start with. This happens if a particular cluster function, which does not appear in the final potential, cannot be a geometrical common part of the two parent clusters, either of the same type or different types, no matter how those parent clusters are translated and rotated with respect to each other in the given lattice of the physical system.

We can work with many different approximations; for example, the one-site, two-site (pair), triangular, square, tetrahedron, octahedron, etc. approximations. This formulation is called the *cluster variation method* in applications of equilibrium statistical mechanics. The cluster variation method was originally introduced by Kikuchi (1951); however, the formulation presented here is due to Morita (1957).

In the formulation of the cluster variation method, it is crucially important to decide on the types of parent clusters to be included in the variational potential, to express the variational potential in terms of the cumulant functions, and then to translate the cumulant functions into the cluster functions. One cannot work directly with the cluster function representation from the outset.

Up to this point, the formulation is entirely algebraic and no variational calculation is involved. The number of times each cluster function appears in the variational potential depends uniquely upon the types of parent clusters included in the variational potential and the crystallographic structure of the physical system under investigation. A mathematical formulation pertaining to this aspect of the cluster variation method is fully developed by Morita (1994).

6.4 The mean-field approximation

Let us, in this section, examine possibly the simplest, or crudest, approximation. This is the *mean-field approximation*. From the point of view of the cluster variation method, the mean-field approximation can be given a simple mathematical characterization, i.e., all the cumulant functions are neglected except $g^{(1)}$:

$$g^{(n)} = 0, \quad n = 2, 3, \ldots, N. \tag{6.17}$$

In particular, $g^{(2)}(i, j) = 0$ is equivalent to

$$\rho_t^{(2)}(i, j) = \rho_t^{(1)}(i)\rho_t^{(1)}(j). \tag{6.18}$$

For this reason, the mean-field approximation can be given another characterization, i.e., an approximation in which all the *statistical correlation effects* are ignored.

Let us take the nearest-neighbor interacting Ising ferromagnet as a working system. The Hamiltonian is

$$H = -2J \sum_{\langle i,j \rangle} S_i^z S_j^z, \quad (S_i^z = \pm\tfrac{1}{2}, i = 1, 2, \ldots, N), \tag{6.19}$$

where $\langle i, j \rangle$ means that the sites i and j are the nearest neighbors of each other and the summation is taken throughout the lattice over the nearest-neighbor pairs.

The variational potential is, then, given by[†]

$$F = -2J \sum_{\langle i,j \rangle} \mathrm{tr}\big[\rho_{\mathrm{t}}^{(1)}(i)S_i^z\big]\mathrm{tr}\big[\rho_{\mathrm{t}}^{(1)}(j)S_j^z\big]$$
$$+ kT \sum_i \mathrm{tr}\big[\rho_{\mathrm{t}}^{(1)}(i) \log \rho_{\mathrm{t}}^{(1)}(i)\big] - \sum_i \big[(f_i + kT)(\mathrm{tr}\rho_{\mathrm{t}}^{(1)}(i) - 1)\big],$$

$$(6.20)$$

where f_i is the Lagrange multiplier for the normalization of the density matrix $\rho_{\mathrm{t}}^{(1)}(i)$.

When F is minimized with respect to $\rho_{\mathrm{t}}^{(1)}(i)$, we obtain

$$\rho^{(1)}(i) = \exp\big(2\beta z J \langle S^z \rangle S_i^z\big)/Z(\beta),$$

$$(6.21)$$

where $\langle S^z \rangle = \mathrm{tr}[S_j^z \rho^{(1)}(j)]$, $\beta = 1/kT$, and $Z(\beta)$ is the normalization factor

$$Z(\beta) = \exp(-\beta f_i) = \mathrm{tr}\big[\exp\big(2\beta z J \langle S^z \rangle S_i^z\big)\big]$$
$$= 2\cosh(\beta z J \langle S^z \rangle).$$

$$(6.22)$$

The factor z, in the above equations, is the number of the nearest-neighbor lattice sites of a given lattice point; $z = 6, 8, 12$ for the simple cubic, body center cubic, and face center cubic lattices, respectively. In the derivation of (6.21) it is assumed that the lattice has a translational symmetry and hence the average $\langle s \rangle$ is independent of the lattice site.

If (6.21) and (6.22) are substituted into the equation

$$\langle S^z \rangle = \mathrm{tr}\big[S_i^z \rho^{(1)}(i)\big],$$

$$(6.23)$$

a transcendental equation for $\langle S^z \rangle$,

$$\langle S^z \rangle = \tfrac{1}{2}\tanh[\beta z J \langle S^z \rangle]$$

$$(6.24)$$

results. By inspection it is found that $\langle S^z \rangle = 0$ is a solution at all temperatures. This is a trivial solution.

In order to find a nontrivial solution, we can set up a parametric representation of (6.24) as follows:

$$\langle S^z \rangle = \tfrac{1}{2}\tanh x,$$

$$(6.25)$$

$$\langle S^z \rangle = \frac{kT}{zJ}x.$$

$$(6.26)$$

The solution of (6.24) is now found as an intersection of two curves represented by the above two equations (see Fig. 6.1). Equation (6.25) is a curve which approaches

[†] Quantum mechanics terminologies are used because the formulations developed in this section are readily applicable to quantum systems, even though the Ising ferromagnet is a classical system.

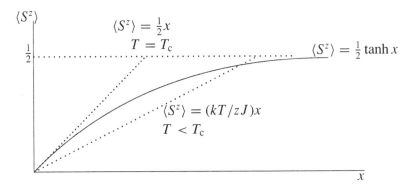

Fig. 6.1. Solution of Equation (6.24).

$\frac{1}{2}$ asymptotically at large values of x; (6.26) is a straight line through the origin and intersects (6.25) at $\langle S^z \rangle \neq 0$, $x \neq 0$ if the slope is smaller than $\frac{1}{2}$. In terms of the temperature, if T is higher than $zJ/2k$, $\langle S^z \rangle = 0$ is the only solution, and if T is lower than $zJ/2k$, then there is a nonzero solution; for this reason, $T_c = zJ/2k$ is called the critical temperature. In this way, it is possible to predict a phase transition from the *paramagnetic phase*, $\langle S^z \rangle = 0$, to a phase in which there is a *spontaneous magnetization*, $\langle S^z \rangle \neq 0$.

It would be appropriate to give some explanation for the name *mean-field* approximation. From the Hamiltonian (6.19) we see that any one of the spins in the lattice, say the ith spin, interacts with the nearest-neighbor spins, $j \neq i$, through the exchange interaction of strength J. Since any correlation between the spins in the lattice is ignored in the mean-field approximation, the effect of the neighboring spins upon the ith spin may be replaced by a mean-field; i.e., an effective Hamiltonian for the ith spin would be given by

$$H_i = -2J \sum_{j=1,2,\ldots,z} \langle S_j^z \rangle S_i^z = -2z \langle S^z \rangle J S_i^z, \tag{6.27}$$

where all the mean-fields of the neighboring spins are assumed to be the same. The value of the mean-field $\langle s \rangle$ is not known at this moment. From the form of the approximate Hamiltonian (6.27), however, the statistical probability that the ith spin is found in the state S_i^z will be given also by (6.21), although the mathematical derivation of it is different from the one which led to (6.21).

The final assumption is that the statistical average of the spin variable S_i^z is also equal to $\langle S^z \rangle$, i.e., a consistency equation results:

$$\langle S^z \rangle = \text{tr}\left[S_i^z \rho_1(i) \right] = \tfrac{1}{2}\tanh[\beta z J \langle S^z \rangle]. \tag{6.28}$$

This equation turns out to be identical to (6.24).

A nonzero solution of (6.24) exists for T smaller than T_c given by

$$kT_c = zJ/2, \qquad [\beta_c zJ/2 = 1]. \tag{6.29}$$

The term *Weiss' molecular field* may be used instead of the *mean-field* in some literature; this refers to the pioneering work of P. Weiss (1907). Historically this approximation was introduced in the theory of the order–disorder phase transformation in β-brass by Bragg & Williams (1934), and hence the approximation is also called the *Bragg–Williams approximation*. However, it has been called the *mean-field* approximation in recent years.

6.5 The Bethe approximation

A one step higher approximation in the cluster variation method is naturally the two-site or pair approximation. The variational potential without subsidiary conditions, in this case, is given from (6.13) by

$$F = -2J \sum_{\langle i,j \rangle} \mathrm{tr}\big[S_i^z S_j^z \rho_t^{(2)}(i,\,j)\big] + kT \left[\sum_i g^{(1)}(i) + \sum_{\langle i,j \rangle} g^{(2)}(i,\,j)\right].$$

$$\tag{6.30}$$

It should be pointed out that there is a significant difference between the above potential and the one given by (6.13).

Because of the statistical correlations among interacting spins, the range or the spatial distance over which the two-site correlation extends is not limited, in general, to the range of the interaction potential in the Hamiltonian. For this reason, the summation of $g^{(2)}(i,\,j)$ in (6.13) extends over all possible distances between i and j. The summation over $\langle i,\,j \rangle$ in (6.30), however, extends only over the nearest-neighbor pairs in the lattice. The two-site cumulant function $g^{(2)}(i,\,j)$ for $(i,\,j)$ distances beyond the range of the interaction potential vanishes identically as a consequence of the pair approximation [Exercise 6.2].

When $g^{(2)}(i,\,j)$ is expressed in terms of the density matrices from (6.14) and some rearrangement of terms is carried out, we find

$$F = -2J \sum_{\langle i,j \rangle} \mathrm{tr}_{i,j}\big[S_i^z S_j^z \rho_t^{(2)}(i,\,j)\big]$$

$$+ kT \left\{ \sum_i (1-z)\mathrm{tr}_i\big[\rho_t^{(1)}(i) \log \rho_t^{(1)}(i)\big] + \sum_{\langle i,j \rangle} \mathrm{tr}_{i,j}\big[\rho_t^{(2)}(i,\,j) \log \rho_t^{(2)}(i,\,j)\big] \right\}.$$

$$\tag{6.31}$$

In the next step, various normalization conditions and the reducibility conditions must be incorporated into the variational function by means of the Lagrange
multipliers. The added terms are:

$$+\sum_{\langle i,j \rangle} \Big\{ \mathrm{tr}_i \big([\mathrm{tr}_j \rho_t^{(2)}(i,j) - \rho_t^{(1)}(i)][\lambda(i) - kT \log \rho^{(1)}(i)] \big)$$

$$+ \mathrm{tr}_j \big([\mathrm{tr}_i \rho_t^{(2)}(i,j) - \rho_t^{(1)}(j)][\lambda(j) - kT \log \rho^{(1)}(j)] \big) \Big\}$$

$$- \sum_{\langle i,j \rangle} [\mathrm{tr}_{i,j} \rho_t^{(2)}(i,j) - 1][f_2 - 2f_1 + kT] - \sum_i [\mathrm{tr}_i \rho_t^{(1)}(i) - 1][f_1 + kT].$$

$$(6.32)$$

The first line represents the reducibility from $\rho_t^{(2)}(i,j)$ to $\rho_t^{(1)}(i)$,

$$\mathrm{tr}_j \rho_t^{(2)}(i,j) = \rho_t^{(1)}(i). \qquad (6.33)$$

Since this is a matrix equation, the corresponding Lagrange multiplier must also
be a matrix. In (6.32), the matrix multiplier has been chosen to be

$$\big[\lambda(i) - kT \log \rho^{(1)}(i) \big], \qquad (6.34)$$

where $\lambda(i)$ is a 2×2 matrix which is linearly independent of the unit matrix and it
should also have the dimensions of energy. We can further specify this to be

$$\lambda(i) = 2J\lambda s_i, \qquad (6.35)$$

where λ is a scalar multiplier which should be determined later by the reducibility
condition. The term $-kT \log \rho^{(1)}(i)$ in (6.32) is added for the sake of convenience.
This $\rho^{(1)}(i)$ is not the trial density matrix, but rather it is the equilibrium density
matrix which will be found later as a result of the variational calculation, hence it
is not a variational parameter. The addition of any known quantity, $-kT \log \rho^{(1)}(i)$,
to $\lambda(i)$ does not change the indeterminate nature of the multiplier.

It should be noted that the reducibility condition as given in the first line of (6.32)
has a somewhat important implication. If the $\lambda(i)$ is distributed to both $\rho_t^{(2)}(i,j)$
and $\rho_t^{(1)}(i)$, the condition becomes

$$\mathrm{tr}_i \mathrm{tr}_j \lambda(i) \rho_t^{(2)}(i,j) = \mathrm{tr}_i \lambda(i) \rho_t^{(1)}(i). \qquad (6.36)$$

This means that the thermal average of any one-site quantity must be evaluated
consistently by the two-site density matrix as well as by the one-site density matrix.
This is the reason why $\lambda(i)$ should not be a constant matrix which simply leads to
successive normalization.

The third and fourth lines in (6.32) are for the normalization of $\rho_t^{(2)}(i,j)$ and
$\rho_t^{(1)}(i)$, respectively. f_2 and f_1 are the corresponding multipliers, and other factors

are added for the sake of convenience. Because of these nonessential modifications of the multipliers, the final forms of $\rho^{(2)}(i, j)$ and $\rho^{(1)}(i)$ will be found in concise forms. After some manipulation, we find the density matrices in their final forms [Exercise 6.3]:

$$\rho^{(1)}(i) = \exp \beta \left[f_1 + 2zJ\lambda S_i^z \right],$$
$$\rho^{(2)}(i, j) = \exp \beta \left[f_2 + 2(z - 1)J\lambda \left(S_i^z + S_j^z \right) + 2J S_i^z S_j^z \right]. \tag{6.37}$$

These expressions have simple interpretations. The Lagrange multiplier λ plays the role of a molecular field. The Ising spin S_i^z is surrounded by the z nearest neighbors, and hence the strength of the molecular field is z times $2J\lambda$.

For the nearest-neighbor pair, (i, j), each of the spins, say S_i^z, sees the molecular field produced by its $z - 1$ neighbors but not that of S_j^z, because the interaction with S_j^z is explicitly taken into account through the exchange interaction. It is found that $\rho^{(1)}(i)$ in (6.37) is the same as that in (6.21) except λ replaces $\langle S^z \rangle$.

Equations (6.37) may be brought forth without resorting to the cluster variation method; this was exactly what H. Bethe proposed in his original paper (Bethe, 1935).

The normalization factors are fixed immediately:

$$1 = \text{tr}_i \rho^{(1)}(i) = 2 \exp(\beta f_1) \cosh(\beta z J \lambda),$$
$$1 = \text{tr}_{i,j} \rho^{(2)}(i, j) = 2 \exp(\beta f_2)[\exp(\beta J/2) \cosh 2\beta(z - 1)J\lambda + \exp(-\beta J/2)]. \tag{6.38}$$

In order to determine λ, the consistency condition (6.36) will be employed [Exercise 6.4]:

$$\langle S^z \rangle = \tfrac{1}{2} \tanh(\beta z J \lambda)$$
$$= \frac{\tfrac{1}{2}\exp(\beta J/2) \sinh 2\beta(z - 1)J\lambda}{\exp(\beta J/2) \cosh 2\beta(z - 1)J\lambda + \exp(-\beta J/2)}. \tag{6.39}$$

Once λ is determined as a nontrivial solution of the above equation, the spontaneous magnetization or the long range order, $\langle S^z \rangle$, is determined. Since λ is not identical to $\langle S^z \rangle$, some improvement in the spontaneous magnetization compared with that in the mean-field approximation should be expected. The reciprocal critical temperature, β_c, at which the long range order, $\langle S^z \rangle$ and hence λ, vanish is found from (6.39) by taking the limit as λ approaches zero:

$$\frac{z}{2(z - 1)} = \frac{\exp(\beta_c J/2)}{\exp(\beta_c J/2) + \exp(-\beta_c J/2)}, \tag{6.40}$$

or solving for $\beta_c J$, or for $\beta_c z J/2$ in comparison with (6.28),

$$\beta_c z J/2 = -\frac{z}{2}\log\left(1 - \frac{2}{z}\right). \tag{6.41}$$

If the logarithm is expanded into an infinite series, we obtain

$$\beta_c J z/2 = 1 + \tfrac{1}{2}(2/z) + \tfrac{1}{3}(2/z)^2 + \cdots. \tag{6.42}$$

This expansion suggests that the mean-field approximation becomes exact in the limit as z approaches infinity, i.e., if one retains only the first term of the expansion one finds the mean-field result:

$$kT_c = 1/\beta_c = zJ/2. \tag{6.43}$$

6.6 Four-site approximation

In this section a four-site approximation will be developed, but at the same time the formulation is generalized to the spin $= \tfrac{1}{2}$ Heisenberg ferromagnet. The results for the classical case of the Ising model can be found easily from the results developed in this section by setting all the off-diagonal elements of the density matrices equal to zero. The four-site approximation developed in this section will be applied to the two-dimensional square lattice, three-dimensional simple cubic, and face center cubic lattices with only minor modifications.

The Heisenberg Hamiltonian in the presence of an external magnetic field B is given by

$$H = -2J\sum_{\langle i,j\rangle} \mathbf{S}_i \cdot \mathbf{S}_j - B\sum_i S_i^z. \tag{6.44}$$

The variational potential is

$$F = -B\sum_1 \mathrm{tr}_1 S_1^z \rho_t^{(1)}(1) - 2J\sum_{\langle 1,2\rangle} \mathrm{tr}_{1,2}\left[\mathbf{S}_1 \cdot \mathbf{S}_2 \rho_t^{(2)}(1,2)\right]$$
$$+ kT\left[\sum_1 g^{(1)}(1) + \sum_{\langle 1,2\rangle} g^{(2)}(1,2) + \sum_{\langle 1,3\rangle} g^{(2)}(1,3)\right.$$
$$\left. + \sum_{1,2,3} g^{(3)}(1,2,3) + \sum_{1,2,3,4} g^{(4)}(1,2,3,4)\right]. \tag{6.45}$$

In this approximation, the parent, largest, cluster is the square, and all the subclusters of the square are included in the cumulant expansion (Fig. 6.2).

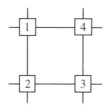

Fig. 6.2. Square cluster.

The next step is to express all the cumulant functions, g, in terms of the cluster functions, G:

$$G^{(1)}(1) = \mathrm{tr}\rho_t^{(1)}(1) \log \rho_t^{(1)}(1) = g^{(1)}(1),$$

$$G^{(2)}(1, 2) = \mathrm{tr}\rho_t^{(2)}(1, 2) \log \rho_t^{(2)}(1, 2)$$
$$= g^{(1)}(1) + g^{(1)}(2) + g^{(2)}(1, 2),$$

$$G^{(3)}(1, 2, 3) = \mathrm{tr}\rho_t^{(3)}(1, 2, 3) \log \rho_t^{(3)}(1, 2, 3)$$
$$= g^{(1)}(1) + g^{(1)}(2) + g^{(1)}(3)$$
$$+ g^{(2)}(1, 2) + g^{(2)}(1, 3)$$
$$+ g^{(2)}(2, 3) + g^{(3)}(1, 2, 3),$$

$$G^{(4)}(1, 2, 3, 4) = \mathrm{tr}\rho_t^{(4)}(1, 2, 3, 4) \log \rho_t^{(4)}(1, 2, 3, 4)$$
$$= g^{(1)}(1) + g^{(1)}(2) + g^{(1)}(3) + g^{(1)}(4)$$
$$+ g^{(2)}(1, 2) + g^{(2)}(2, 3) + g^{(2)}(3, 4)$$
$$+ g^{(2)}(1, 4) + g^{(2)}(1, 3) + g^{(2)}(2, 4)$$
$$+ g^{(3)}(1, 2, 3) + g^{(3)}(2, 3, 4) + g^{(3)}(3, 4, 1)$$
$$+ g^{(3)}(4, 1, 2) + g^{(4)}(1, 2, 3, 4). \tag{6.46}$$

When these equations are solved for g in terms of G and substituted in (6.45), we have

$$F = -B \sum_1 \mathrm{tr}_1\left[S_1^z \rho_t^{(1)}(1)\right] - 2J \sum_{\langle 1,2 \rangle} \mathrm{tr}_{1,2}\left[S_1 \cdot S_2 \rho_t^{(2)}(1, 2)\right]$$

$$+ kT\left[\sum_1 (1 + z_2 - z)G^{(1)}(1) + \sum_{\langle 1,2 \rangle} (1 - z_4)G^{(2)}(1, 2)\right.$$

$$\left. + \sum_{1,2,3,4} G^{(4)}(1, 2, 3, 4)\right], \tag{6.47}$$

where z and z_2 are the numbers of the nearest and second neighbors in the lattice, respectively, z_4 is the number of neighboring squares which share one side of the square. For the square lattice $z = 4$, $z_2 = 4$, $z_3 = 4$, $z_4 = 2$ and $z = 6$, $z_2 = 12$, $z_4 = 4$ for the simple cubic lattice.

It is interesting to note that neither $G^{(3)}(1, 2, 3)$ nor $G^{(2)}(1, 3)$ appears in the final potential. The reason is that neither of these clusters can be a common part of two squares, the parent clusters, no matter how they are oriented themselves with respect to each other in the lattice space.

The Helmholtz potential F/N now contains three trial density matrices $\rho_t^{(1)}$, $\rho_t^{(2)}$, and $\rho_t^{(4)}$, and they must be determined in such a way that the potential is a minimum. Since they must satisfy the reducibility and normalization conditions, they are not statistically independent.

In the variational formulation, it is a standard procedure to incorporate the subsidiary conditions into the variational function by means of the Lagrange undetermined multipliers. The expression which should be added to the potential may be given in the following form:

$$+\sum_{1,2,3,4} \mathrm{tr}_{1,2}\big[\mathrm{tr}_{3,4}\rho_t^{(4)}(1, 2, 3, 4) - \rho_t^{(2)}(1, 2)\big]\big[\lambda_{3,4}(1, 2) - kT \log \rho_t^{(2)}(1, 2)\big],$$

$$(6.48)$$

$$+\sum_{1,2,3,4} \mathrm{tr}_1\big[\mathrm{tr}_{2,3,4}\rho_t^{(4)}(1, 2, 3, 4) - \rho_t^{(1)}(1)\big]\big[\lambda_{2,3,4}(1) - kT \log \rho_t^{(1)}(1)\big], \quad (6.49)$$

$$+\sum_{1,2} \mathrm{tr}_1\big[\mathrm{tr}_2\rho_t^{(2)}(1, 2) - \rho_t^{(1)}(1)\big]\big[\lambda_2(1) - kT \log \rho_t^{(1)}(1)\big], \quad (6.50)$$

$$-\sum_{1,2,3,4} f_1^{(4)}\mathrm{tr}_{1,2,3,4}\big[\rho_t^{(4)}(1, 2, 3, 4) - 1\big] - \sum_{1,2} f_1^{(2)}\mathrm{tr}_{1,2}\big[\rho_t^{(2)}(1, 2) - 1\big]$$

$$-\sum_1 f_1^{(1)}\mathrm{tr}_1\big[\rho_t^{(1)}(1) - 1\big], \quad (6.51)$$

where $f_1^{(1)}$, $f_1^{(2)}$, and $f_1^{(4)}$ are the parameters for the normalizations and the λ's are the matrix Lagrange multipliers for the reducibility conditions.

In the above expression, several Lagrange multipliers, λ, have been introduced. $\lambda_{3,4}(1, 2)$ is an operator defined in the two-spin subspace and is regarded as a molecular field acting upon spins (1,2) due to spins (3,4). Similarly, $\lambda_{2,3,4}(1)$ is another operator defined in one-spin subspace and is regarded as a molecular field acting upon spin (1) due to three spins (2,3,4). Still another, $\lambda_2(1)$ is defined in one-spin subspace and is regarded as a molecular field acting upon spin (1) due to spin (2). The molecular field $\lambda_{2,3,4}(1)$ is often ignored in some formulations, for no good reason except that the reducibility (6.49) from four-site density matrix to one-site density matrix may appear to be redundant. It should be pointed out, however, that the redundancy statement is questionable and that the significance of this molecular field should be examined.

The molecular field $\lambda_{3,4}(1, 2)$ is usually expressed as

$$\lambda_{3,4}(1, 2) = -\lambda_3 S_1 \cdot S_2 - \tfrac{1}{2}\lambda_2\big(S_1^z + S_2^z\big), \tag{6.52}$$

where λ_3 and λ_2 are the Lagrange multipliers, which are scalars, and $\lambda_{2,3,4}(1)$ may be expressed as

$$\lambda_{2,3,4}(1) = \lambda_4 S_1^z, \tag{6.53}$$

where λ_4 is another scalar multiplier. Similarly,

$$\lambda_2(1) = \lambda_1 S_1^z. \tag{6.54}$$

The $\rho^{(2)}(1, 2)$ and $\rho^{(1)}(1)$ appearing in (6.48), (6.49), and (6.50), which do not bear the subscript t, are the density matrices which will be found eventually as results of variational calculations, and hence they should not be varied during the calculation.

When the variations with respect to $\rho_t^{(1)}$, $\rho_t^{(2)}$, and $\rho_t^{(4)}$ are individually set equal to zero, one finds the final forms of these density matrices. Rather tedious but straightforward manipulations are left for Exercise 6.6.

$$\rho^{(1)}(1) = \exp\beta\big[f^{(1)} + (B + z\lambda_1 + z_2\lambda_4)S_1^z\big], \tag{6.55}$$

$$\begin{aligned}
\rho^{(2)}(1, 2) = \exp\beta\big\{ &f^{(2)} + [B + (z - 1)\lambda_1 + (z_4/2)\lambda_2 + z_2\lambda_4]\big(S_1^z + S_2^z\big) \\
&+ (2J + z_4\lambda_3)S_1 \cdot S_2\big\},
\end{aligned} \tag{6.56}$$

$$\begin{aligned}
\rho^{(4)}(1, 2, 3, 4) = \exp\beta\big\{ &f^{(4)} + [B + (z - 2)\lambda_1 + (z_4 - 1)\lambda_2 + (z_2 - 1)\lambda_4] \\
&\times \big(S_1^z + S_2^z + S_3^z + S_4^z\big) \\
&+ [2J + (z_4 - 1)\lambda_3] \\
&\times (S_1 \cdot S_2 + S_2 \cdot S_3 + S_3 \cdot S_4 + S_4 \cdot S_1)\big\},
\end{aligned} \tag{6.57}$$

where $\beta = 1/kT$. $f^{(1)}$, $f^{(2)}$, and $f^{(4)}$ are the normalization constants. They are different from the previously introduced $f_1^{(1)}$, $f_1^{(2)}$, and $f_1^{(4)}$ by nonessential additive constants.

After having determined the structure of the reduced density matrices, the potential and all the necessary thermodynamical quantities must be evaluated. Since the reduced density matrices are all given in the form of exponential functions, the cumulant functions take rather simple expressions. The Helmholtz potential per lattice site may be given in an extremely simple form [Exercise 6.7]:

$$F/N = \frac{z_2}{4}f^{(4)} + \left(\frac{z}{2} - z_2\right)f^{(2)} + (1 - z + z_2)f^{(1)}. \tag{6.58}$$

The first factor, $\frac{z_2}{4}$, is easily understood, but interpretation of the factors $(z/2 - z_2)$ and $(1 - z + z_2)$ is not as clear. There are z_2 squares meeting at one lattice site; however, each square contributes four times to single lattice sites through its four vertices, and hence the contribution should be divided by four. Contributions of the pairs and singles are, respectively, $z/2$ and unity; however, there must be some over-counting.

There is a sum rule. If we assume that there is no interaction and no correlation effect in the lattice, then

$$\rho^{(2)}(1, 2) = \rho^{(1)}(1)\rho^{(1)}(2),$$
$$\rho^{(4)}(1, 2, 3, 4) = \rho^{(1)}(1)\rho^{(1)}(2)\rho^{(1)}(3)\rho^{(1)}(4). \tag{6.59}$$

Then,

$$f^{(2)} = 2f^{(1)}, \quad f^{(4)} = 4f^{(1)}, \tag{6.60}$$

$$F/N = \frac{z_2}{4} \times 4f^{(1)} + \left(\frac{z}{2} - z_2\right) \times 2f^{(1)} + (1 - z + z_2)f^{(1)}$$

$$= f^{(1)}. \tag{6.61}$$

This justifies the foregoing distribution of the four-, two-, and one-site contributions.

6.7 Simplified cluster variation methods

As demonstrated in the previous section, many different molecular fields appear in the reduced density matrices as a result of the successive reducibility conditions, none of which may be ignored. In the literature, however, two-site, three-site, and four-site reduced density matrices are derived by some intuitive arguments rather than by the accurate variational principle, and hence some of the molecular fields have been unjustly omitted. Such a formulation has been called the *simplified cluster variation method* abbreviated as SCVM.

Let us in the following examine some of these approximations in connection with the four-site approximation.

λ_2 is the molecular field arising from the four-site to two-site reduction, and λ_4 is from the four-site to one-site reduction which was thought to be redundant. The situation seems to be the other way around. In Fig. 6.3, site 1 is surrounded by four bonds, $\langle 1, g \rangle$, $\langle 1, 1' \rangle$, $\langle 1, 2 \rangle$, and $\langle 1, 4 \rangle$, and is also surrounded by four squares, [A], [B], [C], and [D].

The one-site density matrix (6.55) contains the bond molecular field, λ_1, z times and the square molecular field, λ_4, z_2 times. The molecular field λ_4 pertaining to the reducibility from four sites to one site is presumably more important than λ_2, which does not appear in the one-site density matrix. This is true with the four-site density matrix (6.57) where the effect of the four outermost squares, [D], [E], [F], and [G], upon the central square appears through the molecular field λ_4. If we look at the structure of the two-site density matrix (6.56), the molecular field

$$(z_4/2)\lambda_2\left(S_1^z + S_2^z\right) \tag{6.62}$$

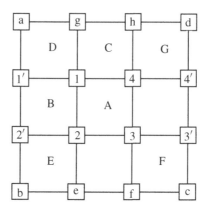

Fig. 6.3.

is seen to have been introduced in connection with the four-site to two-site reducibility, and the corresponding operator is supposed to be a two-spin operator.
Expression (6.62) is not a genuine two-spin operator, but a pseudo-two-spin operator, and a similar field should also be introduced in connection with the four-site
to one-site reducibility, as is seen in (6.57).

It is, therefore, more appropriate to ignore λ_2 rather than λ_4. In the formulation
by Tanaka & Kimura (1994), the molecular field, λ_4, is totally ignored in favor of
λ_2. Numerical results based on this assumption will be discussed in Sec. 6.10.

At this moment it would be desirable to set some guidelines by which we can
choose the most appropriate set of molecular field parameters.

Let us, for this purpose, work with the Heisenberg ferromagnet in the eight-
site, cubic approximation on the simple cubic lattice in the full molecular field
formulation [Exercise 6.7]. The molecular fields arising from the eight-site to four-
site (eight-to-four site) reducibility and appearing in the four-site density matrix
are given by

$$
\begin{aligned}
(\lambda_5/4)&\left(S_1^z + S_2^z + S_3^z + S_4^z\right) \\
&+ (\lambda_6/4)(S_1 \cdot S_2 + S_2 \cdot S_3 + S_3 \cdot S_4 + S_4 \cdot S_1) \\
&+ (\lambda_7/2)(S_1 \cdot S_3 + S_2 \cdot S_4) \\
&+ (\lambda_8/2)\left[S_1 \cdot S_2\left(S_3^z + S_4^z\right) + S_2 \cdot S_3\left(S_4^z + S_1^z\right)\right. \\
&\left.+ S_3 \cdot S_4\left(S_1^z + S_2^z\right) + S_4 \cdot S_1\left(S_2^z + S_3^z\right)\right] \\
&+ (\lambda_9/2)\left[S_1 \cdot S_3\left(S_2^z + S_4^z\right) + S_2 \cdot S_4\left(S_1^z + S_3^z\right)\right] \\
&+ \lambda_{10}[(S_1 \cdot S_2)(S_3 \cdot S_4) + (S_2 \cdot S_3)(S_4 \cdot S_1)] \\
&+ \lambda_{11}(S_1 \cdot S_3)(S_2 \cdot S_4), \qquad\qquad\qquad (6.63)
\end{aligned}
$$

where the factor $\frac{1}{4}$ in front of the first term is to make it a pseudo-four-spin operator

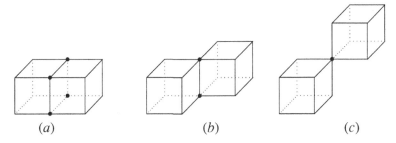

Fig. 6.4. (*a*) Eight-to-four, (*b*) eight-to-two and (*c*) eight-to-one reducibilities.

by adding four one-spin operators. Other factors have the same implication. The field due to the eight-to-two reducibility is

$$(\lambda_{12}/2)(S_1^z + S_2^z). \tag{6.64}$$

Finally the field due to the eight-to-one reducibility is

$$\lambda_{13} S_1^z. \tag{6.65}$$

The Oguchi–Kitatani formulation (Oguchi & Kitatani 1988) can be interpreted as a simplified cluster variation method (SCVM) if:

(i) only λ_{13} is retained in the one–eight combination;
(ii) only λ_{12} is retained in the two–eight combination; and
(iii) only λ_5 is retained in the four–eight combination.

The approximation may be improved further by retaining two or all of the molecular field parameters simultaneously.

By observing the crystallographic structure of the underlying lattice, we can easily gain some insight into estimating how many times each molecular field should appear in different reduced density matrices upon which particular reducibility conditions are imposed. For instance, in the eight-to-one reducibility, any one-site quantity must be calculated consistently either by the single-site density matrix or by the eight-site density matrix. The pertaining molecular field parameter will then be fixed by means of this consistency condition.

Similar calculations should be carried out for all of the possible reducibility conditions. When all of the unknown mean-field parameters are fixed in this way, we have the complete solution of the cluster variation method. In doing so, it is easy to forget the very important fact that we are working with the cluster variation formulation. For this reason, the formulation may be called the *method of self-consistent fields*.

6.8 Correlation function formulation

The molecular field formulation of the cluster variation method has been fully developed in the preceding sections. It was seen that each time a new cumulant function was added to the variational potential, the introduction of a new set of molecular field parameters was required. These molecular field parameters are then fixed either by the consistency conditions or by minimizing the potential with respect to those parameters.

Since the cluster variation method is most suitable for localized systems, any cumulant function appearing in the cumulant expansion is defined in conjunction with a particular set of lattice sites. For this reason the reduced density matrices should satisfy the crystallographic symmetry as well as the physical symmetry imposed by the Hamiltonian under consideration. By physical symmetry we mean that a certain physical quantity is conserved because of the nature of the Hamiltonian. In the problem of the Heisenberg ferromagnet, for instance, the total magnetization is said to be a good quantum number, implying that it is a quantity which should be conserved. There is no term in the Hamiltonian of the Heisenberg ferromagnet which changes the total magnetization [Exercise 6.8]. All the reduced density matrices employed in the cluster variation method, therefore, should be constructed in such a way that all the matrix elements which do not conserve the magnetization are set equal to zero.

In this section, it will be shown that there is an entirely different formulation of the cluster variation method in which all the reduced density matrices are constructed in such a way that they satisfy the reducibility and normalization conditions from the outset. Anticipating such a formulation, we have constructed the necessary reduced density matrices in Chapter 5. These density matrices are exact as far as the dependence of their elements upon the correlation functions is concerned. Numerical values of the correlation functions are unknown at the time of the construction of the density matrices. The formulation presented in the remainder of this chapter is based on the treatment by Aggarwal & Tanaka (1977).

For the sake of simplicity only the spin $\frac{1}{2}$ Heisenberg ferromagnet and the corresponding Ising model will be treated unless otherwise stated explicitly.

6.8.1 One-site density matrix

The one-site density matrix is a 2×2 matrix, because there are only two states, either up or down, of each spin. The matrix is diagonal because of the conservation of the total magnetization, and the two diagonal elements are available from (5.33):

$$R_1(1) = \tfrac{1}{2}(1 + x_1),$$
$$R_1(2) = \tfrac{1}{2}(1 - x_1),$$
$$x_1 = 2\langle S_i^z \rangle. \tag{6.66}$$

According to the Heisenberg exchange Hamiltonian, these are of the exact form of the two diagonal elements of the one-site reduced density matrix which will be used as one of the ingredients of the cluster variation method regardless of the degree of approximation; in either one-site, two-site, or even eight-site approximation.

6.8.2 *Two-site density matrix*

The two-site density matrix, in its diagonalized form, is readily available from (5.40):

$$R_2(1) = N_2(1 + 2x_1 + x_2),$$
$$R_2(2) = N_2(1 - x_2 + 4y_1),$$
$$R_2(3) = N_2(1 - x_2 - 4y_1),$$
$$R_2(4) = N_2(1 - 2x_1 + x_2), \tag{6.67}$$

where $x_1 = 2\langle S_i^z \rangle = 2\langle S_j^z \rangle$, $x_2 = 4\langle S_i^z S_j^z \rangle$, and $y_1 = \langle S_i^- S_j^+ \rangle$.

This is again the exact form of the two-site reduced density matrix for any distance between the two sites (i, j) and for any crystallographic structure of the underlying lattice.

By virtue of the variational principle, these correlation functions are now determined one by one as a function of temperature by minimizing the variational potential with respect to the unknown correlation functions, rather than with respect to the unknown reduced density matrices. This alternative method is henceforth called the *correlation function formulation* (CFF) of the cluster variation method.

6.9 The point and pair approximations in the CFF

If the energy and temperature are measured in units of $zJ/2$, the variational potential in the mean-field approximation is given by

$$\beta F/N = -\tfrac{1}{2}\beta x_1^2 + g_1, \tag{6.68}$$

where g_1 is the one-site cumulant function expressed as

$$g_1 = G_1 = \mathrm{tr}\rho^{(1)}(1) \log \rho^{(1)}(1) = R(1) \log R_1 + R(2) \log R_2,$$
$$R_1(1) = \tfrac{1}{2}(1 + x_1),$$
$$R_1(2) = \tfrac{1}{2}(1 - x_1),$$
$$x_1 = 2\langle S_i^z \rangle. \tag{6.69}$$

The potential is now minimized with respect to x_1:

$$-\beta x_1 + \tfrac{1}{2} \log \frac{(1 + x_1)}{(1 - x_1)} = 0, \tag{6.70}$$

or

$$x_1 = \tanh \beta x_1. \tag{6.71}$$

This is the same as (6.24) if the temperature renormalization is recognized.

Let us next consider the pair approximation. The variational potential is, from (6.31),

$$\beta F/N = -\beta \mathrm{tr}_{1,2}\left[2S_1 \cdot S_2\right]\rho^{(2)}(1,2)\right] + \left[(1-z)G^{(1)} + (z/2)G^{(2)}\right],$$
$$= -(\beta/2)x_2 - 2\beta y_1 + \left[(1-z)G^{(1)} + (z/2)G^{(2)}\right], \tag{6.72}$$

except that the Ising interaction is now replaced by the Heisenberg exchange interaction, the temperature is normalized: $\beta = zJ/2kT$, $x_2 = 4\langle S_1^z S_2^z \rangle$, and $y_1 = \langle S_1^+ S_2^- \rangle$. The one-site and two-site cluster functions are given by (6.66) and (6.67);

$$G^{(1)} = \tfrac{1}{2}[(1+x_1)\log(1+x_1) + (1-x_1)\log(1-x_1)],$$
$$G^{(2)} = \sum_{i=1}^{4}[R_2(i)\log R_2(i)]. \tag{6.73}$$

Now the variations of the potential with respect to x_1, x_2, and y_1 are performed:

- x_1 equation:

$$\frac{1-z}{2}\log\frac{1+x_1}{1-x_1} + \frac{z}{4}\log\frac{1+2x_1+x_2}{1-2x_1+x_2} = 0; \tag{6.74}$$

- x_2 equation:

$$-\frac{\beta}{2} + \frac{z}{8}\log\frac{(1+2x_1+x_2)(1-2x_1+x_2)}{(1-x_2+4y_1)(1-x_2-4y_1)} = 0; \tag{6.75}$$

- y_1 equation:

$$-2\beta + \frac{z}{2}\log\frac{1-x_2+4y_1}{1-x_2-4y_1} = 0. \tag{6.76}$$

In the close vicinity of the transition temperature, $x_1 \to 0$. So, if the terms in (6.74) are expanded in powers of x_1, retaining only the terms of the first power of x_1, we obtain

$$x_2 = \frac{1}{z-1}. \tag{6.77}$$

If (6.75) and (6.76) are compared at $x_1 = 0$, we see that $x_2 = 2y_1$, and hence

$$x_2 = 2y_1 = \frac{1}{z-1}. \tag{6.78}$$

Finally, from (6.75) the reciprocal transition temperature β_c will be found:

$$\beta_c = \frac{z}{4} \log \frac{z}{z-4} = -\frac{z}{4} \log \left(1 - \frac{4}{z}\right)$$
$$= 1 + \tfrac{1}{2}(4/z) + \tfrac{1}{3}(4/z)^2 + \cdots . \tag{6.79}$$

It is rather interesting to compare this result with (6.41) for the Ising ferro-magnet. The transition temperature is lowered because of the quantum effect of the Heisenberg exchange interaction, i.e., the off-diagonal terms $S_1^+ S_2^-$ act like a kinetic energy and hence raise the temperature.

If we set $y_1 = 0$ in (6.75), (6.74) is not changed, but is the transition temperature from (6.75), given by

$$\beta_c = \frac{z}{2} \log \frac{z}{z-2} = -\frac{z}{2} \log \left(1 - \frac{2}{z}\right), \tag{6.80}$$

which is exactly the Ising result.

6.10 The tetrahedron approximation in the CFF

The four-site approximation is once again examined; this time, however, in the correlation function formulation.[†] The energy and temperature will be measured in units of the mean-field critical temperature, $zJ/2$.

The variational potential is given by (see (6.45))

$$\beta F/N = - (\beta/2)x_2 - 2\beta y_1$$
$$+ z_4 g^{(4)}(1, 2, 3, 4) + z_3 g^{(3)}(1, 2, 3)$$
$$+ (z/2)g^{(2)}(1, 2) + (z_2/2)g^{(2)}(1, 3) + g^{(1)}(1), \tag{6.81}$$

where z and z_2 are the numbers of the nearest and second neighbors, and z_3 and z_4 are the numbers of triangles and the four-site clusters per lattice site.

- Two-dimensional square
  $z = 4, \quad z_2 = 4, \quad z_3 = 1, \quad z_4 = 1$;
- three-dimensional simple cubic (square cluster)
  $z = 6, \quad z_2 = 12, \quad z_3 = 12, \quad z_4 = 3$;
- face center cubic (tetrahedron)
  $z = 12, \quad z_2 = 0, \quad z_3 = 8, \quad z_4 = 2$.

When the cumulant functions, g, are replaced by the cluster functions, G, we find the variational potential for two-dimensional and three-dimensional

---

[†] Formulation in this section was developed by Tanaka & Libelo (1975).

squares as

$$\beta F/N = -\beta H - (\beta/2)x_2 - 2\beta y_1 + z_4 G^{(4)}(1, 2, 3, 4)$$
$$+ \left(\frac{z}{2} - 4z_4\right) G^{(2)}(1, 2) + (1 + 4z_4 - z)G^{(1)}(1), \tag{6.82}$$

and for face center cubic (tetrahedron):

$$\beta F/N = -(\beta/2)x_2 - 2\beta y_1 + z_4 G^{(4)}(1, 2, 3, 4)$$
$$+ \left(\frac{z}{2} - 6z_4\right) G^{(2)}(1, 2) + (1 + 8z_4 - z)G^{(1)}(1). \tag{6.83}$$

Since all the reduced density matrices, $\rho^{(1)}$, $\rho^{(2)}$, and $\rho^{(4)}$, are available from Chapter 5, the variational potential can be readily minimized with respect to the correlation functions.

For the sake of illustration, however, only the tetrahedron approximation in the face center cubic lattice will be worked out in this section.

Let us reconfirm the definitions of the variation parameters appearing in the tetrahedron cluster. They are, from (5.52),

$$x_1 = 2\langle S_1^z \rangle, \quad x_2 = 4\langle S_1^z S_2^z \rangle, \quad x_3 = 8\langle S_1^z S_2^z S_3^z \rangle,$$
$$x_4 = 16\langle S_1^z S_2^z S_3^z S_4^z \rangle,$$
$$y_1 = \langle S_1^+ S_2^- \rangle, \quad y_2 = 2\langle S_1^z S_2^+ S_3^- \rangle,$$
$$y_3 = 4\langle S_1^z S_2^z S_3^+ S_4^- \rangle, \quad y_4 = \langle S_1^+ S_2^- S_3^+ S_4^- \rangle. \tag{6.84}$$

Again the significance of the numerical coefficients is obvious.

Sixteen elements of the diagonalized tetrahedral density matrix are, from (5.64),

$$R_4(1) = N_4(1 + 4x_1 + 6x_2 + 4x_3 + x_4),$$
$$R_4(2) = N_4(1 - 4x_1 + 6x_2 - 4x_3 + x_4),$$
$$R_4(3) = N_4(1 + 2x_1 - 2x_3 - x_4 + 12y_1 + 24y_2 + 12y_3),$$
$$R_4(4) = N_4(1 - 2x_1 + 2x_3 - x_4 + 12y_1 - 24y_2 + 12y_3),$$
$$R_4(5) = N_4(1 + 2x_1 - 2x_3 - x_4 - 4y_1 - 8y_2 - 4y_3), \quad g = 3,$$
$$R_4(6) = N_4(1 - 2x_1 + 2x_3 - x_4 - 4y_1 + 8y_2 - 4y_3), \quad g = 3,$$
$$R_4(7) = N_4(1 - 2x_2 + x_4 + 16y_1 - 16y_3 + 16y_4),$$
$$R_4(8) = N_4(1 - 2x_2 + x_4 - 8y_1 + 8y_3 + 16y_4), \quad g = 2,$$
$$R_4(9) = N_4(1 - 2x_2 + x_4 - 16y_4), \quad g = 3. \tag{6.85}$$

Once the density matrices are diagonalized the cumulant functions are easily found and the variational potential is minimized with respect to the correlation functions x_1 through x_4 and y_1 through y_4.

In the case of the face center cubic lattice one can obtain some information without solving the numerical problem. The variational potential is given by

$$\beta F/N = -\tfrac{1}{2}\beta x_2 - 2\beta y_1 + 5G^{(1)} - 6G^{(2)} + 2G^{(4)}. \qquad (6.86)$$

Variations with respect to the eight correlation functions yield the following equations:

- x_1 equation:

$$0 = \tfrac{5}{2}\log\frac{1+x_1}{1-x_1} - 3\log\frac{1+2x_1+x_2}{1-2x_1+x_2}$$

$$+ \tfrac{1}{2}\log\frac{1+4x_1+6x_2+4x_3+x_4}{1-4x_1+6x_2-4x_3+x_4}$$

$$+ \tfrac{1}{4}\log\frac{1+2x_1-2x_3-x_4+12y_1+24y_2+12y_3}{1-2x_1+2x_3-x_4+12y_1-24y_2+12y_3}$$

$$+ \tfrac{3}{4}\log\frac{1+2x_1-2x_3-x_4-4y_1-8y_2-4y_3}{1-2x_1+2x_3-x_4-4y_1+8y_2-4y_3}; \qquad (6.87)$$

- x_2 equation:

$$0 = -\tfrac{3}{2}\log\frac{(1+2x_1+x_2)(1-2x_1+x_2)}{(1-x_2+4y_1)(1-x_2-4y_1)}$$

$$+ \tfrac{3}{4}\log(1+4x_1+6x_2+4x_3+x_4)(1-4x_1+6x_2-4x_3+x_4)$$

$$- \tfrac{1}{4}\log(1-2x_2+x_4+16y_1-16y_3+16y_4)$$

$$- \tfrac{2}{4}\log(1-2x_2+x_4-8y_1+8y_3+16y_4)$$

$$- \tfrac{3}{4}\log(1-2x_2+x_4-16y_4) - \frac{1}{2}\beta; \qquad (6.88)$$

- x_3 equation:

$$0 = \tfrac{1}{2}\log\frac{1+4x_1+6x_2+4x_3+x_4}{1-4x_1+6x_2-4x_3+x_4}$$

$$- \tfrac{1}{4}\log\frac{1+2x_1-2x_3-x_4+12y_1+24y_2+12y_3}{1-2x_1+2x_3-x_4+12y_1-24y_2+12y_3}$$

$$- \tfrac{3}{4}\log\frac{1+2x_1-2x_3-x_4-4y_1-8y_2-4y_3}{1-2x_1+2x_3-x_4-4y_1+8y_2-4y_3}; \qquad (6.89)$$

- x_4 equation:

$$0 = \frac{1}{8}\left[\log\frac{1 + 4x_1 + 6x_2 + 4x_3 + x_4}{1 + 2x_1 - 2x_3 - x_4 + 12y_1 + 24y_2 + 12y_3}\right.$$
$$\left. + \log\frac{1 - 4x_1 + 6x_2 - 4x_3 + x_4}{1 - 2x_1 + 2x_3 - x_4 + 12y_1 - 24y_2 + 12y_3}\right]$$
$$- \frac{3}{8}[\log(1 + 2x_1 - 2x_3 - x_4 - 4y_1 - 8y_2 - 4y_3)$$
$$+ \log(1 - 2x_1 + 2x_3 - x_4 - 4y_1 + 8y_2 - 4y_3)]$$
$$+ \frac{1}{8}\log(1 - 2x_2 + x_4 + 16y_1 - 16y_3 + 16y_4)$$
$$+ \frac{2}{8}\log(1 - 2x_2 + x_4 - 8y_1 + 8y_3 + 16y_4)$$
$$+ \frac{3}{8}\log(1 - 2x_2 + x_4 - 16y_4); \tag{6.90}$$

- y_1 equation:

$$0 = -6\log\frac{1 - x_2 + 4y_1}{1 - x_2 - 4y_1}$$
$$+ \frac{3}{2}\log\frac{1 + 2x_1 - 2x_3 - x_4 + 12y_1 + 24y_2 + 12y_3}{1 + 2x_1 - 2x_3 - x_4 - 4y_1 - 8y_2 - 4y_3}$$
$$+ \frac{3}{2}\log\frac{1 - 2x_1 + 2x_3 - x_4 + 12y_1 - 24y_2 + 12y_3}{1 - 2x_1 + 2x_3 - x_4 - 4y_1 + 8y_2 - 4y_3}$$
$$+ 2\log\frac{1 - 2x_2 + x_4 + 16y_1 - 16y_3 + 16y_4}{1 - 2x_2 + x_4 - 8y_1 + 8y_3 + 16y_4} - 2\beta; \tag{6.91}$$

- y_2 equation:

$$0 = \log\frac{1 + 2x_1 - 2x_3 - x_4 + 12y_1 + 24y_2 + 12y_3}{1 - 2x_1 + 2x_3 - x_4 + 12y_1 - 24y_2 + 12y_3}$$
$$+ \log\frac{1 - 2x_1 + 2x_3 - x_4 - 4y_1 + 8y_2 - 4y_3}{1 + 2x_1 - 2x_3 - x_4 - 4y_1 - 8y_2 - 4y_3}; \tag{6.92}$$

- y_3 equation:

$$0 = \frac{3}{2}\left[\log\frac{1 + 2x_1 - 2x_3 - x_4 + 12y_1 + 24y_2 + 12y_3}{1 + 2x_1 - 2x_3 - x_4 - 4y_1 - 8y_2 - 4y_3}\right.$$
$$\left. + \log\frac{1 - 2x_1 + 2x_3 - x_4 + 12y_1 - 24y_2 + 12y_3}{1 - 2x_1 + 2x_3 - x_4 - 4y_1 + 8y_2 - 4y_3}\right]$$
$$- 2\log\frac{1 - 2x_2 + x_4 + 16y_1 - 16y_3 + 16y_4}{1 - 2x_1 + x_4 - 8y_1 + 8y_3 + 16y_4}; \tag{6.93}$$

- y_4 equation:

$$0 = 2\log(1 - 2x_2 + x_4 + 16y_1 - 16y_3 + 16y_4)$$
$$+ 4\log(1 - 2x_2 + x_4 - 8y_1 + 8y_3 + 16y_4)$$
$$- 6\log(1 - 2x_2 + x_4 - 16y_4). \tag{6.94}$$

Solution of these equations at each temperature is obtained by any computer program designed for a set of nonlinear equations. However, in the immediate neighborhood of the critical temperature where x_1, x_3, and y_2 are vanishingly small, these equations become algebraic, and a few simple relations result [Exercise 6.8]. These are

$$2\langle S_1^z S_2^z S_3^z \rangle = 3\langle S_1^z S_2^+ S_3^- \rangle, \tag{6.95}$$

$$\langle S_1^+ S_2^- \rangle = 2\langle S_1^z S_2^z \rangle = \tfrac{1}{10}. \tag{6.96}$$

x_4 is the real solution of cubic equation

$$27(11 + 5x_4)(1 - 25x_4)^2 = (9 - 25x_4)^3, \tag{6.97}$$

which yields

$$x_4 = 16\langle S_1^z S_2^z S_3^z S_4^z \rangle = 0.0810. \tag{6.98}$$

Also, we have the relations

$$3\langle S_1^+ S_2^+ S_3^- S_4^- \rangle = 12\langle S_1^z S_2^z S_3^+ S_4^- \rangle = 8\langle S_1^z S_2^z S_3^z S_4^z \rangle. \tag{6.99}$$

We can see that

$$\beta_c = \log \frac{1}{27} \left(\frac{5x_4 + 11}{25x_4 - 1} \right)^2 = (0.6735)^{-1}. \tag{6.100}$$

The generalized magnetizations x_1, x_3, and y_2 satisfy the equilibrium conditions (where the left hand side of (6.87) has been replaced by a single term in the external magnetic field). In the linear response regime, these are solved to give

$$x_1 = \frac{(1 + x_2)(1 + 3x_2)}{(1 - 3x_2)(1 - 5x_2)} \beta H,$$

$$x_3 = \frac{(1 + x_2)(3x_2 + x_4)}{(1 - 3x_2)(1 - 5x_2)} \beta H, \quad y_2 = 6x_3. \tag{6.101}$$

These quantities have a singularity at $x_2 = \tfrac{1}{5}$ and are expressed in terms of x_2 and x_4. The quantities x_2 and x_4 satisfy (6.88) and (6.90), and, in turn, they are expressible as infinite series in $K = \beta z J / 2$:

$$\chi_{4\text{site}} = 6K \left(1 + 6K + 30K^2 + 138K^3 + \frac{2428}{4} K^4 + \cdots \right). \tag{6.102}$$

Finally, the exact infinite series is given by[†]

$$\chi_{\text{exact}} = 6K \left(1 + 6K + 30K^2 + 138K^3 + \frac{2445}{4} K^4 \right.$$

$$\left. + \frac{53171}{20} K^5 + \frac{914601}{80} K^6 + \cdots \right). \tag{6.103}$$

[†] See Chapter 7 for a more detailed analysis.

Exercises

6.1 Derive the canonical density matrix (6.3) by varying the potential (6.5) with respect to ρ_t. Examine the significance of an extra term of kT in the Lagrange multiplier α.

6.2 Evaluate the normalization constants $f^{(1)}$, $f^{(2)}$, and $f^{(4)}$ for the Heisenberg ferromagnet.

6.3 Prove, in the pair approximation, that the two-site cumulant function $g^{(2)}(i, j)$ for (i, j) distances beyond the range of the interaction potential vanishes identically.

[**Hint**]

Because of the lack of an interaction term for the distances (i, j) larger than the nearest neighbor, the equation determining $\rho^{(2)}(i, j)$ contains only single-site terms. This leads to the form

$$\rho^{(2)}(i, j) = \rho^{(1)}(i)\rho^{(1)}(j) \tag{E6.1}$$

and hence

$$g^{(2)}(i, j) = 0. \tag{E6.2}$$

6.4 Show that the final forms of $\rho^{(1)}(i)$ and $\rho^{(2)}(i, j)$ are given by

$$\rho^{(1)}(i) = \exp \beta[f_1 + 2zJ\lambda s_i],$$
$$\rho^{(2)}(i, j) = \exp \beta[f_2 + 2(z-1)J\lambda(s_i + s_j) + 2Js_i s_j]. \tag{E6.3}$$

6.5 Show that the molecular field parameter λ is a solution of the following equations:

$$\langle s \rangle = \tfrac{1}{2} \tanh(\beta z J\lambda)$$
$$= \frac{\tfrac{1}{2}\exp(\beta J/2)\sinh 2\beta(z-1)J\lambda}{\exp(\beta J/2)\cosh 2\beta(z-1)J\lambda + \exp(-\beta J/2)}. \tag{E6.4}$$

6.6 Carry out the variational calculations for the four-site approximation and derive the density matrices (6.55), (6.56), and (6.57).

6.7 Derive the expression for the potential in terms of the normalization parameters as given by (6.58).

6.8 Derive relations (6.95)–(6.100).

7

Infinite-series representations of correlation functions

7.1 Singularity of the correlation functions

Ehrenfest classified the phase transitions occurring in different sytems according to the abruptness of the transition. If the internal energy has a discontinuity, i.e., observed as a finite latent heat, at the transition point, the transition is called the first-order phase transition. If the internal energy itself is continuous but the first derivative of it, i.e., the specific heat, has a discontinuity, the transition is called the second-order transition, and so on. According to this classification, the transition from the solid phase to liquid phase, i.e., melting, or the transition from gas phase to liquid phase, i.e., condensation, are the first-order phase transitions.

Extensive studies have been performed, both experimentally and theoretically, on the nature of the second-order phase transition. It was found that the second-order phase transition is not just characterized by a discontinuity in the specific heat. Many other physical properties, such as the high temperature magnetic susceptibility and the low temperature spontaneous magnetization in ferromagnetic substances, also show discontinuities in the neighborhood of the phase transition temperature. In general, the mathematical nature of the singularity is most accurately represented by a branch-point type singularity expressed as

$$x = A|\epsilon|^{\gamma}, \quad \epsilon = T - T_c, \quad (7.1)$$

where A and γ are called the critical amplitude and critical exponent of the quantity x, respectively. ϵ could be either positive or negative depending upon the definition of the quantity.

From the theoretical point of view it has long been known that a simple statistical treatment cannot predict such a singular behavior accurately. Any of the mean-field type approximations can indeed predict a certain value of the critical exponent γ; however, the numerical value is characteristically incorrect. It has commonly been understood that the situation is not improved by going to more accurate treatments such as the pair, four-site, or even to the six-site approximations.

153

More recently, some success in the theoretical investigation of the critical behavior in the Ising ferromagnet, Heisenberg ferromagnet, or the lattice gauge model has been accomplished. There are two methods of approach: the Padé approximant method, and the renormalization group technique.

According to the method of Padé approximant (Sec. 7.4), the critical exponent for the spontaneous magnetization for the three-dimensional Ising model has been found to be

$$\gamma = \tfrac{1}{3}, \tag{7.2}$$

and the Onsager (1949) exact value for the two-dimensional Ising model is well known (see Chapter 11),

$$\gamma = \tfrac{1}{8}. \tag{7.3}$$

Similarly, for the magnetic susceptibility above the transition temperature

$$\gamma = -\tfrac{5}{4}, \tag{7.4}$$

and the exact value for the two-dimensional Ising model is

$$\gamma = -\tfrac{7}{4}. \tag{7.5}$$

It will be interesting, therefore, to find the value of the critical exponent using an approximate theory of the second-order phase transition.

7.2 The classical values of the critical exponent

It has been widely acknowledged that the mean-field type approximations give an unacceptable value, $\gamma = \tfrac{1}{2}$, of the critical exponent for the spontaneous magnetization.

Let us examine the way in which the critical exponent is calculated in the mean-field type approximations.

7.2.1 The mean-field approximation

From (6.71),

$$x_1 = 2\langle\mu\rangle = \tanh \beta x_1. \tag{7.6}$$

In the immediate neighborhood of the critical temperature, from below, x_1 is vanishingly small but not equal to zero, and hence $\tanh \beta x_1$ can be expanded into an infinite series, retaining only the two lowest order terms on the right hand side:

$$x_1 = \beta x_1 - \tfrac{1}{3}(\beta x_1)^3. \tag{7.7}$$

Dividing both sides of the equation by x_1 and solving for x_1^2, we find

$$x_1^2 = \frac{3}{\beta^3}(\beta - 1). \tag{7.8}$$

This equation shows that the dimensionless magnetization x_1 vanishes at the critical temperature $\beta_c = 1$, and near the transition temperature it behaves as

$$x_1 = \sqrt{3}\epsilon^{\frac{1}{2}}, \quad \epsilon = \beta - 1, \quad \gamma = \tfrac{1}{2}, \quad \text{and} \quad A = \sqrt{3}. \tag{7.9}$$

7.2.2 The pair approximation

From (6.74) and (6.75), with $y_1 = 0$ for the Ising model,

- x_1 equation:

$$\frac{1-z}{2} \log \frac{1+x_1}{1-x_1} + \frac{z}{4} \log \frac{1+2x_1+x_2}{1-2x_1+x_2} = 0; \tag{7.10}$$

- x_2 equation:

$$-\frac{\beta}{2} + \frac{z}{8} \log \frac{(1+2x_1+x_2)(1-2x_1+x_2)}{(1-x_2)^2} = 0. \tag{7.11}$$

When (7.10) and (7.11) are expanded for small x_1, we find

$$(1-z)\left(x_1 + \frac{1}{3}x_1^3\right) + \frac{z}{2}\left[\frac{2x_1}{1+x_2} + \frac{1}{3}\left(\frac{2x_1}{1+x_2}\right)^3\right] = 0, \tag{7.12}$$

$$-\frac{\beta}{2} + \frac{z}{8}\left[\log\left(\frac{1+x_2}{1-x_2}\right)^2 - \left(\frac{2x_1}{1+x_2}\right)^2\right] = 0. \tag{7.13}$$

x_2 and β are also expanded in the neighborhood of their values at the critical point as

$$x_2 = x_{2c} + \mu, \quad \beta = \beta_c + \epsilon, \tag{7.14}$$

where ϵ is the reciprocal temperature shift and hence proportional to the temperature shift from the critical point. These are then introduced into the above equations whence we retain only up to the lowest orders in μ, ϵ and x_1:

$$\frac{1+x_2}{1-x_2} = \frac{1+x_{2c}}{1-x_{2c}}\left(1 + \frac{2\mu}{1-x_{2c}^2}\right). \tag{7.15}$$

After some manipulations we obtain

$$\mu = \frac{1}{3}\frac{(3z-2)(z-2)}{z(z-1)}x_1^2,$$

$$x_1^2 = \frac{3z}{(z-1)(2z-1)}\epsilon. \tag{7.16}$$

The critical exponent for the spontaneous magnetization is, thus, seen to be $\frac{1}{2}$, which is called the classical value.

7.3 An infinite-series representation of the partition function

In the history of the development of statistical mechanics, a great deal of effort was devoted in order to derive exact series expansions for some thermodynamic quantities, such as the partition function, spontaneous magnetization, specific heat, and magnetic susceptibility for the Ising and Heisenberg ferromagnets (Bloch, 1930; Domb, 1960; Domb & Sykes, 1956; Rushbrooke & Wakefield, 1949; Tanaka *et al.*, 1951; Wakefield, 1951a,b). The general graphical procedures employed for this purpose are called the *combinatorial formulation*.

By way of introduction, let us take the Ising Hamiltonian and derive an infinite-series representation of the partition function, $Z(\beta)$:

$$Z(\beta) = \exp(-\beta F) = \text{Tr} \exp(-\beta H), \tag{7.17}$$

where H is the Hamiltonian given by

$$H = -B \sum_i s_i - J \sum_{\langle i,j \rangle} s_i s_j, \quad s_i = \pm 1. \tag{7.18}$$

In the above equation, B is the external magnetic field in suitable units, and J is the exchange integral, which is positive, indicating that the interaction is ferromagnetic.

Let us first define two parameters which appear in the series expansion:

$$\alpha = \exp(-mH/kT), \quad \text{and} \quad v = \exp(-J/kT). \tag{7.19}$$

These parameters are small at low temperatures ($\alpha \to 0$, $v \to 0$ as $T \to 0$).

The most elementary definition of the partition function is considered, i.e., the summation in the partition function is performed according to the energy of excitation and the number of states for the given energy are included, starting from the ground state in which all the spins are parallel to the external magnetic field:

$$f^N = (\alpha^{-1})^N (v^{-1})^{\frac{zN}{2}} \sum_{N_\downarrow, N_{\uparrow\downarrow}} g(N; N_\downarrow, N_{\uparrow\downarrow}) \alpha^{2N_\downarrow} v^{2N_{\uparrow\downarrow}}, \tag{7.20}$$

where f is the partition function per lattice site, $N_\downarrow$ is the number of spins antiparallel to H, and $N_{\uparrow\downarrow}$ is the number of antiparallel spin pairs of neighbors. $g(N; N_\downarrow, N_{\uparrow\downarrow})$ denotes the number of ways of creating $N_\downarrow$ down spins and $N_{\uparrow\downarrow}$ antiparallel spin pairs of neighbors in the lattice.

For the face centered cubic lattice the first leading terms are given by

$$f^N = (\alpha^{-1} v^{-6})^N \cdot Z, \tag{7.21}$$

where

$$Z = 1 + N\alpha^2 v^{24} + \alpha^4 \left[6N v^{46} + \left(\tfrac{1}{2}N(N-1) - 6N \right) v^{48} \right]$$
$$+ \alpha^6 \left[8N v^{60} + 42N v^{64} + (6N^2 - 120N)v^{68} + \frac{N}{6}(N^2 - 39N + 422)v^{72} \right]$$
$$+ \alpha^8 [2N v^{72} + 24N v^{76} + 123N v^{80} + (8N^2 + 126N)v^{84} + \cdots] + \cdots.$$
$$(7.22)$$

The leading term '1', in the above equation, is the contribution from the ground state, i.e., the state in which all the spins are lined up parallel to the external magnetic field.

The term of second order in α is the contribution of the state in which one spin is flipped down against the external field. The excitation energy of this state is $24J$ because 12 parallel pairs are destroyed and 12 antiparallel pairs are created. The number of possibilities of flipping one spin is N. The terms of fourth order in α are the contribution of the states in which two spins are flipped down. There are $6N$ ways in which two spins are nearest neighbors of each other. In this configuration $24 - 1 = 23$ parallel pairs are destroyed and 23 antiparallel pairs are created, resulting in the excitation energy of $46J$. There are $N(N-1)/2 - 6N$ ways in which two spins are not nearest neighbors of each other. In this configuration 24 parallel pairs are destroyed and 24 antiparallel pairs are created, resulting in the excitation energy of $48J$. The terms of the sixth power of α come from the configurations in which three spins are flipped down, and the terms of the eighth power of α come from the states in which four spins are flipped down.

When the Nth root of the above expression is calculated, we find an infinite series for f, i.e., the partition function per single site (Tanaka *et al.*, 1951),

$$f = \alpha^{-1} v^{-6} [1 + \alpha^2 v^{24} + \alpha^4 (6v^{44} - 6v^{48})$$
$$+ \alpha^6 (8v^{60} + 42v^{64} - 114v^{68} + 64v^{72})$$
$$+ \alpha^8 (2v^{72} + 24v^{76} + 123v^{80} + 134v^{84} + \cdots) + \cdots]. \quad (7.23)$$

In the absence of the external magnetic field, the above expression reduces to

$$f = v^{-6}[1 + v^{24} + 6v^{44} - 6v^{48} + 8v^{60} + 42v^{64}$$
$$- 114v^{68} + 8v^{72} + 24v^{76} + 123v^{80} + 134v^{84} + \cdots]. \quad (7.24)$$

From the thermodynamic relation (3.45), the internal energy per single site is given by

$$-6J\lambda_2 = -\frac{\partial \log f}{\partial \beta} = 6Jv \frac{\partial \log f}{\partial v}, \quad (7.25)$$

where x_2 is the nearest-neighbor correlation function, and hence the equation yields an infinite-series representation of x_2 as follows:

$$x_2 = 1 - 4v^{24} - 44v^{44} + 52v^{48} - 80v^{60} - 448v^{64} + 1360v^{68}$$
$$- 868v^{72} - 304v^{76} - 1640v^{80} - 1764v^{84} + \cdots. \tag{7.26}$$

Since x_2 is directly proportional to the internal energy, the specific heat can be found immediately, i.e.,

$$\left(\frac{kT}{J}\right)^2 c_v = -2uz\frac{dx_2}{du}, \tag{7.27}$$

where

$$u = \exp(-4J/kT) = v^4 \tag{7.28}$$

is used for convenience. From the series in the above one finds

$$(kT/J)^2 c_v = 576u^6 + 11\,616u^{11} - 14\,976u^{12} + 28\,800u^{15}$$
$$+ 172\,032u^{16} - 55\,488u^{17} + 374\,976u^{18} + 138\,624u^{19}$$
$$+ 787\,200u^{20} + 889\,056u^{21} + \cdots. \tag{7.29}$$

The high temperature series representation of the partition function and other thermodynamic quantities is developed in Domb (1960), Domb & Sykes (1956, 1962) and Sykes (1961); see Sec. 11.1.

7.4 The method of Padé approximants

As mentioned earlier, a great deal of effort has been devoted to the development of the series representations of the partition function, the low temperature specific heat, and the high temperature magnetic susceptibility for the three-dimensional Ising model.

The motivation for this effort was not clear in the beginning of the development, except for the fact that anything exact was of interest in its own right. However, Baker and colleagues (Baker, 1961, 1963, 1964, 1965; Baker & Gammel, 1961; Baker, Gammel & Wills, 1961) proposed and introduced a significant method of utilizing the exact series for the thermodynamic functions in order to predict singular behaviors of the series which had important implications for the second-order phase transition. The mathematical procedure utilized in their analysis is referred to as the *method of Padé approximants*.

The original idea of the method was to represent the infinite series $f(z)$ as the ratio of two polynomials $P_M(z)$ and $Q_N(z)$ in such a way that

$$f(z)Q_N(z) - P_M(z) = Az^{M+N+1} + Bz^{M+N+2} + \cdots, \tag{7.30}$$

where

$$Q_N(z) = \sum_{n=0}^{N} q_n z^n, \quad P_M(z) = \sum_{n=0}^{M}, \quad q_0 = 1.0 \qquad (7.31)$$

are polynomials of degree N and M, respectively, and $P_M(z)/Q_N(z)$ is called the $[N, M]$ *Padé approximant* to $f(z)$.

In the method of Padé approximants the singularity of the thermodynamic quantity $f(z)$ is expected to be found only as a simple pole on the real z-axis, i.e., as the smallest zero of the denominator Q_N. In reality, however, from both theoretical and experimental observations, it is suggested that the singularity is more like a branch-point with a nonintegral critical exponent.

For this reason, Baker (1961) proposed to employ Padé approximants to the logarithmic derivative of the thermodynamic function. In this way, the singularity becomes a simple pole and the critical exponent will be found as the residue at the pole. By employing a high temperature series representation for the magnetic susceptibility of the face center Ising model developed by Domb & Sykes (1962), the logarithmic derivative is calculated:

$$\frac{d \log \chi_{\text{fcc}}}{dw} = 12 + 120w + 1\,189w^2 + 11\,664w^3 + 114\,492w^4$$
$$+ 1\,124\,856w^5 + 11\,057\,268w^6 + 108\,689\,568w^7 + \cdots, \qquad (7.32)$$

where

$$w = \tanh J/kT. \qquad (7.33)$$

In order to see the stability of the convergence, a series of $[N, N], N = 1, 2, 3, \ldots$ approximants is examined. In the following only the face centered cubic result will be reviewed.

The results in Table 7.1 indicate that the convergence is not quite satisfactory, but gives the critical exponent in the neighborhood of $-1.25 = -\frac{5}{4}$, which confirms the previous observation of Domb & Sykes (1962).

In order to improve the convergence in the location of the singularity, the Padé approximants to $[d \log \chi/dw]^m$, $m = \frac{4}{5}$ are examined in Table 7.2. By looking at

Table 7.1. *Padé approximants to* d log χ/dw.

N	Location w_c	Residue
1	0.10101010	-1.2243648
2	0.10187683	-1.2564708
3	0.10171078	-1.2451572

Table 7.2. *Padé approximants to* $[\mathrm{d}\log\chi/\mathrm{d}w]^{\frac{4}{5}}$.

N	Location w_c	Residue
1	0.10204082	−0.099583490
2	0.10174095	−0.099113327
3	0.10177220	−0.099263130
4	0.10176345	−0.099202123

the results it is concluded that the Padé approximants have converged to the known value for the location of the critical point within a few parts in the sixth decimal place. Furthermore the approximants seem to show an oscillatory pattern of convergence, which enables one to estimate the error. Using the interpolation method the critical point is determined

$$\text{fcc:} \quad w_c = 0.101767, \quad \text{residue} = 0.09923.$$

In order to improve further the convergence of the critical exponent, Padé approximants to $(w - w_c)\,\mathrm{d}\log\chi/\mathrm{d}w$ are evaluated at w_c (Table 7.3). This expression does not have a singularity at w_c and hence is suitable to evaluate the residue. This analysis yielded a more reliable value of the critical exponent for the susceptibility.

The Padé approximant method is not without its difficulties. The method is most suitable for the function which has a branch-point singularity with a nonzero exponent, as was seen in the case of the high temperature magnetic susceptibilty or the spontaneous magnetization at low temperature.

Baker (1963) calculated the $[N, N]$ Padé approximants ($N = 3, \ldots, 7$) to the low temperature, simple cubic lattice series for $(kT/J)^2 c_v(u)$, where $u = \exp(-4J/kT)$; however, convergence was not obtained, but the value at the critical point, as expected, seemed to be tending to infinity. Baker next computed the $[N, N]$ Padé approximants ($N = 2, \ldots, 5$) to the logarithmic derivative of $(kT/J)^2 c_v(u)$.

If c_v were proportional to some power of $(1 - T/T_c)$, then the logarithmic derivative would have a simple pole at T_c. No such simple pole appeared. Therefore, c_v probably does not behave like $(1 - T/T_c)^{-\alpha}$ for any α. After some more careful

Table 7.3. *Padé approximants to* $(w - w_c)\,\mathrm{d}\log\chi/\mathrm{d}w$.

N	Residue
1	−1.2591654
2	−1.2497759
3	−1.2497511

investigation of the behavior of $(kT/J)^2 c_v(u)$ near the critical point, it was concluded that the specific heat goes logarithmically to infinity, i.e., it is proportional to $-\log(1 - T/T_c)$.

7.5 Infinite-series solutions of the cluster variation method

The preceding analysis of the method of Padé approximants strongly suggests that any method of statistical mechanics which is capable of producing the first ten or more exact terms of an infinite-series representation of a thermodynamic quantity could be used to predict a critical singularity of the second-order phase transition. Since the cluster variation method developed in Chapter 6 is capable of predicting the second-order phase transition in the ferromagnetic Ising model and in the Heisenberg model to different orders of accuracy by going from the mean-field, Bethe (pair), tetrahedron, and tetrahedron-plus-octahedron approximations, it is rather tempting to examine the applicability of the infinite-series method to the cluster variation method. For this reason the power-series solutions of the cluster variation method applied to the Ising model will be developed.

A commonly held belief about the mean-field type approximation, which is understood to be the nature of solutions in the cluster variation method, is that it almost always predicts physically unacceptable critical behaviors. This was the conclusion reached in the earlier sections of this chapter.

In the following sections, however, an entirely different approach will be employed in order to analyse the results obtained by the cluster variation method, i.e., the infinite-series representation of the correlation functions. It turns out that the series solution by this method is almost as accurate as the exact series solution obtained by direct counting of the numbers of ways of fitting different clusters into the lattice. If the power-series solution thus obtained is combined with the Padé approximant method, one is able to find nonclassical critical indices of the correlation functions near the second-order transition temperature, in contradiction to the generally accepted notion that the mean-field-type approximations yield only the classical indices.

In order to accomplish this goal, let us summarize the results of variational calculations which were developed in Chapter 6.

7.5.1 Mean-field approximation

The variational potential per lattice point, divided by kT, is

$$F/NkT = -(zJ/2kT)x_1^2 + G_1, \tag{7.34}$$

where $y_1 = \langle s_1 \rangle$ and G_1 is the one-site cluster function given by

$$G_1 = R_{11} \log R_{11} + R_{12} \log R_{12},$$
$$R_{11} = \tfrac{1}{2}(1 + x_1), \quad R_{12} = \tfrac{1}{2}(1 - x_1). \tag{7.35}$$

The minimization condition for x_1 is

$$0 = -(zJ/kT)x_1 + \tfrac{1}{2} \log \frac{1 + x_1}{1 - x_1}. \tag{7.36}$$

7.5.2 Pair approximation

$$F/NkT = -(zJ/2kT)x_2 + \tfrac{1}{2}zG_2(1, 2) - (z - 1)G_1, \tag{7.37}$$

where $G_2(1, 2)$ is the two-site cluster function which is defined by (Chapter 6)

$$G_2(1, 2) = \sum_{i=1}^{3} w_2(i) R_{2i} \log R_{2i}, \tag{7.38}$$

where

$$R_{21} = \tfrac{1}{4}(1 + 2x_1 + x_2), \quad w_2(1) = 1,$$
$$R_{22} = \tfrac{1}{4}(1 - 2x_1 + x_2), \quad w_2(2) = 1,$$
$$R_{23} = \tfrac{1}{4}(1 - x_2), \quad w_2(3) = 2. \tag{7.39}$$

The minimization conditions for x_1 and x_2 are

$$x_1 : 0 = \frac{z}{4} \log \frac{1 + 2x_1 + x_2}{1 - 2x_1 + x_2} - \tfrac{1}{2}(z - 1) \log \frac{1 + x_1}{1 - x_1}, \tag{7.40a}$$

$$x_2 : \frac{zJ}{2kT} = \frac{z}{8} \log \left[\frac{(1 + 2x_1 + x_2)(1 - 2x_1 + x_2)}{(1 - x_2)^2} \right]. \tag{7.40b}$$

7.5.3 Tetrahedron approximation

The variational potential, (6.83), is

$$F/NkT = -(zJ/2kT)x_2 + 2G_4(1, 2, 3, 4) - 6G_2(1, 2) + 5G_1, \tag{7.41}$$

where $G_4(1, 2, 3, 4)$ is the tetrahedron cluster function, which is given by

$$G_4(1, 2, 3, 4) = \sum_{i=1}^{5} w_4(i) R_{4i} \log R_{4i},$$
$$R_{41} = N_4(1 + 4x_1 + 6x_2 + 4x_3 + x_4), \quad w_4(1) = 1,$$
$$R_{42} = N_4(1 - 4x_1 + 6x_2 - 4x_3 + x_4), \quad w_4(2) = 1,$$
$$R_{43} = N_4(1 + 2x_1 - 2x_3 - x_4), \quad w_4(3) = 4,$$
$$R_{44} = N_4(1 - 2x_1 + 2x_3 - x_4), \quad w_4(4) = 4,$$
$$R_{45} = N_4(1 - 2x_2 + x_4), \quad w_4(5) = 6, \tag{7.42}$$

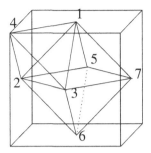

Fig. 7.1. Tetrahedron-plus-octahedron approximation.

where $N_4 = \frac{1}{16}$ and $w_4(i)$ are the multiplicities. The new correlation functions are

$$x_3 = \langle \mu_1 \mu_2 \mu_3 \rangle, \quad \text{and} \quad x_4 = \langle \mu_1 \mu_2 \mu_3 \mu_4 \rangle. \tag{7.43}$$

The minimization conditions for the variational potential are

$$x_1 : 0 = \tfrac{1}{2} \log \left(\frac{R_{41}}{R_{42}} \right) + \log \left(\frac{R_{43}}{R_{44}} \right) - 3 \log \left(\frac{R_{41}}{R_{42}} \right) + \tfrac{5}{2} \log \left(\frac{R_{41}}{R_{42}} \right),$$

$$x_2 : \frac{6J}{kT} = \tfrac{3}{4} \log \left(\frac{R_{41} R_{42}}{R_{45}^2} \right) - \tfrac{3}{2} \log \left(\frac{R_{21} R_{22}}{R_{23}^2} \right),$$

$$x_3 : 0 = \log \left(\frac{R_{41}}{R_{42}} \right) - 2 \log \left(\frac{R_{43}}{R_{44}} \right),$$

$$x_4 : 0 = \tfrac{1}{8} \log \left(\frac{R_{41} R_{42} R_{45}^6}{(R_{43} R_{44})^4} \right). \tag{7.44}$$

7.5.4 Tetrahedron-plus-octahedron approximation

The variational potential per lattice site divided by kT is given by[†]

$$F/NkT = -(6J/kT)x_2 - G_1 + 6G_2(1, 2) - 8G_3(1, 2, 3)$$
$$+ 2G_4(1, 2, 3, 4) + G_6(1, 2, 3, 5, 6, 7). \tag{7.45}$$

$G_3(1, 2, 3)$ and $G_6(1, 2, 3, 5, 6, 7)$ are the triangle and tetrahedron cluster functions, respectively, given by

$$G_3(1, 2, 3) = \sum_{i=1}^{4} w_3(i) R_{3i} \log R_{3i},$$

$$G_6(1, 2, 3, 5, 6, 7) = \sum_{i=1}^{10} w_6(i) R_{6i} \log R_{6i}. \tag{7.46}$$

[†] The formulation presented in this subsection was developed by Aggarwal & Tanaka (1977).

The elements of the three-site functions are

$$
\begin{aligned}
R_{31} &= N_3(1 + 3x_1 + 3x_2 + x_3), \quad w_3(1) = 1, \\
R_{32} &= N_3(1 - 3x_1 + 3x_2 - x_3), \quad w_3(2) = 1, \\
R_{33} &= N_3(1 + x_1 - x_2 - x_3), \quad w_3(3) = 3, \\
R_{34} &= N_3(1 - x_1 - x_2 + x_3), \quad w_3(4) = 3, \\
N_3 &= 1/2^3.
\end{aligned}
\tag{7.47}
$$

The four-site functions are already given by (7.42), and the six-site functions are

$$
\begin{aligned}
R_{61} &= N_6(S_{12} + L_{12}), \quad w_6(1) = 1, \\
R_{62} &= N_6(S_{12} - L_{12}), \quad w_6(2) = 1, \\
R_{63} &= N_6(S_{34} + L_{34}), \quad w_6(3) = 6, \\
R_{64} &= N_6(S_{34} - L_{34}), \quad w_6(4) = 6, \\
R_{65} &= N_6(S_{56} + L_{56}), \quad w_6(5) = 12, \\
R_{66} &= N_6(S_{56} - L_{56}), \quad w_6(6) = 12, \\
R_{67} &= N_6(S_{78} + L_{78}), \quad w_6(7) = 3, \\
R_{68} &= N_6(S_{78} - L_{78}), \quad w_6(8) = 3, \\
R_{69} &= N_6(1 - 3x_3 + 3x_7 - x_{10}), \quad w_6(9) = 8, \\
R_{60} &= N_6(1 - 4x_1 + x_3 - x_7 + 4x_9 + x_{10}), \quad w_6(10) = 12,
\end{aligned}
\tag{7.48}
$$

where $N_6 = 1/2^6$, and

$$
\begin{aligned}
S_{12} &= 1 + 12x_1 + 3x_6 + 12x_7 + x_9 + 3x_{10}, \\
L_{12} &= 6x_1 + 8x_3 + 12x_5 + 6x_8, \\
S_{34} &= 1 + 4x_1 - x_6 - 4x_7 - x_9 + x_{10}, \\
L_{34} &= 4x_1 - 4x_8, \\
S_{56} &= 1 - x_6 + x_9 - x_{10}, \\
L_{56} &= 2x_1 - 4x_5 + 2x_8, \\
S_{78} &= 1 - 4x_1 + 3x_6 - 3x_7 + x_9 + 3x_{10}, \\
L_{78} &= 2x_1 - 8x_3 + 4x_5 + 2x_8.
\end{aligned}
\tag{7.49}
$$

Newly introduced correlation functions are

$$
\begin{aligned}
x_6 &= \langle \mu_2 \mu_3 \mu_5 \mu_7 \rangle, \quad x_7 = \langle \mu_3 \mu_5 \mu_6 \mu_7 \rangle, \\
x_9 &= \langle \mu_1 \mu_2 \mu_3 \mu_5 \mu_6 \mu_7 \rangle, \quad x_{10} = \langle \mu_1 \mu_6 \rangle, \\
x_5 &= \langle \mu_1 \mu_2 \mu_7 \rangle, \quad x_8 = \langle \mu_2 \mu_3 \mu_5 \mu_6 \mu_7 \rangle.
\end{aligned}
\tag{7.50}
$$

Minimization of F, (7.45), leads to the following set of ten conditions:

$$x_1 : 0 = \tfrac{3}{32} \log \frac{R_{61} R_{67}}{R_{62} R_{68}} + \tfrac{3}{8} \log \frac{R_{63} R_{65}}{R_{64} R_{66}} + \tfrac{1}{2} \log \frac{R_{41}}{R_{42}} + \log \frac{R_{43}}{R_{44}}$$
$$- 3 \log \frac{R_{31} R_{33}}{R_{32} R_{34}} + 3 \log \frac{R_{21}}{R_{22}} - \tfrac{1}{2} \log R_{11} R_{12},$$

$$x_2 : \frac{6J}{kT} = \tfrac{3}{16} \log \frac{R_{61} R_{62}}{R_{67} R_{68}} + \tfrac{1}{2} \log \frac{R_{63} R_{64}}{R_{60}^2} + \tfrac{3}{4} \log \frac{R_{41} R_{42}}{R_{45}^2}$$
$$- 3 \log \frac{R_{31} R_{32}}{R_{33} R_{34}} + \tfrac{3}{2} \log \frac{R_{21} R_{22}}{R_{23}^2},$$

$$x_3 : 0 = \tfrac{1}{8} \log \frac{R_{61}}{R_{62}} - \tfrac{3}{8} \log \frac{R_{67}}{R_{68}} + \tfrac{1}{2} \log \frac{R_{41}}{R_{42}} - \log \frac{R_{43}}{R_{44}} - \log \frac{R_{31}}{R_{32}} + 3 \log \frac{R_{33}}{R_{34}},$$

$$x_4 : 0 = \tfrac{1}{8} \log(R_{41} R_{42}) - \tfrac{1}{2} \log(R_{43} R_{44}) + \tfrac{3}{4} \log R_{45},$$

$$x_5 : 0 = \tfrac{3}{16} \log \frac{R_{61}}{R_{62}} - \tfrac{3}{4} \log \frac{R_{65}}{R_{66}} + \tfrac{3}{16} \log \frac{R_{67}}{R_{68}},$$

$$x_6 : 0 = \tfrac{3}{64} \log(R_{61} R_{62}) - \tfrac{3}{32} \log(R_{63} R_{64}) - \tfrac{3}{16} \log(R_{65} R_{66}) + \tfrac{3}{8} \log R_{69}$$
$$+ \tfrac{9}{64} \log(R_{67} R_{68}) - \tfrac{3}{16} \log R_{60},$$

$$x_7 : 0 = \tfrac{3}{16} \log \frac{R_{61} R_{62}}{R_{67} R_{68}} - \tfrac{3}{8} \log \frac{R_{63} R_{64}}{R_{60}^2},$$

$$x_8 : 0 = \tfrac{1}{64} \log \frac{R_{61}}{R_{62}} - \tfrac{3}{8} \log \frac{R_{63} R_{66}}{R_{64} R_{65}} + \tfrac{3}{32} \log \frac{R_{67}}{R_{68}},$$

$$x_9 : 0 = \tfrac{1}{64} \log(R_{61} R_{62}) - \tfrac{3}{32} \log(R_{63} R_{64}) + \tfrac{3}{16} \log(R_{65} R_{66})$$
$$+ \tfrac{3}{64} \log(R_{67} R_{68}) - \tfrac{1}{8} \log R_{69} - \tfrac{3}{16} \log R_{60},$$

$$x_{10} : 0 = \tfrac{3}{64} \log(R_{61} R_{62}) - \tfrac{3}{32} \log(R_{63} R_{64}) + \tfrac{3}{16} \log(R_{65} R_{66})$$
$$+ \tfrac{3}{64} \log(R_{64} R_{68}) - \tfrac{3}{8} \log R_{69} + \tfrac{3}{16} \log R_{60}. \tag{7.51}$$

7.6 High temperature specific heat

The expression for the high temperature specific heat is given by

$$(kT/J)^2 c_v = \tfrac{1}{2} z(1 - w^2) \frac{dx_2}{dw}, \tag{7.52}$$

where x_2 is the nearest-neighbor correlation function, and

$$w = \tanh(J/kT). \tag{7.53}$$

7.6.1 Pair approximation

Since $y_1 = 0$ above the critical temperature, the x_2-equation is reduced to

$$J/kT = \tfrac{1}{2} \log(1 + x_2)/(1 - x_2) = \tanh^{-1} x_2, \tag{7.54}$$

or, solving for x_2,

$$x_2 = \tanh(J/kT) = w, \tag{7.55}$$

and hence (7.52) gives, setting $(z = 12)$,

$$(kT/J)^2 c_v = 6 - 6w^2. \tag{7.56}$$

7.6.2 Tetrahedron approximation

Noting the fact that x_2 and x_4 go to zero as T goes to infinity, we can write

$$x_2 = \sum_{1}^{\infty} b_n w^n \quad \text{and} \quad x_4 = \sum_{1}^{\infty} d_n w^n, \tag{7.57}$$

where b_n and d_n are to be determined. Substituting the above expansions into (7.44), the expressions for x_2 and x_4 are obtained as [Exercise 7.1]

$$\begin{aligned} x_2 &= w + 4w^2 + 20w^3 + 116w^4 + \cdots, \\ x_4 &= 3w^2 + 16w^3 + 87w^4 + \cdots. \end{aligned} \tag{7.58}$$

This leads to the following expression for the specific heat:

$$(kT/J)^2 c_v = 6 + 48w + 354w^2 + \cdots. \tag{7.59}$$

7.6.3 Tetrahedron-plus-octahedron approximation

For $T > T_c$, the equations for x_1, x_3, x_5, and x_8 in (7.51) are satisfied identically with $x_1 = x_3 = x_5 = x_8 = 0$. The remaining equations are manipulated to yield [Exercise 7.2]

$$x_6 = x_7 \quad \text{and} \quad x_9 + x_{10} = 2x_6. \tag{7.60}$$

Since all the above x correlation functions go to zero at infinite temperature, the following form of series expansion can be assumed for those x-functions:

$$x = \sum_{1}^{\infty} a_n w^n. \tag{7.61}$$

Substituting this form of expansion into the x-equations in (7.51) yields [Exercise 7.3]

$$
\begin{aligned}
x_2 &= w + 4w^2 + 22w^3 + 136w^4 + 880w^5 + 5908w^6 + \cdots, \\
x_4 &= 3w^2 + 16w^3 + 108w^4 + 744w^5 + 5180w^6 + \cdots, \\
x_6 &= x_4 = 2w_2 + 16w^3 + 108w^4 + 736w^5 + 5144w^6 + \cdots, \\
x_9 &= 8w^3 + 72w^4 + 528w^5 + 3858w^6 + \cdots, \\
x_{10} &= 4w^2 + 24w^3 + 144w^4 + 944w^5 + 6432w^6 + \cdots.
\end{aligned}
\tag{7.62}
$$

Using the series for x_2 in (7.52), the high temperature specific heat is found:

$$
(kT/J)^2 c_v = 6 + 48w + 390w^2 + 3216w^3 + 26004w^4 + \cdots.
\tag{7.63}
$$

The exact infinite series (Baker, 1963) is

$$
\begin{aligned}
(kT/J)^2 c_v &= 6 + 48w + 390w^2 + 3216w^3 + 26\,448w^4 + 229\,584w^5 \\
&\quad + 2\,006\,736w^6 + 17\,809\,008w^7 + \cdots.
\end{aligned}
\tag{7.64}
$$

Thus the six-spin approximation reproduces the first four of the eight known terms exactly. The main reason for this rather poor agreement of the cluster variation series with the exact one is attributed to the fact that the four y-equations do not play any important part in the determination of the x-functions. The situation is substantially improved in the low temperature expansion.

7.7 High temperature susceptibility

In the presence of a small external magnetic field H in appropriate units, the variational potential will have an extra term

$$
F/Nkt = -(H/kT)x_1 + \cdots,
\tag{7.65}
$$

and hence the x_1-equation is modified. Consequently, all the odd correlation functions can have nonzero values even at temperatures higher than the critical temperature. The reduced susceptibility per spin, which is the deviation from Curie's law, is defined by

$$
\chi = kT(\partial x_1/\partial H)_{T,H=0}.
\tag{7.66}
$$

7.7.1 Mean-field approximation

The variational potential is modified, cf. (7.34),

$$
\frac{F}{NkT} = -\frac{H}{kT}x_1 - \frac{zJ}{2kT}x_1^2 + \tfrac{1}{2}[(1+x_1)\log(1+x_1) + (1-x_1)\log(1-x_1)].
\tag{7.67}
$$

The minimization condition for x_1 is also modified:

$$\frac{H}{kT} = -\frac{zJ}{kT}x_1 + \tfrac{1}{2}\log\frac{1+x_1}{1-x_1}. \tag{7.68}$$

If the logarithm term is expanded into an infinite series and only the terms of the first power of x_1 are retained, we find

$$\frac{H}{kT} = \left(1 - \frac{zJ}{kT}\right)x_1. \tag{7.69}$$

When this is solved for x_1,

$$x_1 = \frac{H}{kT} \cdot \frac{1}{1 - \frac{zJ}{kT}}, \tag{7.70}$$

and hence the reduced susceptibility is given by ($z = 12$)

$$\chi = 1/(1 - zJ/kT) = 1 + 12w + 144w^2 + \cdots. \tag{7.71}$$

7.7.2 Pair approximation

The susceptibility is given by [Exercise 7.4]

$$\begin{aligned}\chi &= (1 + x_1)/[1 - (z - 1)x_1] \\ &= 1 + zx_1 + z(z - 1)x_1 + z(z - 1)^2 x_1^2 + \cdots.\end{aligned} \tag{7.72}$$

Using (7.55) and $z = 12$, this is reduced to

$$\chi = 1 + 12w + 132w^2 + 1452w^3 + \cdots. \tag{7.73}$$

7.7.3 Tetrahedron approximation

In Sec. 6.10, the expression for the induced magnetization was derived as

$$x_1 = \beta H \frac{(1 + x_2)(1 + 3x_2)}{(1 - 3x_2)(1 - 5x_2)}. \tag{7.74}$$

Substituting from (7.62) and expanding the above fraction into an infinite series yields the reduced susceptibility [Exercise 7.5]

$$\chi = 1 + 12w + 132w^2 + 1404w^3 + 14676w^4 + \cdots. \tag{7.75}$$

7.7.4 Tetrahedron-plus-octahedron approximation

In the presence of an external magnetic field, the x_1-equation, (7.51), is modified as

$$x_1 : \frac{H}{kT} = \tfrac{3}{32} \log \frac{R_{61}R_{67}}{R_{62}R_{68}} + \tfrac{3}{8} \log \frac{R_{63}R_{65}}{R_{64}R_{66}} + \tfrac{1}{2} \log \frac{R_{41}}{R_{42}} + \log \frac{R_{43}}{R_{44}}$$

$$- 3 \log \frac{R_{31}R_{33}}{R_{32}R_{34}} + 3 \log \frac{R_{21}}{R_{22}} - \tfrac{1}{2} \log R_{11} R_{12}. \tag{7.76}$$

The x_3-, x_5-, and x_8- equations in (7.51) are not changed, but they are infinitesimally small. The logarithmic terms in these equations are expanded, and we retain only the terms of the first power in these quantities. The resulting set of first-order inhomogeneous equations is solved, only the inhomogeneous term being H/kT, with respect to x_1, x_3, x_5, x_8 in terms of $x_2, x_4, x_6, x_7, x_9, x_{10}$. Then using (7.62), the following expression for the susceptibility is found [Exercise 7.6]:

$$\chi = 1 + 12w + 132w^2 + 1404w^3 + 14\,652w^4$$

$$+ 151\,116w^5 + 1\,546\,668w^6 + \cdots. \tag{7.77}$$

The exact infinite series result (Rapaport, 1974) is

$$\chi = 1 + 12w + 132w^2 + 1404w^3 + 14\,652w^4 + 151\,116w^5 + 1\,546\,332w^6$$

$$+ 15\,734\,460w^7 + 159\,425\,580w^8 + 160\,998\,770w^9 + 16\,215\,457\,188w^{10}$$

$$+ 162\,961\,837\,500w^{11} + 1\,634\,743\,178\,420w^{12}$$

$$+ 16\,373\,484\,437\,340w^{13} + \cdots. \tag{7.78}$$

Thus the six-site approximation reproduces the first six coefficients exactly and the seventh one within 0.02%. The agreement of the high temperature susceptibility from the cluster variation method with the exact series is not satisfactory. The main reason is that the odd order correlations do not play any part in the determination of the even order correlation functions above the transition temperature. The situation is substantially improved in the low temperature series expansion.

7.8 Low temperature specific heat

The low temperature specific heat is given by

$$\left(\frac{kT}{J}\right)^2 c_v = -2uz \frac{dx_2}{du}, \tag{7.79}$$

where

$$u = \exp(-4J/kT) = v^4. \tag{7.80}$$

At zero temperature, all the correlation functions are unity. Consequently, all the correlation functions are expanded in the following infinite series:

$$x = 1 - u^{\alpha} \sum_{0}^{\infty} a_n u^n, \quad a_0 \neq 0, \tag{7.81}$$

where α and a_n are to be determined for each correlation function.

7.8.1 Mean-field approximation

Only one correlation function appears in this case, i.e.,

$$0 = -zJx_1 + \frac{kT}{2} \log(1 + x_1)/(1 - x_1), \tag{7.82}$$

and it is found to be (only the leading term is expected to be correct)

$$\alpha = \tfrac{1}{2}z. \tag{7.83}$$

A further comparison of coefficients of different powers of u in (7.82) leads to

$$x_1 = 1 - 2u^{z/2} + 2u^z + \cdots, \tag{7.84}$$

which on using $x_2 = x_1^2$ and (7.78) yields

$$(kT/J)^2 c_v(u) = 4z^2 u^{z/2} - 16z^2 u^z + \cdots. \tag{7.85}$$

For face centered cubic ($z = 12$),

$$(kT/J)^2 c_v(u) = 576u^6 - 2304u^{12} + \cdots. \tag{7.86}$$

7.8.2 Pair approximation

Since $1 - x_1, 1 - x_2$, and $1 - 2x_1 + x_2$ in (7.40a,b) go to zero in the limit as $u \to 0$, the following expansions are introduced into (7.40a,b):

$$1 - x_1 = 2u^{\alpha_1} \sum_{0}^{\infty} a_n u^n, \quad a_0 \neq 0,$$

$$1 - x_2 = 2u^{\alpha_2} \sum_{0}^{\infty} b_n u^n, \quad b_0 \neq 0,$$

$$1 - 2x_1 + x_2 = 2u^{\alpha_3} \sum_{0}^{\infty} c_n u^n, \quad c_0 \neq 0. \tag{7.87}$$

Comparison of the coefficients of $\log u$ in (7.40a,b) leads to (for $z = 12$)

$$0 = -3\alpha_3 + \tfrac{11}{2}\alpha_1, \quad -1 = \alpha_3 - 2\alpha_2. \tag{7.88}$$

The set (7.88) by itself does not give a unique solution. This inconsistency comes from the fact that three different expansions are introduced for only two correlation functions. If only the first two expansions are used one finds a condition

$$\alpha_3 \geq \min(\alpha_1, \alpha_2), \tag{7.89}$$

with equality holding only when $\alpha_1 = \alpha_2$. If $\alpha_1 = \alpha_2$ (the mean-field case) is assumed,

$$\alpha_1 = 6, \quad \alpha_2 = 6, \quad \alpha_3 = 11. \tag{7.90}$$

Feeding this information back into (7.40a,b) and comparing the coefficients of higher powers of u, we obtain

$$x_1 = 1 - 2u^6(1 + 12u^5 - 13u^6 + 198u^{10} + \cdots),$$
$$x_2 = 1 - 4u^6(1 + 11u^5 - 13u^6 + 176u^{10} + \cdots). \tag{7.91}$$

The expression for the low temperature specific heat is given by

$$(kT/J)^2 c_v(u) = 576u^6 + 11\,616u^{11} - 14\,976u^{12} + 270\,336u^{16} + \cdots. \tag{7.92}$$

7.8.3 Tetrahedron approximation

Following the same procedure as in the pair approximation, an undetermined set of equations similar to (7.88) is obtained. The set can be uniquely solved, if the results (7.90) are assumed to be known. Feeding this information back into (7.44), the following series for the low temperature specific heat is obtained:

$$(kT/J)^2 c_v = 576u^6 + 11\,616u^{11} - 14\,976u^{12} + 28\,800u^{15} + 172\,032u^{16}$$
$$- 55\,488u^{17} + 374\,976u^{18} + 138\,624u^{19} + 768\,000u^{20} + \cdots. \tag{7.93}$$

7.8.4 Tetrahedron-plus-octahedron approximation

In this approximation (Aggarwal & Tanaka, 1977), the series for the specific heat, which reproduces all the terms up to u^{24} given by Baker (1963) and more, is obtained:

$$(kT/J)^2 c_v = 576u^6 + 11\,616u^{11} - 14\,976u^{12} + 28\,800u^{15} + 172\,032u^{16}$$
$$- 55\,488u^{17} + 374\,976u^{18} + 138\,624u^{19} + 787\,200u^{20}$$
$$+ 889\,056u^{21} - 12\,568\,512u^{22} + 20\,465\,952u^{23} - 4\,564\,224u^{24}$$
$$+ 8\,220\,000u^{25} - 29\,235\,648u^{26} - 180\,931\,968u^{27}$$
$$+ 633\,948\,672u^{28} - 558\,773\,856u^{29} + 125\,758\,080u^{30} + \cdots. \tag{7.94}$$

7.9 Infinite series for other correlation functions

In the process of obtaining an infinite-series representation of the low temperature specific heat, infinite series for other correlation functions are also obtained. These series are not available in the literature, except for a few correlation functions.

The cluster variation method may, therefore, be used as a means of producing a number of terms of exact infinite series for the correlation functions without using the combinatorial formulation. The infinite series for ten correlation functions appearing in the tetrahedron-plus-octahedron approximation are listed in Tables 7.4–7.7 (Aggarwal & Tanaka, 1977). For this reason, the cluster variation method may be regarded as another way of finding exact critical exponents for the Ising ferromagnet when combined with the Padé approximant method.

Table 7.4. *Coefficients of infinite series for correlation functions.* $\langle 123 \rangle = \langle \mu_1 \mu_2 \mu_3 \rangle$, etc.

Correlation function	u^0	u^6	u^{11}	u^{12}	u^{15}
$\langle 1 \rangle$	1	−2	−24	26	−48
$\langle 12 \rangle$	1	−4	−44	52	−80
$\langle 16 \rangle$	1	−4	−48	56	−96
$\langle 123 \rangle$	1	−6	−60	78	−104
$\langle 124 \rangle$	1	−6	−64	82	−112
$\langle 1234 \rangle$	1	−8	−72	104	−128
$\langle 2357 \rangle$	1	−8	−80	112	−128
$\langle 3567 \rangle$	1	−8	−76	108	−128
$\langle 23567 \rangle$	1	−10	−88	138	−144
$\langle 123567 \rangle$	1	−12	−96	168	−160

Table 7.5. *Continuation from Table 7.4 for u^{16} to u^{20}.*

Correlation function	u^{16}	u^{17}	u^{18}	u^{19}	u^{20}
$\langle 1 \rangle$	−252	720	−438	−192	−984
$\langle 12 \rangle$	−448	1360	−868	−304	−1640
$\langle 16 \rangle$	−488	1504	−964	−368	−1896
$\langle 123 \rangle$	−588	1920	−1298	−384	−2088
$\langle 124 \rangle$	−636	2064	−1386	−416	−2264
$\langle 1234 \rangle$	−672	2400	−1720	−480	−2448
$\langle 2357 \rangle$	−784	2688	−1896	−480	−2624
$\langle 3567 \rangle$	−728	2544	−1816	−464	−2544
$\langle 23567 \rangle$	−828	3088	−2334	−512	−2856
$\langle 123567 \rangle$	−888	3312	−2860	−528	−3144

Table 7.6. *For u^{21} to u^{25}.*

Correlation function	u^{21}	u^{22}	u^{23}	u^{24}	u^{25}
$\langle 1 \rangle$	-1008	$12\,924$	$-19\,536$	3062	-8280
$\langle 12 \rangle$	-1764	$23\,804$	$-37\,076$	7924	$-13\,700$
$\langle 16 \rangle$	-1664	$26\,312$	$-41\,648$	8660	$-15\,072$
$\langle 123 \rangle$	-2068	$32\,496$	$-52\,932$	$13\,818$	$-16\,500$
$\langle 124 \rangle$	-2296	$35\,300$	$-57\,256$	$14\,786$	$-18\,112$
$\langle 1234 \rangle$	-1720	$39\,048$	$-67\,224$	$19\,528$	$-16\,632$
$\langle 2357 \rangle$	-2832	$45\,088$	$-75\,568$	$22\,520$	$-20\,240$
$\langle 3567 \rangle$	-2276	$42\,020$	$-71\,428$	$21\,480$	$-19\,108$
$\langle 23567 \rangle$	-2288	$49\,788$	$-88\,240$	$29\,822$	$-20\,472$
$\langle 123567 \rangle$	-1904	$55\,752$	$-103\,488$	$38\,956$	$-22\,224$

Table 7.7. *For u^{26} to u^{30}.*

Correlation function	u^{26}	u^{27}	u^{28}	u^{29}	u^{30}
$\langle 1 \rangle$	$26\,694$	$153\,536$	$-507\,948$	$406\,056$	$-78\,972$
$\langle 12 \rangle$	$46\,852$	$279\,216$	$-943\,376$	$802\,836$	$-174\,664$
$\langle 16 \rangle$	$57\,484$	$302\,048$	$-1\,057\,024$	$908\,432$	$-184\,296$
$\langle 123 \rangle$	$65\,010$	$369\,136$	$-1\,308\,588$	$1\,185\,348$	$-268\,050$
$\langle 124 \rangle$	$70\,274$	$404\,368$	$-1\,425\,780$	$1\,286\,640$	$-291\,684$
$\langle 1234 \rangle$	$86\,856$	$418\,528$	$-1\,609\,440$	$1\,541\,112$	$-318\,704$
$\langle 2357 \rangle$	$85\,144$	$511\,424$	$-1\,847\,296$	$1\,746\,736$	$-413\,872$
$\langle 3567 \rangle$	$86\,048$	$464\,624$	$-1\,726\,960$	$1\,649\,564$	$-387\,488$
$\langle 23567 \rangle$	$103\,838$	$536\,416$	$-2\,086\,524$	$2\,090\,952$	$-515\,116$
$\langle 123567 \rangle$	$123\,876$	$579\,232$	$-2\,387\,904$	$2\,511\,840$	$-669\,456$

Exercises

7.1 Derive relations (7.58):

$$x_2 = w + 4w^2 + 20w^3 + 116w^4 + \cdots,$$
$$x_4 = 3w^2 + 16w^3 + 87w^4 + \cdots.$$

7.2 Derive relations (7.60):

$$x_6 = x_7 \quad \text{and} \quad x_9 + x_{10} = 2x_6.$$

7.3 Derive relations (7.62):

$$x_2 = w + 4w^2 + 22w^3 + 136w^4 + 880w^5 + 5908w^6 + \cdots,$$
$$x_4 = 3w^2 + 16w^3 + 108w^4 + 744w^5 + 5180w^6 + \cdots,$$
$$x_6 = x_4 = 2w_2 + 16w^3 + 108w^4 + 736w^5 + 5144w^6 + \cdots,$$
$$x_9 = 8w^3 + 72w^4 + 528w^5 + 3858w^6 + \cdots,$$
$$x_{10} = 4w^2 + 24w^3 + 144w^4 + 944w^5 + 6432w^6 + \cdots.$$

7.4 Derive (7.72):

$$\chi = (1 + x_1)/[1 - (z - 1)x_1]$$
$$= 1 + zx_1 + z(z - 1)x_1 + z(z - 1)^2 x_1^2 + \cdots.$$

7.5 Derive (7.75):

$$\chi = 1 + 12w + 132w^2 + 1404w^3 + 14676w^4 + \cdots.$$

7.6 Derive (7.77):

$$\chi = 1 + 12w + 132w^2 + 1404w^3 + 14\,652w^4$$
$$+ 151\,116w^5 + 1\,546\,668w^6 + \cdots.$$

7.7 Derive (7.84) and (7.85):

$$x_1 = 1 - 2u^{z/2} + 2u^z + \cdots,$$

$$(kT/J)^2 c_v(u) = 4z^2 u^{z/2} - 16z^2 u^z + \cdots.$$

8

The extended mean-field approximation

8.1 The Wentzel criterion

G. Wentzel (1960) presented a formulation called the *method of thermodynamically equivalent Hamiltonian*. He proved that a certain type of Hamiltonian can be treated in the mean-field approximation, in an extended sense,[†] in order to obtain an exact result for the potential. The reduced Hamiltonian of the Bardeen–Cooper–Schrieffer (BCS) theory of superconductivity is one which falls under the category satisfying the *Wentzel criterion* (Bardeen *et al.*, 1957).

In the BCS reduced Hamiltonian, only two types of operators occur, namely the occupation number and the pair absorption (and creation) operators:

$$a_k^* a_k + a_{-k}^* a_{-k} = b_{k1}, \quad a_{-k} a_k = b_{k2}. \tag{8.1}$$

A general expression for the BCS Hamiltonian is then

$$H = \sum_{k\lambda} E_{k\lambda} b_{k\lambda} + \frac{1}{V} \sum_{k\lambda, k'\lambda'} J_{k\lambda, k'\lambda'} b_{k\lambda}^* b_{k'\lambda'}, \tag{8.2}$$

where $E_{k2} = 0$. One distinct feature of this Hamiltonian is that the two-particle interaction term is expressed as a double summation instead of the usual triple summation indicating a momentum transfer as a mechanism of interaction. More generally, we will admit Hamiltonians of the form (8.2) where the operators b_k are specified bilinear combinations of the a and a^* operators, not necessarily those of the BCS theory. One can also allow the a and a^* operators to refer to either fermions or bosons, since the procedure can be described without specifying the commutation relations.

The given Hamiltonian is transformed into the so-called thermodynamically equivalent Hamiltonian by means of the following transformation. We define new

---

[†] Physical properties are defined in the momentum space and all the correlation functions containing two or more momenta will be decomposed as products of individual averages.

operators

$$B_{k\lambda} = b_{k\lambda} - \eta_{k\lambda}, \tag{8.3}$$

where $\eta_{k\lambda}(\lambda = 1, 2)$ are trial functions (c numbers, not necessarily real), and we subtract from the Hamiltonian (8.2) a perturbation defined as

$$H' = \frac{1}{V} \sum_{k\lambda,k'\lambda'} J_{k\lambda,k'\lambda'} B_{k\lambda}^* B_{k'\lambda'}. \tag{8.4}$$

The remaining *unperturbed* Hamiltonian may be written as

$$H^{(0)} = H - H' = U + \sum_{k\lambda}(E_{k\lambda}b_{k\lambda} + G_{k\lambda}b_{k\lambda} + G_{k\lambda}^* b_{k\lambda}^*), \tag{8.5}$$

where

$$U = -\frac{1}{V} \sum_{k\lambda,k'\lambda'} J_{k\lambda,k'\lambda'} \eta_{k\lambda}^* \eta_{k'\lambda'}, \tag{8.6}$$

$$G_{k\lambda} = \frac{1}{V} \sum_{k'\lambda'} J_{k\lambda,k'\lambda'} \eta_{k'\lambda'}^*. \tag{8.7}$$

$H^{(0)}$, being bilinear in the operators a and a^*, can be diagonalized in closed form, e.g., by a Bogoliubov transformation (Bogoliubov, 1958), for any given set of functions $\eta_{k\lambda}$.

The Helmholtz potential $F^{(0)}$ of the corresponding canonical ensemble, defined by

$$\text{Tr} \exp\left[\beta\left(F^{(0)} - H^{(0)}\right)\right] = 1 \tag{8.8}$$

can now be minimized with respect to the trial functions:

$$\partial F^{(0)}/\partial \eta_{k\lambda} = \partial F^{(0)}/\partial \eta_{k\lambda}^* = 0. \tag{8.9}$$

Note that $F^{(0)}$ can be written as

$$F^{(0)} = U + F_1(G_{k\lambda}, G_{k'\lambda'}^*), \tag{8.10}$$

where U, $G_{k\lambda}$, and $G_{k\lambda}^*$ depend on $\eta_{k\lambda}$ according to (8.6) and (8.7). It is then easily seen that (8.9) is satisfied by setting [Exercise 8.1]

$$\eta_{k\lambda} = \partial F_1/\partial G_{k\lambda}. \tag{8.11}$$

Substituting this back into (8.7) leads to a set of coupled integral equations[†] for the functions $G_{k\lambda}$, with solutions depending on the given coefficients $E_{k\lambda}$ and $J_{k\lambda,k'\lambda'}$, and on the temperature β^{-1}.

[†] When the $\lim V \to \infty$ is considered, then $V^{-1} \sum_k$ becomes a k-space integral.

The essential point is to prove that this variational solution is in some way rigorous. For this purpose it should be noted that the thermal average of $b_{k\lambda}$, with the unperturbed density matrix, is given by

$$\langle b_{k\lambda} \rangle = \text{Tr}\left[b_{k\lambda} \exp \beta \left(F^{(0)} - H^{(0)} \right) \right] = \partial F_1 / \partial G_{k\lambda}, \qquad (8.12)$$

as is easily seen by substituting (8.10) and (8.5) into (8.8) and then differentiating with respect to $G_{k\lambda}$ [Exercise 8.2]. From (8.11) and (8.12) it follows that the thermal averages of the quantities (8.3), and also of their Hermitian conjugates, vanish:

$$\langle B_{k\lambda} \rangle = \langle B_{k\lambda}^* \rangle = 0. \qquad (8.13)$$

Having rigorously determined the potential $F^{(0)}$ using (8.6), (8.7), and (8.11), we now investigate how the perturbation H', (8.4), affects the potential F of the system $H = H^{(0)} + H'$. We can use the perturbation expansion developed in Sec. 5.3. Writing down the nth-order correction to the partition function, $\text{Tr} \exp[-\beta(H^{(0)} + H')]$, with H' given by (8.4), yields expressions of the following type; since each of the terms in H' is a product of $B_{k\lambda}^*$ and $B_{k'\lambda'}$,

$$\frac{1}{V^m} \text{Tr} \left\{ \exp \left(-\beta H^{(0)} \right) \prod_{i=1,\ldots,2m} \left[\exp \left(u_i H^{(0)} \right) C_{ki} \exp \left(-u_i H^{(0)} \right) \right] \right\}, \qquad (8.14)$$

where each C_{ki} stands for one of the operators $B_{k\lambda}$ or $B_{k\lambda}^*$. (The u_i are the integration variables.) At this point, we can assume that the commutation properties of the $b_{k\lambda}^{(*)}$ are such that, when $H^{(0)}$ is written as $U + \sum_k H_k$,

$$[H_k, H_{k'}] = 0, \quad [H_k, C_{k'}] = 0 \quad \text{for} \quad k \neq k'. \qquad (8.15)$$

It is then easily seen that the trace (8.14) vanishes, because of (8.13), if one momentum k_i, say k_1, is different from all the other momenta, $k_2, \ldots, k_{2m}$, occurring in the product. In order to obtain a nonvanishing term, there have to be m pairs of equal k_i' so that after multiplying with the appropriate factors $J_{k\lambda,k'\lambda'}$ and then summing over $k_1, \ldots, k_{2m}$, the sum runs over only m independent k vectors. If we then divide by $\text{Tr} \exp(-\beta H^{(0)})$ and finally write the sums as integrals, the factors V^{-m} and V^m cancel out and the result becomes volume independent. Hence, to all orders in H' (assuming convergence),

$$\lim_{V \to \infty} \frac{\text{Tr} \exp \left[-\beta \left(H^{(0)} + H' \right) \right]}{\text{Tr} \exp \left(-\beta H^{(0)} \right)} = \text{finite}, \qquad (8.16)$$

and

$$\lim_{V \to \infty} \frac{1}{V} \left(F - F^{(0)} \right) = 0. \qquad (8.17)$$

We can conclude that H' does not affect the volume-proportional part of the Helmholtz potential. $H^{(0)}$ alone determines the thermodynamics of the system H

exactly. $H^{(0)}$ can be called the thermodynamically equivalent Hamiltonian. $G_{k,\lambda}$ defined by (8.5), being a c number, acts as a mean-field in an extended sense, in the thermodynamic limit.

This result may be stated slightly differently as follows. The variational formulation based on the thermodynamically equivalent Hamiltonian is called the *extended mean-field approximation*. Since H' does not affect the Helmholtz potential, all the correlation effects vanish, and hence the extended mean-field approximation becomes exact.

The criterion by which the extended mean-field approximation gives an exact result is inherent in the original Hamiltonian given by (8.2), i.e., the two-body interaction term is given in double summation in contrast with the general cases of the two-body interaction term being given in triple summation. If a three-body interaction term is expressed in triple summation the extended mean-field approximation still yields an exact result, and so forth. This criterion may be called the *Wentzel criterion*.

8.2 The BCS Hamiltonian

After having established the fact that the BCS Hamiltonian satisfies the Wentzel criterion for the thermodynamically equivalent Hamiltonian giving an exact result, we may evaluate all the thermodynamical quantities in an extended mean-field approximation without destroying the exact nature of the physical properties represented by the Hamiltonian. Keeping this in mind, let us carry out the actual evaluation of various physical quantities as an exercise by the cluster variation method in the extended mean-field approximation. In doing so it is not even necessary to replace the original Hamiltonian by the thermodynamically equivalent Hamiltonian.

The Hamiltonian is given by (8.2); however, it can be given a more specific form for the purposes of this section:

$$H = \sum_{k,\sigma} \epsilon_k n_{k,\sigma} - \frac{1}{V} \sum_{k,k'} J_{kk'} a^*_{k'\uparrow} a^*_{-k'\downarrow} a_{-k\downarrow} a_{k\uparrow}, \tag{8.18}$$

where $n_{k,\sigma} = a^*_{k\sigma} a_{k\sigma}$, and the interaction constant $J_{kk'}$ is assumed to be a positive quantity so that the attractive nature of the interaction is explicitly represented by the minus sign in front of the summation sign.

Let us, once again, confirm the nature of the Wentzel criterion, i.e., $J_{kk'}$ is a quantity of the order of unity and the interaction term is given as a double summation containing only two momenta. It means, then, that the exact effect of the interaction terms can be evaluated by means of the approximation

$$\langle a_{k'\uparrow}^* a_{-k'\downarrow}^* a_{-k\downarrow} a_{k\uparrow} \rangle = \langle a_{k'\uparrow}^* a_{-k'\downarrow}^* \rangle \langle a_{-k\downarrow} a_{k\uparrow} \rangle. \tag{8.19}$$

It is evident that the evaluation of either $\langle a_{k'\uparrow}^* a_{-k'\downarrow}^* \rangle$ or $\langle a_{-k\downarrow} a_{k\uparrow} \rangle$ requires us to find the correlation between two particles in two different one-particle states $(k, \uparrow)$ and $(-k, \downarrow)$, and hence these quantities are two-particle properties. This may appear to be inconsistent with the concept of the mean-field approximation which is genuinely a one-body approximation. The interpretation of the situation is that if one employs a suitable unitary transformation, a Bogoliubov transformation in this case, the two-particle properties are indeed transformed into a single-particle property: the thermal average of the Cooper pair creation or annihilation operator. This is one of the purposes of this section: to show that we can get away with this mathematically involved transformation if we employ the cluster variation formulation (Bose *et al.*, 1973).

The smallest reduced density matrix which contains these correlations is the two-particle density matrix given by

$$\rho^{(2)}(k, -k) = \begin{bmatrix} A_0 & 0 & 0 & E_0 \\ 0 & B_0 & 0 & 0 \\ 0 & 0 & C & 0 \\ F_0 & 0 & 0 & D \end{bmatrix}, \tag{8.20}$$

where

$$\begin{aligned} A_0 &= \langle n_k n_{-k} \rangle, \\ B_0 &= \langle n_k (1 - n_{-k}) \rangle, \\ C &= \langle (1 - n_k) n_{-k} \rangle, \\ D &= \langle (1 - n_k)(1 - n_{-k}) \rangle, \\ E_0 &= \langle a_{-k} a_k \rangle, \\ F_0 &= \langle a_k^* a_{-k}^* \rangle. \end{aligned} \tag{8.21}$$

From now on the spin symbols may be suppressed as long as one remembers that k and $-k$ are associated with up and down spins, respectively.

In this approximation the variational potential is given by

$$\begin{aligned} F = \sum_k \epsilon_k [\langle n_{k\uparrow} \rangle + \langle n_{-k\downarrow} \rangle] - \frac{1}{V} \sum_{k,k'} J_{kk'} \langle a_k^* a_{-k}^* \rangle \langle a_{-k} a_k \rangle \\ + k_B T \sum_k [g_1(k \uparrow) + g_1(k \downarrow) + g_2(k \uparrow, -k \downarrow)]. \end{aligned} \tag{8.22}$$

If, however, the cumulant g functions are replaced by the cluster functions G, the complete cancellation of the g_1 terms follows, and the variational function is found in the following form:

$$F = \sum_k \epsilon_k [\langle n_{k\uparrow}\rangle + \langle n_{-k\downarrow}\rangle] - \frac{1}{V}\sum_{k,k'} J_{kk'}\langle a_k^* a_{-k'}^*\rangle\langle a_{-k}a_k\rangle$$
$$+ k_\mathrm{B}T\sum_k \mathrm{tr}[\rho_2(k,-k)\log\rho_2(k,-k)]. \tag{8.23}$$

After carrying out the usual diagonalization of $\rho_2(k,-k)$, we can perform the trace operation in the variational potential and obtain F explicitly in terms of the unknown expectation values as

$$F = \sum_k \epsilon_k(n_k + n_{-k}) - \frac{1}{V}\sum_{k,k'} J_{kk'}\Gamma_{k'}^*\Gamma_k$$
$$+ k_\mathrm{B}T\sum_k [(A+B)\log(A+B) + (n_k - N_k)\log(n_k - N_k)$$
$$+ (n_{-k} - N_k)\log(n_{-k} - N_k) + (A-B)\log(A-B)], \tag{8.24}$$

where

$$A = \tfrac{1}{2}(1 - n_k - n_{-k}) + N_k,$$
$$B = \tfrac{1}{2}[(1 - n_k - n_{-k})^2 + 4\Gamma_k\Gamma_k^*]^{\frac{1}{2}}, \tag{8.25}$$

and

$$n_k = \langle n_k\rangle, \quad \Gamma_k = \langle a_{-k}a_k\rangle, \quad N_k = \langle n_k n_{-k}\rangle \tag{8.26}$$

are the unknown expectation values.

These expectation values can now be determined by minimizing F with respect to N_k, Γ_k^*, and n_k, i.e.,

$$k_\mathrm{B}T\log\frac{(A+B)(A-B)}{(n_k - N_k)(n_{-k} - N_k)} = 0, \tag{8.27}$$

$$k_\mathrm{B}T\frac{\Gamma_k}{4B}\log\frac{(A+B)}{(A-B)} = \frac{1}{V}\sum_{k'} J_{kk'}\Gamma_{k'}, \tag{8.28}$$

and

$$\epsilon_k - k_\mathrm{B}T\left\{\frac{1}{2}\left[1 + \frac{1 - n_k - n_{-k}}{4B}\right][1 + \log(A+B)] - [1 + \log(n_k - N_k)]\right.$$
$$\left. + \frac{1}{2}\left[1 - \frac{1 - n_k - n_{-k}}{4B}\right][1 + \log(A-B)]\right\} = 0. \tag{8.29}$$

Since (8.27) holds for all temperatures,

$$(A + B)(A - B) = (n_k - N_k)(n_{-k} - N_k). \tag{8.30}$$

Substituting (8.25) into (8.27) yields

$$N_k = n_k n_{-k} + \Gamma_k \Gamma_k^*. \tag{8.31}$$

It is apparent from the symmetry of the problem that $n_k = n_{-k}$ (this can be verified by inspection of the variational potential or by direct calculation), and hence (8.29) is simplified as

$$\epsilon_k + \frac{k_B T}{2} \left[\log \frac{(n_k - N_k)(n_{-k} - N_k)}{(A + B)(A - B)} - \frac{(1 - 2n_k)}{4B} \log \frac{(A + B)}{(A - B)} \right] = 0. \tag{8.32}$$

The second term in this equation, however, vanishes because of (8.27), and hence the equation is simplified:

$$\epsilon_k - \frac{k_B T}{2} \frac{(1 - 2n_k)}{4B} \log \frac{(A + B)}{(A - B)} = 0. \tag{8.33}$$

Now substituting (8.28) into (8.33) yields

$$\frac{2\epsilon_k}{(1 - 2n_k)} = \frac{1}{V} \frac{\sum_{k'} J_{kk'} \Gamma_{k'}}{\Gamma_k}. \tag{8.34}$$

Let us now look at the special case of $T = 0$. As $T \to 0$, (8.28) can be satisfied if

$$\log \frac{(A + B)}{(A - B)} \to \infty, \tag{8.35}$$

since the right hand side of (8.28) remains finite in this limit. This can be true only if $A = B$ or

$$(1 - n_k - n_{-k} + 2N_k)^2 = (1 - n_k - n_{-k})^2 + 4\Gamma_k \Gamma_k^*. \tag{8.36}$$

Using (8.31) in (8.36) yields

$$\Gamma_k = \pm[n_k(1 - n_k)]^{\frac{1}{2}}; \tag{8.37}$$

(8.37) can be substituted into the integral equation (8.34).

Defining $n_k(T = 0) = h_k$, we have

$$\frac{2\epsilon_k}{(1 - 2h_k)} = \frac{1}{V} \frac{\sum_{k'} J_{kk'} [h_{k'}(1 - h_{k'})]^{\frac{1}{2}}}{[h_k(1 - h_k)]^{\frac{1}{2}}}, \tag{8.38}$$

which can be solved for h_k,

$$h_k = \frac{1}{2} \left[1 - \frac{\epsilon_k}{E_k} \right], \tag{8.39}$$

where

$$E_k = \left(\epsilon_k^2 + \epsilon_0^2\right)^{\frac{1}{2}}, \quad \text{and} \quad \epsilon_0 = \frac{1}{V} \sum_{k'} J_{kk'}[h_{k'}(1 - h_{k'})]^{\frac{1}{2}}. \tag{8.40}$$

If $J_{kk'}$ is taken to be a constant, ϵ_0 can be found from (8.38)

$$\epsilon_0 = \frac{\hbar\omega_c}{\sinh\left(\frac{V}{N(0)J}\right)}, \tag{8.41}$$

where ω_c is the Debye frequency.

Returning to the finite temperature case, we observe that we must first evaluate Γ_k in order to solve the integral equation (8.34). Defining

$$\epsilon_{0k}(T) = \frac{1}{V} \sum_{k'} J_{kk'} \Gamma_{k'}, \tag{8.42}$$

we have from (8.34)

$$n_k = \frac{1}{2}\left[1 - \frac{2\epsilon_k}{\epsilon_{0k}(T)}\Gamma_k\right]. \tag{8.43}$$

Using (8.43) and (8.31) in (8.25) yields

$$A = \frac{1}{4} + \frac{E_k^2}{\epsilon_0^2}\Gamma_k^2, \tag{8.44}$$

and

$$B = \frac{E_k}{\epsilon_0}\Gamma_k, \tag{8.45}$$

where

$$E_k = \left(\epsilon_k^2 + \epsilon_0^2\right)^{\frac{1}{2}} \quad \text{and} \quad \epsilon_0 = \epsilon_{0k}(T). \tag{8.46}$$

After substituting (8.44) and (8.45) into (8.33) and carrying out some straightforward algebra, we obtain the following quadratic equation for Γ_k:

$$\Gamma_k^2 - \frac{\Gamma_k \epsilon_0}{E} \coth \beta E_k + \frac{\epsilon_0^2}{4E_k^2} = 0 \tag{8.47}$$

of which solutions are

$$\Gamma_k = \frac{\epsilon_0}{2E_k}\coth\frac{\beta E_k}{2} \quad (+\,\text{sign}) \quad \text{or} \quad \frac{\epsilon_0}{2E_k}\tanh\frac{\beta E_k}{2} \quad (-\,\text{sign}). \tag{8.48}$$

Both solutions have the correct behavior as

$$\beta = \frac{1}{k_B T} \to \infty \quad (T \to 0), \tag{8.49}$$

but only the solution corresponding to the negative sign gives the proper physical behavior as $\beta \to 0$ $(T \to \infty)$. Therefore, only the second solution is acceptable. Hence

$$\Gamma_k = \frac{\epsilon_0}{2E_k} \tanh \frac{\beta E_k}{2}. \tag{8.50}$$

Note that in the limit $\beta \to \infty$

$$\Gamma_k \to \frac{\epsilon_0}{2E_k} = \pm[h_k(1-h_k)]^{\frac{1}{2}}, \tag{8.51}$$

which was found earlier.

Substituting (8.50) into (8.43) yields

$$n_k(T) = \frac{1}{2}\left[1 - \frac{\epsilon_k}{E_k}\tanh\frac{\beta E_k}{2}\right], \tag{8.52}$$

which is exactly the BCS result. Also, substituting (8.50) into (8.42), we obtain the finite temperature BCS gap equation

$$\epsilon_{0k}(T) = \frac{1}{V}\sum_{k'} J_{kk'} \frac{\epsilon_{0k'(T)}}{2E'_k}\tanh\frac{\beta E_{k'}}{2}. \tag{8.53}$$

As shown by BCS, for a constant $J_{kk'}$, a temperature dependence of the energy gap $\epsilon_0(T)$ can be obtained by solving (8.53) numerically. Also, in the weak-coupling limit, (8.53) gives the transition temperature as

$$k_{\rm B}T_{\rm c} = 1.14\omega_{\rm c}\exp\left[-\frac{V}{N(0)J}\right]. \tag{8.54}$$

The next important quantity to calculate by the cluster variation method is the entropy of the superconducting state; this will allow us to derive the various thermodynamic properties of the superconducting state. The entropy term from the expression of the potential (8.24) is seen to be

$$-TS = k_{\rm B}T\sum_k[(A+B)\log(A+B) + 2(n_k - N_k)\log(n_k - N_k)$$
$$+ (A-B)\log(A-B)]. \tag{8.55}$$

After substituting (8.25), (8.31), and (8.37) into (8.55) and simplifying, we obtain

$$-TS = \frac{1}{2}k_{\rm B}T\sum_k\left\{\frac{1}{2}\left[1+\tanh\frac{\beta E_k}{2}\right]^2\log\frac{1}{4}\left[1+\tanh\frac{\beta E_k}{2}\right]^2\right.$$
$$+ {\rm sech}^2\frac{\beta E_k}{2}\log\left[\frac{1}{4}{\rm sech}^2\frac{\beta E_k}{2}\right]$$
$$\left.+\frac{1}{2}\left[1-\tanh\frac{\beta E_k}{2}\right]^2\log\frac{1}{4}\left[1-\tanh\frac{\beta E_k}{2}\right]^2\right\}, \tag{8.56}$$

which can be simplified even further to obtain

$$-TS = 2k_{\mathrm{B}}T \sum_k \left\{ \log\left[\tfrac{1}{2}\mathrm{sech}\left(\frac{\beta E_k}{2} \right) \right] + \tfrac{1}{2}\beta E_k \tanh\left(\frac{\beta E_k}{2} \right) \right\}. \quad (8.57)$$

Defining

$$f(E_k) = \frac{1}{\exp(\beta E_k) + 1}, \quad (8.58)$$

it follows that (8.57) may be written as

$$-TS = 2k_{\mathrm{B}}T \sum_k \{ f(E_k) \log f(E_k) + [1 - f(E_k)] \log[1 - f(E_k)] \}, \quad (8.59)$$

which is exactly the same as the BCS expression for the entropy and shows that the entropy is that of the mean-field approximation.

In this section the cluster variation method is applied in the extended mean-field approximation to the BCS Hamiltonian. In this formulation we do not have to worry about the explicit evaluation of the ground state of the superconducting system because we are considering the direct determination of the unknown expectation values of operators in the superconducting ground state. We utilize the important property that the nonparticle conserving averages $\langle a_{k\sigma}^* a_{-k-\sigma}^* \rangle$ and $\langle a_{k\sigma} a_{-k-\sigma} \rangle$, which create or destroy a pair of electrons of opposite spin and momenta, do not vanish in the superconducting ground state. Both the zero temperature and finite temperature results in this theory follow in a very natural way from the minimization of variational potential in the appropriate temperature limits.

8.3 The s–d interaction

The system under consideration is described by the standard s–d Hamiltonian:[†]

$$H = \sum_{k\sigma} \epsilon_k n_{k\sigma} - (J/2N) \sum_k \sum_{k'} [(a_{k\uparrow}^* a_{k'\uparrow} - a_{k\downarrow}^* a_{k'\downarrow})S^z$$
$$+ a_{k\uparrow}^* a_{k'\downarrow}S^- + a_{k\downarrow}^* a_{k'\uparrow}S^+]. \quad (8.60)$$

The quantity ϵ_k is the kinetic energy of the conduction electron (measured from the Fermi surface); J is the strength of the exchange coupling between conduction electrons and the impurity spin; N is the total number of particles in the system; S^z, S^-, and S^+ are the spin operators for the localized magnetic impurity (for simplicity, the case of $S = \tfrac{1}{2}$ is considered); and the operators $a_{k\sigma}^*$ and $a_{k\sigma}$ ($\sigma = \uparrow$ or $\downarrow$), respectively, create and destroy a conduction electron of momentum k and spin σ. The system has been the subject of intensive investigation in connection with the resistivity minimum at low temperatures due to low density magnetic impurities,

---

[†] The subscripts k and k' and the summation parameters should be bold face; however, scalar k and k' are used for the sake of notational simplicity.

known as the Kondo effect. The exchange coupling, J, can either be ferromagnetic or antiferromagnetic. In this section we are concerned with a nonperturbative nature of the ground state, which would be an interesting system to study by the cluster variation method (Halow et al.,1968).

In order to evaluate the ground state energy without decoupling any interaction term, it is necessary to retain up to three-particle cumulant functions. Since each variable has two internal degrees of freedom, three-particle density matrices have the dimensionality of 8×8. However, if only transitions that conserve total spin and total number of particles are allowed, most of the off-diagonal elements are zero. A further simplification occurs because of the rotational invariance of the Hamiltonian. In the absence of an external magnetic field, the expectation values are related to each other as follows:

$$\langle a_{k\uparrow}^* a_{k'\downarrow} S^- \rangle = \langle a_{k\downarrow}^* a_{k'\uparrow} S^+ \rangle$$
$$= 2 \langle a_{k\uparrow}^* a_{k'\uparrow} S^z \rangle$$
$$= -2 \langle a_{k\downarrow}^* a_{k'\downarrow} S^z \rangle$$
$$= \gamma_{kk'},$$
$$\langle a_{k\uparrow}^* a_{k'\uparrow} \rangle = \langle a_{k\downarrow}^* a_{k'\downarrow} \rangle$$
$$= \nu_{kk'},$$
$$\langle S^z \rangle = 0. \tag{8.61}$$

These relations hold for all k, k', so that for $k = k'$,

$$\langle a_{k\uparrow}^* a_{k\downarrow} S^- \rangle = \langle a_{k\downarrow}^* a_{k\uparrow} S^+ \rangle$$
$$= 2 \langle a_{k\uparrow}^* a_{k\uparrow} S^z \rangle = 2 \langle n_{k\uparrow} S^z \rangle$$
$$= -2 \langle a_{k\downarrow}^* a_{k\downarrow} S^z \rangle = -2 \langle n_{k\downarrow} S^z \rangle$$
$$= \gamma_k,$$
$$\langle a_{k\uparrow}^* a_{k\uparrow} \rangle = \langle n_{k\uparrow} \rangle$$
$$= \langle a_{k\downarrow}^* a_{k\downarrow} \rangle$$
$$= \langle n_{k\downarrow} \rangle$$
$$= \nu_k. \tag{8.62}$$

The first step of the calculation is to construct the required density matrices. To do this, it is necessary to choose a particular representation for the operators appearing in F. It is convenient to choose the standard ones

$$a_{k\sigma} = \begin{bmatrix} 0 & 0 \\ 1 & 0 \end{bmatrix}, \quad a_{k\sigma}^* = \begin{bmatrix} 0 & 1 \\ 0 & 0 \end{bmatrix},$$
$$n_{k\sigma} = \begin{bmatrix} 1 & 0 \\ 0 & 0 \end{bmatrix}, \quad 1 = \begin{bmatrix} 1 & 0 \\ 0 & 1 \end{bmatrix}. \tag{8.63}$$

The convention for the operators is fixed so that an operator containing the variable $k\sigma$ always stands to the left and an operator containing the variable S always stands to the right. For example, the operator $a_{k\uparrow}^* a_{k'\uparrow} S^z$ has the matrix representation

$$a_{k\uparrow}^* \times a_{k'\uparrow} \times S^z = \begin{bmatrix} 0 & 1 \\ 0 & 0 \end{bmatrix} \times \begin{bmatrix} 0 & 0 \\ 1 & 0 \end{bmatrix} \times \begin{bmatrix} \frac{1}{2} & 0 \\ 0 & -\frac{1}{2} \end{bmatrix}$$

$$= \begin{bmatrix} \cdot & \cdot & \cdot & \cdot & \cdot & \cdot & \cdot & \cdot \\ \cdot & \cdot & \cdot & \cdot & \cdot & \cdot & \cdot & \cdot \\ \cdot & \cdot & \cdot & \frac{1}{2} & \cdot & \cdot & \cdot & \cdot \\ \cdot & \cdot & \cdot & \cdot & -\frac{1}{2} & \cdot & \cdot & \cdot \\ \cdot & \cdot & \cdot & \cdot & \cdot & \cdot & \cdot & \cdot \\ \cdot & \cdot & \cdot & \cdot & \cdot & \cdot & \cdot & \cdot \\ \cdot & \cdot & \cdot & \cdot & \cdot & \cdot & \cdot & \cdot \\ \cdot & \cdot & \cdot & \cdot & \cdot & \cdot & \cdot & \cdot \end{bmatrix}. \tag{8.64}$$

In the above 8×8 matrix, only the nonzero elements are indicated explicitly, and the dots are the zero element.

Having established the above convention for the operators, it is now possible to construct the density matrices explicitly in terms of unknown correlation functions.

The one-particle density matrices are given in the following forms:

$$\rho^{(1)}(k\sigma) = \begin{bmatrix} \langle n_{k\sigma} \rangle & 0 \\ 0 & 1 - \langle n_{k\sigma} \rangle \end{bmatrix}, \tag{8.65}$$

where σ is either $\uparrow$ or $\downarrow$, and

$$\rho^{(1)}(S) = \begin{bmatrix} \frac{1}{2} + \langle S^z \rangle & \langle S^- \rangle \\ \langle S^+ \rangle & \frac{1}{2} - \langle S^z \rangle \end{bmatrix}. \tag{8.66}$$

Because of symmetry and conservation laws, however,

$$\langle S^z \rangle = \langle S^- \rangle = \langle S^+ \rangle = 0, \tag{8.67}$$

so that $\rho^{(1)}(S)$ takes a trivial form:

$$\rho^{(1)}(S) = \begin{bmatrix} \frac{1}{2} & 0 \\ 0 & \frac{1}{2} \end{bmatrix}. \tag{8.68}$$

The two-particle density matrices are found in a similar way. For example, the matrix $\rho^{(2)}(k_\uparrow, S)$ is

$$\begin{bmatrix} \langle n_{k\uparrow} A \rangle & \cdot & \cdot & \cdot \\ \cdot & \langle n_{k\uparrow} B \rangle & \cdot & \cdot \\ \cdot & \cdot & \langle (1 - n_{k\uparrow}) A \rangle & \cdot \\ \cdot & \cdot & \cdot & \langle (1 - n_{k\uparrow}) B \rangle \end{bmatrix},$$

$$A = \tfrac{1}{2} + S^z, \qquad B = \tfrac{1}{2} - S^z. \tag{8.69}$$

The reducibility conditions

$$\text{tr}_{k\uparrow}\rho^{(2)}(\boldsymbol{k}_{\uparrow}, S) = \rho^{(1)}(S),$$
$$\text{tr}_{S}\rho^{(2)}(\boldsymbol{k}_{\uparrow}, S) = \rho^{(1)}(\boldsymbol{k}_{\uparrow}), \tag{8.70}$$

can easily be tested by explicitly carrying out the required partial trace operations.

Other higher-order matrices are found in the same way. The three-particle density matrices that contain the expectation value of the Hamiltonian are $\rho^{(3)}(\boldsymbol{k}_{\uparrow}, \boldsymbol{k}'_{\uparrow}, S)$, $\rho^{(3)}(\boldsymbol{k}_{\downarrow}, \boldsymbol{k}'_{\downarrow}, S)$, and $\rho^{(3)}(\boldsymbol{k}_{\uparrow}, \boldsymbol{k}'_{\downarrow}, S)$. Actually, only one of these is needed for the ground state calculation. This is due to the symmetry of the system, expressed by (8.61). Since the matrix $\rho^{(3)}(\boldsymbol{k}_{\uparrow}, \boldsymbol{k}'_{\downarrow}, S)$ has the simplest form, it is advantageous to choose it for the calculation. This density matrix has the following form:

$$\rho^{(3)}(\boldsymbol{k}_{\uparrow}, \boldsymbol{k}'_{\downarrow}, S) = \begin{bmatrix} \rho_{11} & \cdot & \cdot & \cdot & & \cdot & & \cdot \\ \cdot & \rho_{22} & \cdot & \cdot & & \cdot & & \cdot \\ \cdot & \cdot & \rho_{33} & \cdot & & \cdot & & \cdot \\ \cdot & \cdot & \cdot & \rho_{44} & \rho_{45} & \cdot & & \cdot \\ \cdot & \cdot & \cdot & \rho_{54} & \rho_{55} & \cdot & & \cdot \\ \cdot & \cdot & \cdot & \cdot & & \rho_{66} & \cdot & \cdot \\ \cdot & \cdot & \cdot & \cdot & & \cdot & \rho_{77} & \cdot \\ \cdot & \cdot & \cdot & \cdot & & \cdot & \cdot & \rho_{88} \end{bmatrix}, \tag{8.71}$$

where

$$\rho_{11} = \langle n_{k\uparrow} n_{k'\downarrow}(\tfrac{1}{2} + S^{z})\rangle,$$
$$\rho_{22} = \langle n_{k\uparrow} n_{k'\downarrow}(\tfrac{1}{2} - S^{z})\rangle,$$
$$\rho_{33} = \langle n_{k\uparrow}(1 - n_{k'\downarrow})(\tfrac{1}{2} + S^{z})\rangle,$$
$$\rho_{44} = \langle n_{k\uparrow}(1 - n_{k'\downarrow})(\tfrac{1}{2} - S^{z})\rangle,$$
$$\rho_{45} = \langle a^{*}_{k'\downarrow} a_{k\uparrow} S^{+}\rangle,$$
$$\rho_{54} = \langle a^{*}_{k\uparrow} a_{k'\downarrow} S^{-}\rangle,$$
$$\rho_{55} = \langle (1 - n_{k\uparrow}) n_{k'\downarrow}(\tfrac{1}{2} + S^{z})\rangle,$$
$$\rho_{66} = \langle (1 - n_{k\uparrow}) n_{k'\downarrow}(\tfrac{1}{2} - S^{z})\rangle,$$
$$\rho_{77} = \langle (1 - n_{k\uparrow})(1 - n_{k'\downarrow})(\tfrac{1}{2} + S^{z})\rangle,$$
$$\rho_{88} = \langle (1 - n_{k\uparrow})(1 - n_{k'\downarrow})(\tfrac{1}{2} - S^{z})\rangle. \tag{8.72}$$

The center block of $\rho^{(3)}(\boldsymbol{k}_{\uparrow}, \boldsymbol{k}'_{\downarrow}, S)$ is diagonalized by a unitary transformation

$$U^{-} \begin{bmatrix} \rho_{44} & \rho_{45} \\ \rho_{54} & \rho_{55} \end{bmatrix} U = \begin{bmatrix} A + B & 0 \\ 0 & A - B \end{bmatrix}, \tag{8.73}$$

where A and B are given by

$$A = \tfrac{1}{4}(\langle n_{k\uparrow}\rangle + \langle n_{k'\downarrow}\rangle - 2\langle n_{k\uparrow}S^z\rangle + 2\langle n_{k'\downarrow}S^z\rangle - 4\langle n_{k\uparrow}n_{k'\downarrow}\rangle),$$

$$B = \tfrac{1}{4}[(\langle n_{k\uparrow}\rangle - \langle n_{k'\downarrow}\rangle - 2\langle n_{k\uparrow}S^z\rangle - 2\langle n_{k'\downarrow}S^z\rangle + 4\langle n_{k\uparrow}n_{k'\downarrow}S^z\rangle)^2$$
$$+ 16\langle a_{k\uparrow}^* a_{k'\downarrow}S^-\rangle^2]^{\frac{1}{2}}. \tag{8.74}$$

In the approximation in which all the cumulant functions up to three-particle terms are included, the Helmholtz potential is given by the following equation:

$$F = \sum_{k\sigma}\epsilon_k\langle n_{k\sigma}\rangle - (J/2N)\sum_k\sum_{k'}[\langle(a_{k\uparrow}^*a_{k'\uparrow} - a_{k\downarrow}^*a_{k'\downarrow})S^z\rangle$$
$$+ \langle a_{k\uparrow}^*a_{k'\downarrow}S^-\rangle + \langle a_{k\downarrow}^*a_{k'\uparrow}S^+\rangle]$$
$$+ k_{\mathrm B}T\Big\{g^{(1)}(S) + \sum_{k\sigma}[g^{(1)}(k\sigma) + g^{(2)}(k\sigma, S)]$$
$$+ (1/2N)\sum_{k\sigma}\sum_{k'\sigma'}[g^{(2)}(k\sigma, k'\sigma') + g^{(3)}(k\sigma, k'\sigma', S)]\Big\}. \tag{8.75}$$

When the cumulant functions, g, are replaced by the cluster functions, G, all the smaller cumulant functions are eliminated, and the variational potential F takes a very simple form:

$$F = \sum_{k\sigma}\epsilon_k\langle n_{k\sigma}\rangle - (J/2N)\sum_k\sum_{k'}[\langle(a_{k\uparrow}^*a_{k'\uparrow} - a_{k\downarrow}^*a_{k'\downarrow})S^z\rangle$$
$$+ \langle a_{k\uparrow}^*a_{k'\downarrow}S^-\rangle + \langle a_{k\downarrow}^*a_{k'\uparrow}S^+\rangle] + (k_{\mathrm B}T/2N)\sum_{k\sigma}\sum_{k'\sigma'}G^{(3)}(k\sigma, k'\sigma', S),$$
$$\tag{8.76}$$

where

$$G^{(3)}(k\sigma, k'\sigma', S) = \mathrm{tr}\rho^{(3)}(k\sigma, k'\sigma', S)\log\rho^{(3)}(k\sigma, k'\sigma', S). \tag{8.77}$$

Because of the way in which $\langle a_{k\uparrow}^*a_{k'\downarrow}S^-\rangle$ is included in the variational potential, the variation of F with respect to this quantity yields the following equation:

$$k_{\mathrm B}T\frac{\langle a_{k\uparrow}^*a_{k'\downarrow}S^-\rangle}{B}\ln\frac{A+B}{A-B} = J/2. \tag{8.78}$$

Since $A + B$ and $A - B$ are diagonal elements of a probability matrix, they are both positive and their values lie between zero and unity. Therefore, $1 \le (A + B)/(A - B)$, and $\ln[(A + B)/(A - B)]$ must be positive or zero. Thus, if J is positive, $\langle a_{k\uparrow}^*a_{k'\downarrow}S^-\rangle$ must be positive, and if J is negative, $\langle a_{k\uparrow}^*a_{k'\downarrow}S^-\rangle$ must be negative. At $T = 0$, the only possible solution of the above equation is for the logarithm term to diverge, or for

$$A = B. \tag{8.79}$$

Assuming that, in the ground state, the correlation between conduction electrons and the impurity spin is stronger than the correlation between the conduction electrons of opposite spins, the averages $\langle n_{k\uparrow} n_{k'\downarrow} \rangle$ and $\langle n_{k\uparrow} n_{k'\downarrow} S^z \rangle$ can be decoupled in symmetric fashion:

$$\langle n_{k\uparrow} n_{k'\downarrow} \rangle = \langle n_{k\uparrow} \rangle \langle n_{k'\downarrow} \rangle,$$
$$\langle n_{k\uparrow} n_{k'\downarrow} S^z \rangle = \langle n_{k\uparrow} \rangle \langle n_{k'\downarrow} S^z \rangle + \langle n_{k'\downarrow} \rangle \langle n_{k\uparrow} S^z \rangle. \tag{8.80}$$

The justification for this assumption is as follows. Suppose a conduction electron of spin up is close to the impurity spin. If the exchange coupling is antiferromagnetic, the impurity will have a tendency to orient its spin downward. Then, again due to antiferromagnetic J, it is unlikely that a second conduction electron with spin down would interact very strongly with the impurity. Therefore, the indirect exchange interaction between two conduction electrons becomes important only when their spins are aligned. A similar argument for the case of ferromagnetic coupling leads to the same conclusion. Introducing (8.80) and the symmetry relations (8.61) into (8.79) yields

$$\gamma_{kk'} = \pm \tfrac{1}{2} [v_{k'}(1 - v_k) - \gamma_{k'}(1 - v_k) - v_{k'}\gamma_k]^{\frac{1}{2}}$$
$$\times [v_k(1 - v_{k'}) - \gamma_k(1 - v_{k'}) - v_k\gamma_{k'}]^{\frac{1}{2}}, \tag{8.81}$$

where the plus sign corresponds to the case $J > 0$, and the minus sign to the case $J < 0$. For $k = k'$, $\gamma_{kk'} = \gamma_k$, and (8.62) becomes

$$\gamma_k = \mu v_k(1 - v_k), \quad \mu = \frac{1}{3}, \quad J > 0$$
$$= -1, \quad J < 0. \tag{8.82}$$

Substituting this expression for γ_k back into (8.81) leads to the result

$$\gamma_{kk'} = \mu [v_k(1 - v_k)v_{k'}(1 - v_{k'})]^{\frac{1}{2}} [1 - (1/4)(v_k - v_{k'})^2]^{\frac{1}{2}}. \tag{8.83}$$

On taking the exchange constant J to be nonzero only in a band of width $2D$ centered on the Fermi surface, the quantity $\frac{1}{4}(v_k - v_{k'})^2$ can be neglected with respect to unity, and the $T = 0$ variational function becomes

$$F_{T=0} = 2 \sum_k \epsilon_k v_k - (3\mu J/2N) \sum_k \sum_{k'} [v_k(1 - v_k)v_{k'}(1 - v_{k'})]^{\frac{1}{2}}. \tag{8.84}$$

Except for the numerical constant in front of the double summation in (8.84), this expression is exactly the BCS variational function. The quantity μJ is positive, so that the energy is lowered for both ferromagnetic and antiferromagnetic J. On minimizing $F_{T=0}$ with respect to v_k, we obtain a self-consistent expression for v_k:

$$\frac{[v_k(1 - v_k)]^{\frac{1}{2}}}{1 - 2v_k} = \frac{(3\mu J/2N) \sum_{k'} [v_{k'}(1 - v_{k'})]^{\frac{1}{2}}}{2\epsilon_k}. \tag{8.85}$$

On solving this for v_k and performing the necessary integration, the ground state energy lowering is found to be

$$W_0 = -2N(0)D^2 \exp{-[4N/3\mu J N(0)]} \tag{8.86}$$

for weak coupling ($J N(0) \ll 1$), where $N(0)$ is the density of states at the Fermi surface.

8.4 The ground state of the Anderson model

The Anderson model (Anderson, 1961) was introduced in order to study the stability of the nonvanishing spin magnetic moment arising from the d-orbitals in a metallic ferromagnet such as iron.

In the case of insulating ferromagnets, such as EuO and EuS, the magnetic moment of f-orbitals in each Eu ion is spatially well shielded by the outer s-orbitals. For this reason, the interaction between the f-orbital magnetic spins of neighboring Eu ions is quite accurately represented by the Heisenberg exchange interaction.

In the case of the metallic ferromagnet, however, the energy of the d-orbital is somewhere in the middle of the conduction s-band so that the d-orbital electrons join with the s-band electrons in the conduction band and vice versa. Since electrons in the d-orbitals are not localized, the Heisenberg exchange interaction is totally inadequate in explaining the occurrence of the ferromagnetism. A crucial question is the stability of the magnetic moment arising from the d-orbitals, i.e., if there is a single d-orbital of which the energy is in the middle of the s-band, the d-orbital will be occupied by two electrons of opposite spins and hence there is no net magnetic moment. If the energy of the d-orbital is above the Fermi level of the s-band, the d-orbit will be empty and hence there is no magnetic moment either. In order to resolve the problem of stability of a nonvanishing moment, Anderson introduced a repulsive interaction between the two electrons in the same d-orbital arising from the Coulomb interaction and investigated conditions under which the nonvanishing magnetic moment is stable in the ground state. The Anderson model gives a rather convincing mechanism for the occurrence of the stable magnetic moment on the d-orbital.

Since the study of the Anderson model, the idea has been amplified to the case of many d-orbitals, i.e., the entire lattice is made up of the Anderson d-orbitals. This system is called the Anderson lattice, which has become the subject of intensive investigation as a mechanism of the metallic ferromagnet.

In this section the ground state of the Anderson model is studied by the cluster variation method in its lowest order approximation (Bose & Tanaka, 1968). In order to accomplish this, the Helmholtz potential at finite temperature is calculated, and then an extrapolation to the absolute zero of temperature is used to find the ground state of the system.

The Hamiltonian of the system is defined as follows:

$$H = \sum_{k,\sigma} \epsilon_k a_{k\sigma}^* a_{k\sigma} + \sum_{\sigma} \epsilon_d a_{d\sigma}^* a_{d\sigma}$$

$$+ U a_{d\sigma}^* a_{d\sigma} a_{d-\sigma}^* a_{d-\sigma} + V \sum_{k,\sigma} (a_{k\sigma}^* a_{d\sigma} + a_{d\sigma}^* a_{k\sigma}), \qquad (8.87)$$

where $a_{k\sigma}^* (a_{k\sigma})$ and $a_{d\sigma}^* (a_{d\sigma})$ are the creation (annihilation) operators for the conduction and the localized d-electrons, respectively. Hence the first term on the right hand side represents the total energy of the conduction band, of which the width is assumed to be $2D$; the second term is the energy of the d-orbital; the third term gives the Coulomb repulsion between spin up and spin down electrons in the d-level; and the last term represents the energy due to mixing between the conduction electron states and the d-orbital.

The variational potential of the system can be written as

$$F = \sum_{k,\sigma} \epsilon_k \langle a_{k\sigma}^* a_{k\sigma} \rangle + \sum_{\sigma} \epsilon_d \langle a_{d\sigma}^* a_{d\sigma} \rangle$$

$$+ U \langle a_{d\sigma}^* a_{d\sigma} \rangle \langle a_{d-\sigma}^* a_{d-\sigma} \rangle + V \sum_{k,\sigma} (\xi_{kd}^{\sigma} + \xi_{kd}^{\sigma*})$$

$$+ kT \sum_{k,\sigma} \mathrm{tr}\left[\rho^{(2)}(k\sigma, d\sigma) \log \rho^{(2)}(k\sigma, d\sigma) \right], \qquad (8.88)$$

where

$$\rho^{(2)}(k\sigma, d\sigma) = \begin{bmatrix} A & \cdot & \cdot & \cdot \\ \cdot & B & \xi_{kd}^{\sigma} & \cdot \\ \cdot & \xi_{kd}^{\sigma*} & C & \cdot \\ \cdot & \cdot & \cdot & D \end{bmatrix}, \qquad (8.89)$$

and

$$A = \langle a_{k\sigma}^* a_{k\sigma} a_{d\sigma}^* a_{d\sigma} \rangle = \langle n_{k\sigma} n_{d\sigma} \rangle,$$
$$B = \langle n_{k\sigma} (1 - n_{d\sigma}) \rangle,$$
$$C = \langle (1 - n_{k\sigma}) n_{d\sigma} \rangle,$$
$$D = \langle (1 - n_{k\sigma})(1 - n_{d\sigma}) \rangle,$$
$$\xi_{kd}^{\sigma} = \langle a_{k\sigma}^* a_{d\sigma} \rangle. \qquad (8.90)$$

When the center block of $\rho^{(2)}(k\sigma, d\sigma)$ is diagonalized, four eigenvalues of $\rho^{(2)}$ are found as

$$E_1 = \langle n_{k\sigma}\rangle\langle n_{d\sigma}\rangle,$$

$$E_2 = \tfrac{1}{2}[\langle n_{k\sigma}\rangle + \langle n_{d\sigma}\rangle - 2\langle n_{k\sigma}\rangle\langle n_{d\sigma}\rangle + R],$$

$$E_3 = \tfrac{1}{2}[\langle n_{k\sigma}\rangle + \langle n_{d\sigma}\rangle - 2\langle n_{k\sigma}\rangle\langle n_{d\sigma}\rangle - R],$$

$$E_4 = 1 - \langle n_{k\sigma}\rangle - \langle n_{d\sigma}\rangle + \langle n_{k\sigma}\rangle\langle n_{d\sigma}\rangle,$$

$$R = \left[(\langle n_{k\sigma}\rangle - \langle n_{d\sigma}\rangle)^2 + 4\left|\xi_{kd}^\sigma\right|^2\right]^{\frac{1}{2}}. \qquad (8.91)$$

In the above equations, it should be noted that some nonessential decouplings have been introduced, i.e., $\langle n_{k\sigma}n_{d\sigma}\rangle = \langle n_{k\sigma}\rangle\langle n_{d\sigma}\rangle$ and $\langle n_{d\uparrow}n_{d\downarrow}\rangle = \langle n_{d\uparrow}\rangle\langle n_{d\downarrow}\rangle$.

It should be noted that the variational potential F is expressed in terms only of three parameters, i.e., $\langle n_{k\sigma}\rangle$, $\langle n_{d\sigma}\rangle$, and ξ_{kd}. Let us first find ξ_{kd} by the variation. This leads to

$$V + \frac{\xi_{kd}^{\sigma*}kT}{R}\ln\frac{\langle n_{k\sigma}\rangle + \langle n_{d\sigma}\rangle - 2\langle n_{k\sigma}\rangle\langle n_{d\sigma}\rangle + R}{\langle n_{k\sigma}\rangle + \langle n_{d\sigma}\rangle - 2\langle n_{k\sigma}\rangle\langle n_{d\sigma}\rangle - R} = 0. \qquad (8.92)$$

To calculate ξ_{kd}^σ in the ground state, we take the limit in which the temperature T goes to zero. Since the first term V is finite, the factor multiplying T has to be infinite. Since the denominator dividing T cannot be zero, because it is the square root of a positive quantity, the logarithmic term must diverge. Since only the denominator of the logarithmic term can go to zero, V and ξ_{kd}^σ must be of opposite signs so that (8.92) can be satisfied. This amounts to a unique solution for ξ_{kd}^σ in terms of other unknown quantities. We have

$$[\langle n_{k\sigma}\rangle + \langle n_{d\sigma}\rangle - 2\langle n_{k\sigma}\rangle\langle n_{d\sigma}\rangle]^2$$
$$- \left[(\langle n_{k\sigma}\rangle - \langle n_{d\sigma}\rangle)^2 + 4|\xi_{kd}^\sigma|^2\right] = 0, \qquad (8.93)$$

which gives

$$\xi_{kd}^\sigma = \pm[\langle n_{k\sigma}\rangle\langle n_{d\sigma}\rangle(1 - \langle n_{k\sigma}\rangle)(1 - \langle n_{d\sigma}\rangle)]^{\frac{1}{2}}. \qquad (8.94)$$

As mentioned earlier, the sign of ξ_{kd}^σ must be chosen in such a way that the product $V\xi_{kd}^\sigma$ is a negative real number so that (8.92) is satisfied.

Introducing this solution in the variational potential at $T = 0$ yields

$$F_{T=0} = \sum_{k,\sigma}\epsilon_k\langle n_{k\sigma}\rangle + \sum_\sigma\epsilon_k\langle n_{d\sigma}\rangle + U\langle n_{d\sigma}\rangle\langle n_{d-\sigma}\rangle$$
$$- 2|V|\sum_{k,\sigma}[\langle n_{k\sigma}\rangle\langle n_{d\sigma}\rangle(1 - \langle n_{k\sigma}\rangle)(1 - \langle n_{d\sigma}\rangle)]^{\frac{1}{2}}. \qquad (8.95)$$

Minimizing F with respect to $\langle n_{k\sigma}\rangle$ and $\langle n_{d\sigma}\rangle$ leads to the following coupled equations:

$$\epsilon_k - |V|\left[\frac{n_{d\sigma}(1-n_{d\sigma})}{n_{k\sigma}(1-n_{k\sigma})}\right]^{\frac{1}{2}}(1-2n_{k\sigma}) = 0, \tag{8.96}$$

$$\epsilon_d + Un_{d-\sigma} - |V|\sum_k \left[\frac{n_{k\sigma}(1-n_{k\sigma})}{n_{d\sigma}(1-n_{d\sigma})}\right]^{\frac{1}{2}}(1-2n_{d\sigma}) = 0, \tag{8.97}$$

where the bracket notations, $\langle\cdots\rangle$, have been omitted for simplicity. Furthermore, let us introduce more simplifying notations:

$$\epsilon_{d\sigma} = \epsilon_d + Un_{d-\sigma}, \tag{8.98}$$

$$\eta_{d\sigma} = n_{d\sigma}(1-n_{d\sigma}), \tag{8.99}$$

$$\eta_{k\sigma} = n_{k\sigma}(1-n_{k\sigma}), \tag{8.100}$$

$$\epsilon_{0\sigma} = 2|V|\sum_k [n_{k\sigma}(1-n_{k\sigma})]^{\frac{1}{2}}. \tag{8.101}$$

Equations (8.96) and (8.97) then assume the following forms:

$$\epsilon_k - |V|(\eta_{d\sigma}/\eta_{k\sigma})^{\frac{1}{2}}(1-2n_{k\sigma}) = 0, \tag{8.102}$$

$$\epsilon_{d\sigma} - \left(\epsilon_{0\sigma}/2\eta_{d\sigma}^{\frac{1}{2}}\right)(1-2n_{d\sigma}) = 0. \tag{8.103}$$

From (8.102)

$$\eta_{k\sigma} = \eta_{d\sigma}|V|^2/(\epsilon_k^2 + 4\eta_{d\sigma}|V|^2), \tag{8.104}$$

which can be solved for $\eta_{k\sigma}$ to find

$$\eta_{k\sigma} = \tfrac{1}{2}\left[1 - \epsilon_k/(\epsilon_k^2 + 4\eta_{d\sigma}|V|^2)^{\frac{1}{2}}\right]. \tag{8.105}$$

Equation (8.102) gives the distribution of states modified from the Fermi distribution by the presence of the s–d interaction. If (8.105) and (8.101) are combined, one obtains

$$\epsilon_{0\sigma} = 2|V|^2\eta_{d\sigma}^{\frac{1}{2}}\sum_k \left(\epsilon_k^2 + 4\eta_{d\sigma}|V|^2\right)^{-\frac{1}{2}}. \tag{8.106}$$

Replacing the sum over k by an integral over ϵ_k and recalling that the width of the conduction band is $2D$, we see that

$$\epsilon_{0\sigma} = 2N(0)|V|^2\eta_{d\sigma}^{\frac{1}{2}}\int_{-D}^{D}\frac{d\epsilon_k}{\left(\epsilon_k^2 + 4\eta_{d\sigma}|V|^2\right)^{\frac{1}{2}}}$$

$$= 4N(0)|V|^2\eta_{d\sigma}^{\frac{1}{2}}\sinh^{-1}\left(D/2\eta_{d\sigma}^{\frac{1}{2}}|V|\right), \tag{8.107}$$

where the density of states is assumed to be constant and is given by $N(0)$, the density of the conduction states of one spin per unit energy at the Fermi surface.

Combining (8.107), (8.103), and (8.98), yields the following equations:

$$\sinh^{-1} \frac{D}{2[n_{d\uparrow}(1 - n_{d\uparrow})]^{1/2}|V|} = \frac{\epsilon_d + Un_{d\downarrow}}{2N(0)|V|^2(1 - 2n_{d\uparrow})},$$

$$\sinh^{-1} \frac{D}{2[n_{d\downarrow}(1 - n_{d\downarrow})]^{1/2}|V|} = \frac{\epsilon_d + Un_{d\uparrow}}{2N(0)|V|^2(1 - 2n_{d\downarrow})}. \tag{8.108}$$

To understand these two equations, we plot five different cases in Fig. 8.1. In Figs. 8.1(a)–(c), $U = 2|\epsilon_d|$, and in Figs. 8.1(d) and (e), $U = 4|\epsilon_d|$. From Figs. 8.1(a), (b), (d), and (e) it is clear that when $U/N(0)|V|^2$ is large there are three sets of solutions for these equations. The first set is at $n_{d\uparrow} = n_1$ and $n_{d\downarrow} = n_2$; the second set is at $n_{d\uparrow} = n_{d\downarrow}$; and the third is at $n_{d\uparrow} = n_2$ and $n_{d\downarrow} = n_1$. The $n_{d\uparrow} = n_{d\downarrow}$ solution is unstable. The other two solutions are stable energetically, and they correspond to the up and down configurations of a localized spin.

When $U/N(0)|V|^2$ is small, as in Fig. 8.1(c), the set of equations has only one self-consistent solution, $n_{d\uparrow} = n_{d\downarrow}$. In this case, the net spin of the system is zero, and hence this solution is magnetically uninteresting. Since $U = 2|\epsilon_d|$ is the most favorable case for the localized moment, we find that the condition $n_{d\uparrow} = 1 - n_{d\downarrow}$ is satisfied for the two magnetic solutions in Figs. 8.1(a) and (b).

It is interesting to observe that although the self-consistent equations for $n_{d\uparrow}$ and $n_{d\downarrow}$ derived in this section look quite different from Anderson's equation (Anderson, 1961, eq. (27)), they exhibit a similar behavior and have similar self-consistent solutions.

We can now proceed to the evaluation of the ground state energy. The energy lowering of the system due to the presence of a magnetic impurity is given by

$$W = \sum_{k>k_F,\sigma} \epsilon_k n_{k\sigma} + \sum_{k<k_F,\sigma} |\epsilon_k|(1 - n_{k\sigma}) + \sum_\sigma \epsilon_{d\sigma} n_{d\sigma}$$

$$+ Un_{d\sigma}n_{d-\sigma} - \sum_\sigma \epsilon_{0\sigma} \eta_{d\sigma}^{\frac{1}{2}} - \epsilon_d. \tag{8.109}$$

Using (8.105) and (8.101) in (8.109) yields

$$W = \sum_{k>k_F,\sigma} \frac{\epsilon_k}{2} \left\{ 1 - \frac{\epsilon_k}{(\epsilon_k^2 + 4\eta_{d\sigma}|V|^2)^{\frac{1}{2}}} \right\} + \sum_{k<k_F,\sigma} \frac{|\epsilon_k|}{2} \left\{ 1 + \frac{\epsilon_k}{(\epsilon_k^2 + 4\eta_{d\sigma}|V|^2)^{\frac{1}{2}}} \right\}$$

$$+ \epsilon_d(n_{d\uparrow} + n_{d\downarrow} - 1) + Un_{d\uparrow}n_{d\downarrow} - \left[\epsilon_{0\uparrow}(\eta_{d\uparrow})^{\frac{1}{2}} + \epsilon_{0\downarrow}(\eta_{d\downarrow})^{\frac{1}{2}} \right]. \tag{8.110}$$

Converting the sum over k to an integral over ϵ_k and carrying out the sum over the

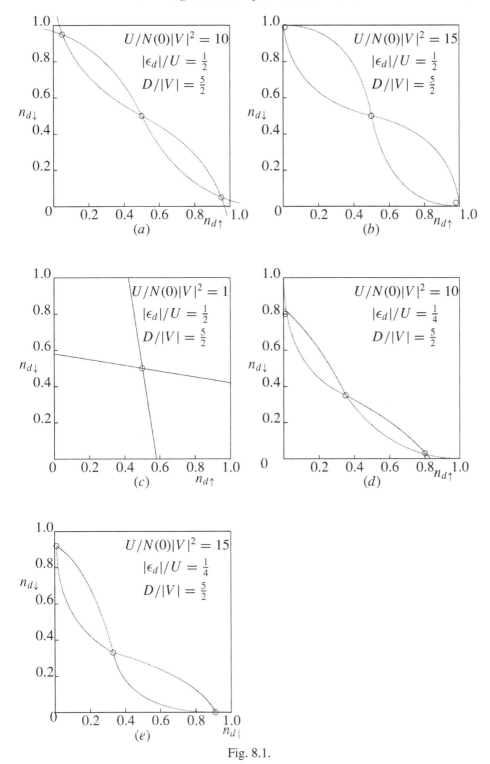

Fig. 8.1.

spin states,

$$W = N(0) \int_0^D \epsilon_k \left\{ 1 - \frac{\epsilon_k}{\left(\epsilon_k^2 + 4\eta_{d\sigma}|V|^2 \right)^{\frac{1}{2}}} \right\} d\epsilon_k$$

$$+ N(0) \int_0^D \epsilon_k \left\{ 1 - \frac{\epsilon_k}{\left(\epsilon_k^2 + 4\eta_{d\sigma}|V|^2 \right)^{\frac{1}{2}}} \right\} d\epsilon_k$$

$$+ \epsilon_d(n_{d\uparrow} + n_{d\downarrow} - 1) + U n_{d\uparrow} n_{d\downarrow} - [\epsilon_{0\uparrow}(\eta_{d\uparrow})^{\frac{1}{2}} + \epsilon_{0\downarrow}(\eta_{d\downarrow})^{\frac{1}{2}}]$$

$$= N(0)D^2 - N(0) \int_0^D \frac{\epsilon_k^2 d\epsilon_k}{\left(\epsilon_k^2 + 4\eta_{d\sigma}|V|^2 \right)^{\frac{1}{2}}} - N(0) \int_0^D \frac{\epsilon_k^2 d\epsilon_k}{\left(\epsilon_k^2 + 4\eta_{d\sigma}|V|^2 \right)^{\frac{1}{2}}}$$

$$+ \epsilon_d(n_{d\uparrow} + n_{d\downarrow} - 1) + U n_{d\uparrow} n_{d\downarrow} - \left[\epsilon_{0\uparrow}(\eta_{d\uparrow})^{\frac{1}{2}} + \epsilon_{0\downarrow}(\eta_{d\downarrow})^{\frac{1}{2}} \right]. \tag{8.111}$$

Carrying out these integrals and using (8.107) yields

$$W = - N(0)|V|^2 \left\{ \eta_{d\uparrow} \left[1 + 2 \sinh^{-1} \left(D/2\eta_{d\uparrow}^{\frac{1}{2}}|V| \right) \right] \right.$$
$$\left. + \eta_{d\downarrow} \left[1 + 2 \sinh^{-1} \left(D/2\eta_{d\downarrow}^{\frac{1}{2}}|V| \right) \right] \right\} + \epsilon_d(n_{d\uparrow} + n_{d\downarrow} - 1) + U n_{d\uparrow} n_{d\downarrow}.$$
$$\tag{8.112}$$

The ground state energy may be calculated in two limiting cases. One is the case of the weak interaction limit, or the Appelbaum (1968) approximation, in which $U/N(0)|V|^2$ tends to infinity. In this limit, the solutions of the self-consistent equations (8.108) are obtained as $n_{d\uparrow} \approx 1$ and $n_{d\downarrow} \approx 0$. Hence, $\eta_{d\uparrow} \approx \eta_{d\downarrow} \approx \eta_d$ and $n_{d\uparrow} + n_{d\downarrow} = 1$. Then from (8.108),

$$U = 4N(0)|V|^2 \sinh^{-1} \left(D/2\eta_d^{\frac{1}{2}}|V| \right); \tag{8.113}$$

and (8.112) reduces to

$$W = -2N(0)|V|^2 \eta_d. \tag{8.114}$$

Again, substituting (8.113) into (8.108) yields

$$\epsilon_d + 2N(0)|V|^2 \sinh^{-1} \left(D/2\eta_d^{\frac{1}{2}}|V| \right) = 0, \tag{8.115}$$

from which we obtain in the weak-coupling limit ($|\epsilon_d| \gg N(0)|V|^2$)

$$\eta_d|V|^2 = D^2 \exp[-|\epsilon_d|/N(0)|V|^2]. \tag{8.116}$$

Substituting (8.116) into (8.114) yields

$$W = -2N(0)|V|^2 \exp[-|\epsilon_d|/N(0)|V|^2]. \tag{8.117}$$

Let us consider the opposite limit of $U = 2|\epsilon_d|$, which is the most favorable case for a localized moment. In this limit, the solutions of the self-consistent equations (8.108) are such that the condition $n_{d\uparrow} + n_{d\downarrow} = 1$ is always satisfied. Hence, $\eta_{d\uparrow} = \eta_{d\downarrow} = \eta_d$. Using these relations in (8.108) yields

$$U = 4N(0)|V|^2 \sinh^{-1}\left(D/2\eta_d^{\frac{1}{2}}|V|\right), \qquad (8.118)$$

which is the same as (8.113). Hence in this limit,

$$W = -2N(0)|V|^2\eta_d, \qquad (8.119)$$

and, as before, for the weak-coupling limit,

$$\eta_d|V|^2 = D^2 \exp[-|\epsilon_d|/N(0)|V|^2], \qquad (8.120)$$

and hence

$$W = -2N(0)|V|^2 \exp[-|\epsilon_d|/N(0)|V|^2], \qquad (8.121)$$

which is the same as (8.117). W is the energy difference between a system with a magnetic impurity and the unperturbed system at absolute zero. Thus, (8.117) and (8.118) show that the ground state energy of the perturbed state is lower than the ground state energy of the unperturbed conduction electrons. The expression for the ground state energy has the same exponential behavior as that obtained by Appelbaum (1968).

8.5 The Hubbard model

The Hubbard model Hamiltonian, which is capable of predicting both ferromagnetism and antiferromagnetism, has been a subject of intensive investigation for many years because it seems also to have, inherently, a capability of producing a superconducting state if the electron density is away from the half-filled value either toward the higher density or the lower density direction.

One of the standard methods of treating the Hubbard Hamiltonian is to use an extended Hartree–Fock approximation after the original Hamiltonian is transformed into a system of quasiparticles by means of a unitary transformation. If one tries to predict both the antiferromagnetism and the superconductivity within one formulation, two types of quasiparticles must be introduced simultaneously, and hence the dimension of the unitary transformation becomes rather large. Since the study of the possibility of high temperature superconductivity based on the Hubbard model has still not been settled, it may not be appropriate to discuss it extensively here. Therefore only a traditional treatment of magnetic properties will be presented. For the sake of consistency, however, we shall develop a formulation based on the cluster variation method. The standard method of unitary transformation will be demonstrated in Appendix 4.

The Hubbard Hamiltonian is given, as usual, by

$$H = -t \sum_{(i,j)} (a_i^* a_j + b_i^* b_j) + U \sum_i a_i^* a_i b_i^* b_i - \mu \sum_i (a_i^* a_i + b_i^* b_i), \quad (8.122)$$

where the summation (i, j) runs over all nearest-neighbor pairs inclusive of both (i, j) and (j, i); t is the hopping energy; U is the repulsive potential when two electrons of opposite spins come to the same lattice site; and μ is the chemical potential which controls the density of electrons in the lattice.

The Hamiltonian is, then, transformed into the momentum space by the plane-wave expansion, as

$$H = \sum_k [\epsilon(k) - \mu](a_k^* a_k + b_k^* b_k) + \frac{U}{N_0} \sum_{k,k',q} a_{k+q}^* b_{k'-q}^* b_{k'} a_k, \quad (8.123)$$

where $\epsilon(k)$ is

$$\epsilon(k) = -2t(\cos k_x + \cos k_y + \cos k_z), \quad (8.124)$$

for a simple cubic lattice in which the lattice constant is chosen to be unity and N_0 is the number of lattice sites in the system.

In order for an antiferromagnetic state to be realizable, the entire lattice must be decomposable into two equivalent sublattices. A two-dimensional square, simple cubic, and body center cubic lattices are the possibilities. In such cases, the entire lattice can be divided into unit cells, each containing one α-site at $r_{i,\alpha}$ and one β-site at $r_{i,\beta}$, where i is the cell number. If the origin site is chosen to be an α-site, then the vector π, in the reciprocal lattice space, has the properties

$$\exp(i\pi \cdot r_{i,\alpha}) = 1, \quad \exp(i\pi \cdot r_{i,\beta}) = -1. \quad (8.125)$$

If the entire α-sublattice is translated by an odd number of steps in any one direction of the primitive translation vectors the entire β-sublattice is reached and vice versa.

The α-sublattice magnetization per lattice site is given by

$$A(\pi) = \frac{1}{N_0} \sum_{i,\text{cell}} (\langle a_{i,\alpha}^* a_{i,\alpha} \rangle - \langle b_{i,\alpha}^* b_{i,\alpha} \rangle). \quad (8.126)$$

However, in the antiferromagnetic spin configuration in the lattice, there exist general relations

$$\langle a_{i,\alpha}^* a_{i,\alpha} \rangle = \langle b_{i,\beta}^* b_{i,\beta} \rangle, \quad \langle b_{i,\alpha}^* b_{i,\alpha} \rangle = \langle a_{i,\beta}^* a_{i,\beta} \rangle, \quad (8.127)$$

and hence $A(\pi)$ is given as

$$A(\pi) = \frac{1}{N_0} \sum_{i,\text{cell}} (\langle a_{i,\alpha}^* a_{i,\alpha} \rangle - \langle a_{i,\beta}^* a_{i,\beta} \rangle)$$

$$= \frac{1}{N_0} \sum_{i,\text{cell}} (\exp(i\pi \cdot \boldsymbol{r}_{i,\alpha}) \langle a_{i,\alpha}^* a_{i,\alpha} \rangle + \exp(i\pi \cdot \boldsymbol{r}_{i,\beta}) \langle a_{i,\beta}^* a_{i,\beta} \rangle)$$

$$= \frac{1}{N_0} \sum_{i,\text{lattice}} \exp(i\pi \cdot \boldsymbol{r}_i) \langle a_i^* a_i \rangle = \frac{1}{N_0} \sum_{k>0} \langle a_{k+\pi}^* a_k \rangle. \tag{8.128}$$

Similarly the β-sublattice magnetization ($= \alpha$-sublattice magnetization) is given by

$$A(\pi) = \frac{1}{N_0} \sum_{i,\text{cell}} (\langle b_{i,\beta}^* b_{i,\beta} \rangle - \langle a_{i,\beta}^* a_{a,\beta} \rangle)$$

$$= \frac{1}{N_0} \sum_{i,\text{cell}} (\langle b_{i,\beta}^* b_{i,\beta} \rangle - \langle b_{i,\alpha}^* b_{a,\alpha} \rangle)$$

$$= -\frac{1}{N_0} \sum_{i,\text{lattice}} \exp(i\pi \cdot \boldsymbol{r}_i) \langle b_i^* b_i \rangle$$

$$= -\frac{1}{N_0} \sum_{k>0} \langle b_{k-\pi}^* b_k \rangle$$

$$= -B(\pi). \tag{8.129}$$

If we look at the Hamiltonian (8.123) we find that there is indeed an interaction between the sublattice magnetizations:

$$\frac{U}{N_0} \sum_{k,k'} a_{k+\pi}^* a_k b_{k'-\pi}^* b_{k'}, \tag{8.130}$$

and hence, in the mean-field approximation, this interaction is expressed as

$$U N_0 A(\pi) B(\pi), \quad A(\pi) = -B(\pi), \tag{8.131}$$

in favor of the antiferromagnetic state, since $A(\pi)$ and $B(\pi)$ have opposite signs due to their phase difference in the spatial oscillations.

The onset of antiferromagnetism is equivalent with the nonvanishing of the thermal average $\langle a_{k+\pi}^* a_k \rangle$. Hence the coupling between the two states $\boldsymbol{k}$ and $\boldsymbol{k} + \pi$ must be explicitly taken into account. This is most easily accomplished by the extended Hartree–Fock approximation, and the Hartree–Fock Hamiltonian is given by

$$H = \sum_{0<k<\pi} \{[\epsilon(\boldsymbol{k}) + Un - \mu] a_k^* a_k + [\epsilon(\boldsymbol{k} + \pi) + Un - \mu] a_{k+\pi}^* a_{k+\pi}$$

$$+ U B (a_{k+\pi}^* a_k + a_k^* a_{k+\pi}) \}, \tag{8.132}$$

where $B = B(\pi) = -A$, and n is one-half of the electron density given by

$$n = \frac{1}{2N_0} \sum_i [\langle a_i^* a_i \rangle + \langle b_i^* b_i \rangle].$$ (8.133)

The appropriate density matrix is a two-particle, 4×4 matrix,

$$\rho_2(k, k') = \begin{bmatrix} a_{11} & 0 & 0 & 0 \\ 0 & a_{22} & a_{23} & 0 \\ 0 & a_{32} & a_{33} & 0 \\ 0 & 0 & 0 & a_{44} \end{bmatrix},$$ (8.134)

where

$$\begin{aligned}
a_{11} &= N_k = \langle a_k^* a_k a_{k+\pi}^* a_{k+\pi} \rangle, \\
a_{22} &= n_k - N_k, \\
a_{33} &= n_{k+\pi} - N_k, \\
a_{44} &= 1 - n_k - n_{k+\pi} + N_k, \\
a_{23} &= \langle a_k^* a_{k+\pi} \rangle = \Gamma_k, \\
a_{32} &= \langle a_{k+\pi}^* a_k \rangle = \Gamma_k^* = \Gamma_k.
\end{aligned}$$ (8.135)

In these equations the variables are expectation values, not operators. The center, 2×2, matrix of (8.134) is easily diagonalized, and the diagonalized matrix elements are

$$\begin{aligned}
a'_{22} &= n_{k+} - N_k + R_k, \\
a'_{33} &= n_{k+} - N_k - R_k,
\end{aligned}$$ (8.136)

where

$$n_{k+} = \tfrac{1}{2}(n_k + n_{k+\pi}), \quad n_{k-} = \tfrac{1}{2}(n_k - n_{k+\pi}), \quad R_k = [n_{k-}^2 + \Gamma_k^2]^{\frac{1}{2}}.$$ (8.137)

The variational potential is given by

$$\begin{aligned}
F = \sum_{0 \le k \le \pi} &\{2(nU - \mu)n_{k+} + 2\epsilon_{k-}n_{k-} + UB(\Gamma_k + \Gamma_k^*) \\
&+ kT[N_k \log N_k + (n_{k+} - N_k + R_k) \log(n_{k+} - N_k + R_k) \\
&+ (n_{k+} - N_k - R_k) \log(n_{k+} - N_k - R_k) \\
&+ (1 - 2n_{k+} + N_k) \log(1 - 2n_{k+} + N_k)]\},
\end{aligned}$$ (8.138)

where

$$\epsilon_{k-} = \tfrac{1}{2}[\epsilon(k) - \epsilon(k + \pi)] = \epsilon(k).$$ (8.139)

We now take variations of the Helmholtz potential with respect to N_k, Γ_k, n_{k-}, and n_{k+}:

$$\log \frac{N_k(1 - 2n_{k+} + N_k)}{(n_{k+} - N_k + R_k)(n_{k+} - N_k - R_k)} = 0, \tag{8.140}$$

$$2\beta U B + \frac{\Gamma_k}{R_k} \log \frac{(n_{k+} - N_k + R_k)}{(n_{k+} - N_k - R_k)} = 0, \tag{8.141}$$

$$2\beta \epsilon(\mathbf{k}) + \frac{n_{k-}}{R_k} \log \frac{(n_{k+} - N_k + R_k)}{(n_{k+} - N_k - R_k)} = 0, \tag{8.142}$$

$$2\beta(nU - \mu) + \log \frac{(n_{k+} - N_k + R_k)(n_{k+} - N_k - R_k)}{(1 - 2n_{k+} + N_k)^2} = 0. \tag{8.143}$$

Equation (8.140) is solved for N_k:

$$N_k = n_{k+}^2 - n_{k-}^2 - \Gamma_k^2 = n_k n_{k+\pi} - \Gamma_k^2. \tag{8.144}$$

This shows that the Hartree–Fock decomposition is precisely a consequence of the cluster variation method.

From (8.141), (8.142), and (8.137),

$$n_{k-} = -\frac{\epsilon(\mathbf{k})\Gamma_k}{UB}, \quad \text{and} \quad R_k = \frac{\Gamma_k E_k}{UA}, \tag{8.145}$$

where

$$E_k = [\epsilon(\mathbf{k})^2 + U^2 A^2]^{\frac{1}{2}}. \tag{8.146}$$

Also,

$$N_k = n_{k+}^2 - R_k^2. \tag{8.147}$$

Again from (8.141),

$$\frac{(n_{k+} - N_k + R_k)}{(n_{k+} - N_k - R_k)} = \exp(2\beta E_k), \tag{8.148}$$

which is solved to give

$$\frac{R_k}{n_{k+} - N_k} = \tanh \beta E_k \tag{8.149}$$

or

$$n_{k+} - N_k = R_k \coth \beta E_k. \tag{8.150}$$

Equations (8.143) with (8.147) and (8.150) will yield

$$\frac{R_k \operatorname{cosech} \beta F_k}{(1 - n_{k+})^2 - R^2} = \exp[-\beta(nU - \mu)], \tag{8.151}$$

and

$$\frac{R_k \operatorname{cosech}\beta E_k}{(1 - n_{k+}) - R_k \coth \beta E_k} = \exp[-\beta(nU - \mu)]. \tag{8.152}$$

The above two equations look similar but they are linearly independent of each other and hence $(1 - n_{k+})$ can be eliminated between the two in order to find R_k. The result is

$$R_k[\exp \beta(nU - \mu) + \exp(\beta E_k)][\exp \beta(nU - \mu) + \exp(-\beta E_k)]$$
$$= \exp \beta(nU - \mu) \sinh \beta E_k. \tag{8.153}$$

When this is solved for R_k, we find

$$2\Gamma_k = \frac{AU}{E_k} \left[\frac{1}{\exp(\beta E_{1k}) + 1} - \frac{1}{\exp(\beta E_{2k}) + 1} \right], \tag{8.154}$$

where

$$E_{1k} = (Un - \mu) - [\epsilon(k)^2 + U^2 B^2]^{\frac{1}{2}},$$
$$E_{2k} = (Un - \mu) + [\epsilon(k)^2 + U^2 B^2]^{\frac{1}{2}}. \tag{8.155}$$

Finally, (8.152) is solved for $(1 - n_{k+})$, and we see that

$$1 - n_{k+} = R_k \frac{\exp \beta(nU - \mu) + \cosh \beta E_k}{\sinh \beta E_k}, \tag{8.156}$$

and, further, if R_k is introduced from (8.153),

$$2n_{k+} = n_k + n_{k+\pi}$$
$$= \frac{1}{\exp(\beta E_{1k}) + 1} + \frac{1}{\exp(\beta E_{2k}) + 1}. \tag{8.157}$$

If both sides of this equation are summed over $\mathbf{k}$ and set equal to the total number of particles in the lattice, what results is the equation which will determine the chemical potential μ as a function of the particle density in the lattice. When the thermodynamic limit is taken, however, the summation over the wave vector is replaced by integrations.

In this way, from (8.154), we obtain the equation for the sublattice magnetization:

$$U \int_0^D \frac{1}{[\epsilon^2 + U^2 A^2]^{\frac{1}{2}}} \left[\frac{1}{\exp(\beta E_1(\epsilon)) + 1} - \frac{1}{\exp(\beta E_2(\epsilon)) + 1} \right] \eta_0(\epsilon) d\epsilon = 1, \tag{8.158}$$

where

$$E_1(\epsilon) = (Un - \mu) - [\epsilon^2 + U^2 B^2]^{\frac{1}{2}},$$
$$E_2(\epsilon) = (Un - \mu) + [\epsilon^2 + U^2 B^2]^{\frac{1}{2}}, \tag{8.159}$$

and D is the half band width ($D = 4t$ for a square lattice and $D = 6t$ for a simple cubic lattice), and $\eta_0(\epsilon)$ is the density of states function.

The sublattice magnetization vanishes ($A = 0$) at the Néel temperature, and hence the reciprocal critical temperature β_c is found as the solution of the integral equation

$$U \int_0^D \frac{1}{\epsilon} \left[\frac{1}{\exp \beta_c(Un - \mu - \epsilon) + 1} - \frac{1}{\exp \beta_c(Un - \mu + \epsilon) + 1} \right] \eta_0(\epsilon) d\epsilon = 1.$$

(8.160)

In order to solve this equation for the reciprocal Néel temperature, β_c, we must know the chemical potential, μ, as a function of both the electron concentration, n, and temperature, β. The chemical potential must therefore be determined as a function of the electron density, n ($n = \frac{1}{2}$ for a half-filled band) [Exercise 8.3]:

$$\frac{1}{N_0} \int_0^D \left[\frac{1}{\exp \beta(Un - \mu - \epsilon) + 1} + \frac{1}{\exp \beta(Un - \mu + \epsilon) + 1} \right] \eta_0(\epsilon) d\epsilon = 1.$$

(8.161)

8.6 The first-order transition in cubic ice[†]

Ice can take many different crystallographic structures at different temperatures and pressures, as shown in Fig. 2.3 (Whalley *et al.*, 1966, 1973; Fletcher, 1970). Above the atmospheric pressure of 20 kbar, ice is found in the body center cubic structure with respect to oxygen ions (Fig. 8.2). At about $0\,°C$ there is a first-order phase transition from the low temperature proton-ordered phase VIII to the proton-disordered phase VII without any appreciable change in the oxygen lattice constant.

Since there are two oxygen ions per cube of the body center structure, there are four protons per cube, and they form a tetrahedrally bonded diamond lattice structure with respect to oxygen ions (see Fig. 8.2). Because there are only four protons per cube, only four out of the eight oxygen ions at the corners of the cube are bonded by the intervening protons with the center oxygen, and the entire crystal of cubic ice is divided into two interpenetrating diamond structures.

The bonding structure of protons is called the ice condition, i.e., each oxygen forms a water molecule with two nearby protons in spite of the fact that there are four hydrogen bonds connecting one oxygen to its four neighboring oxygens. There are two potential minima for each proton to sit in along its hydrogen bond. If one of the two protons forming a water molecule, and hence situated in the potential minimum closer to the center oxygen, moves out to the other potential minimum close to the corner oxygen, one of the protons which is bonded with the two remaining corner oxygens moves into the potential minimum close to the center oxygen so that there

---

[†] This section is mostly reproduced from Tomoyasu Tanaka & Tohru Morita (1999), 'First-order phase transition in cubic ice', *Physica* A, vol. 272, pp. 555–62, with permission from Elsevier Science.

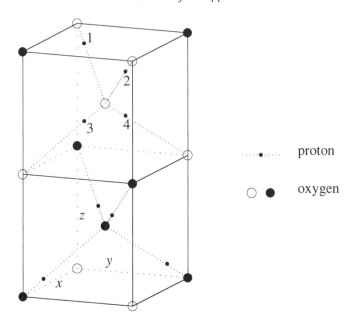

Fig. 8.2. Ordered phase VII of ice. The large circles (representing oxygen ions) form, individually, a diamond structure. Protons (represented by dots) are on the hydrogen bonds of the diamond structure. Arrangements of protons in the ordered phase are such that all the electric dipoles of water molecules in each sublattice are parallel and the two sublattices are arranged in an antiferroelectric configuration along one of the cubic axes, e.g., in the z-direction.

are always two protons at one time around each oxygen. This must be true with all the protons in the lattice, and hence the motions of protons are strongly correlated. All the possible correlated modes of the proton motions have been identified using infrared spectroscopy in the disordered phase VII of ice.

The theory of the first-order phase transition is one of the less developed areas in statistical mechanics, and the mechanism of the transition is not well understood. Haus & Tanaka (1977) presented a theory of the first-order phase transition applied to the ice VII–VIII transition. The mechanism proposed was a simple four-body attractive potential describing the protons around the central oxygen in such a way that the ice condition is favored. The first-order phase transition resulted when the mean-field approximation was applied. The treatment, however, was rather crude, and the stability of the first-order transition with respect to the degree of approximation was not definitive. In the formulation presented in this section, the ice condition is rigorously imposed. This means that the protons can move back and forth between two minima of the potential in their respective hydrogen bonds; however, there are always two protons around each oxygen. An extended mean-field approximation is employed and the first-order transition is concluded. The statistical approximation adopted herein (Tanaka & Morita, 1999) is still rather crude; however, the proposed mechanism of the first-order phase transition seems to be qualitatively justifiable.

Let us introduce dynamical and statistical variables describing the proton states with respect to the central oxygens. Since all the hydrogen bonds in the lattice run diagonally up and down, though not necessarily vertically with respect to the z-axis, the Ising spin variable μ_i can be conveniently employed (see Fig. 8.2). For instance, the up–down positions of the proton along the ith hydrogen bond are represented by

$$\mu_i = \pm 1. \tag{8.162}$$

The projection operators

$$r_i = \tfrac{1}{2}(1 \pm \mu_i) \tag{8.163}$$

have an eigenvalue unity when the proton is in the upper (or lower) minimum, and an eigenvalue 0 when the proton is in the lower (or upper) minimum.

The probabilities of finding the proton in these minima are given by their statistical averages, $s_i(\pm 1)$, respectively, expressed as

$$s_i(\pm) = \tfrac{1}{2}\langle(1 \pm \mu_i)\rangle. \tag{8.164}$$

Configurations of four protons around an oxygen ion are most conveniently represented by the following projection operators, each of which has an eigenvalue unity in its own state and zero in any other states:

$$p_1 = \tfrac{1}{16}(1 + \mu_1)(1 + \mu_2)(1 + \mu_3)(1 + \mu_4), \ [z+],$$

$$p_2 = \tfrac{1}{16}(1 - \mu_1)(1 - \mu_2)(1 - \mu_3)(1 - \mu_4), \ [z-],$$

$$p_3 = \tfrac{1}{16}(1 + \mu_1)(1 - \mu_2)(1 + \mu_3)(1 - \mu_4), \ [x-],$$

$$p_4 = \tfrac{1}{16}(1 - \mu_1)(1 + \mu_2)(1 - \mu_3)(1 + \mu_4), \ [x+],$$

$$p_5 = \tfrac{1}{16}(1 + \mu_1)(1 - \mu_2)(1 - \mu_3)(1 + \mu_4), \ [y-],$$

$$p_6 = \tfrac{1}{16}(1 - \mu_1)(1 + \mu_2)(1 + \mu_3)(1 - \mu_4), \ [y+]. \tag{8.165}$$

In the above definitions, p_1, for example, has eigenvalue unity if all the protons occupy the upper positions ($\mu_1 = \mu_2 = \mu_3 = \mu_4 = +1$), i.e., the dipole is pointing in the positive z-direction (see Fig. 8.2). These projection operators are linearly independent and orthogonal to one another.

The probability, q_1, that the dipole is pointing in the positive z-direction is given by the statistical average of p_1, etc.

Because of the ice condition, no other arrangements of protons around oxygen are permitted. This means that the sum of the six probabilities is equal to unity:

$$\sum_{i=1}^{6} q_i = 1, \tag{8.166}$$

where $q_i = \langle p_i \rangle$.

By looking at (8.165) we obtain the probabilities of the first proton configurations as the condition of reducibility:

$$s_1(+) = q_1 + q_3 + q_5,$$
$$s_1(-) = q_2 + q_4 + q_6. \tag{8.167}$$

The configurations of the system are represented by the proton states on the bonds. In the following, when applying the cluster variation method, the bonds are called sites, $s_i(\pm)$ are one-site probabilities, and q_1–q_6 are the four-site probabilities. It is obvious that both the one-site and four-site probabilities are normalized to unity.

In order to work with the formulation in which the ice condition is rigorously imposed, the smallest clusters which should be included in the variational potential are the four-site clusters and all subclusters of them (Morita, 1994; Tanaka et al., 1994). The crudest approximation employed herein is an extended mean-field approximation. It means that the intersublattice interaction between the neighboring oxygen–proton tetrahedra is estimated in the form of an effective field (mean-field).

The variational potential per oxygen ion, multiplied by the reciprocal temperature, $\beta = 1/T$, is given by

$$\beta F/N = -\tfrac{1}{2}\beta U(q_1 - q_2)^2 + 2g^{(1)} + 6g^{(2)} + 4g^{(3)} + g^{(4)}, \tag{8.168}$$

where $g^{(1)}, \ldots, g^{(4)}$ are the one-, two-, three-, and four-site cumulant functions (Tanaka et al., 1994). The first term on the right hand side of the equation is the interaction term between two sublattices given in the extended mean-field approximation. Originally, it was the statistical average of a product of two dynamical parameters, one from one sublattice and the other from the other sublattice. In the mean-field approximation, however, the statistical correlation between the two sublattices is ignored, and the interaction energy is represented as a product of statistical averages, as given in the above equation, which favors antiferroelectric arrangement of dipoles in different sublattices.

When the cumulant g-functions are represented by the corresponding cluster G-functions, $G^{(2)}$ and $G^{(3)}$ disappear because they are not the common parts of the two neighboring four-site clusters in the lattice, and the variational potential takes the following form:

$$\beta F/N = -\tfrac{1}{2}\beta U(q_1 - q_2)^2 - 2G^{(1)} + G^{(4)}, \tag{8.169}$$

where $G^{(1)}$ and $G^{(4)}$ are given explicitly in terms of the distribution functions, as follows:

$$G^{(1)} = s_1(+) \log s_1(+) + s_1(-) \log s_1(-),$$

$$G^{(4)} = \sum_{i=1}^{6} q_i \log q_i. \tag{8.170}$$

Before minimizing the variational potential with respect to the unknown parameters, q_1 through q_6, a further simplifying condition can be imposed.

In the totally disordered phase, all six orientations of the water dipole are equally probable, and hence it is expected that

$$q_1 = q_2 = q_3 = q_4 = q_5 = q_6 = \tfrac{1}{6}, \quad s_1(+) = s_1(-) = \tfrac{1}{2}. \tag{8.171}$$

The z-axis polarization is created in the ordered phase; however, four horizontal orientations of the dipole are still equivalent. This means that

$$q_3 = q_4 = q_5 = q_6 = \tfrac{1}{4}(1 - q_1 - q_2). \tag{8.172}$$

This condition is seen to be consistent even with the condition in the disordered phase. Hence the formulation becomes a two-parameter problem.

Now parameters q and m, instead of parameters q_1 and q_2, can be used; these are defined by

$$q = \tfrac{1}{2}(q_1 + q_2), \quad m = q_1 - q_2, \tag{8.173}$$

where m is the sublattice polarization per oxygen ion. Then the variational potential takes the following form:

$$\begin{aligned}
\beta F/N = {}&-\tfrac{1}{2}\beta U m^2 + \left(q + \tfrac{1}{2}m\right) \log \left(q + \tfrac{1}{2}m\right) \\
&+ \left(q - \tfrac{1}{2}m\right) \log \left(q - \tfrac{1}{2}m\right) + (1 - 2q) \log \tfrac{1}{4}(1 - 2q) \\
&- (1 + m) \log \tfrac{1}{2}(1 + m) - (1 - m) \log \tfrac{1}{2}(1 - m).
\end{aligned} \tag{8.174}$$

When the Helmholtz potential is minimized with respect to q and m,

$$4(4q^2 - m^2) = (1 - 2q)^2, \tag{8.175}$$

$$-\beta U m + \tfrac{1}{2} \log \frac{2q + m}{2q - m} + \log \frac{1 - m}{1 + m} = 0. \tag{8.176}$$

If (8.175) is solved for q, we obtain

$$q = \tfrac{1}{3}(3m^2 + 1)^{\frac{1}{2}} - \tfrac{1}{6}. \tag{8.177}$$

If q is eliminated between (8.175) and (8.176), we obtain

$$m = \frac{\sinh x}{2 - \cosh x},$$ (8.178)

where $x = \beta U m$, or

$$m = \frac{T}{U}x.$$ (8.179)

The sublattice polarization, m, is found as a solution of two coupled equations (8.178) and (8.179) (Fig. 8.3). This solution corresponds to the local maximum of the Helmholtz potential.

In Fig. 8.4, the Helmholtz potential $f = F/N$ at both the local minimum and maximum, and the sublattice polarization are shown against the reduced temperature, T/U. In the states $m = 0$ and $m = 1$, the entropy is $\log\frac{3}{2}$ and 0, respectively.

Below the phase transition temperature the dipole arrangement is perfect and there is no partially ordered phase. This is likely due to the crudeness of the mean-field approximation; however, the stability of the first-order phase transition with respect to the degree of approximation may very well be maintained. It is interesting to note that the value of the entropy at high temperature, $\log\frac{3}{2}$, is equal to the value estimated by Pauling (1960). The entropy for the ice in which the oxygen ions are situated on a diamond lattice was estimated by Nagle (1966). The value is between $\log(1.50681)$ and $\log(1.50686)$; therefore we can see that the approximation in this section is fairly good.

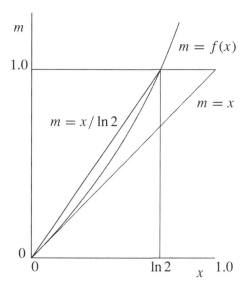

Fig. 8.3. Solution of the transcendental equations, (8.178) and (8.179), exists between two temperatures $T = U$ and $T = U/\ln 2$.

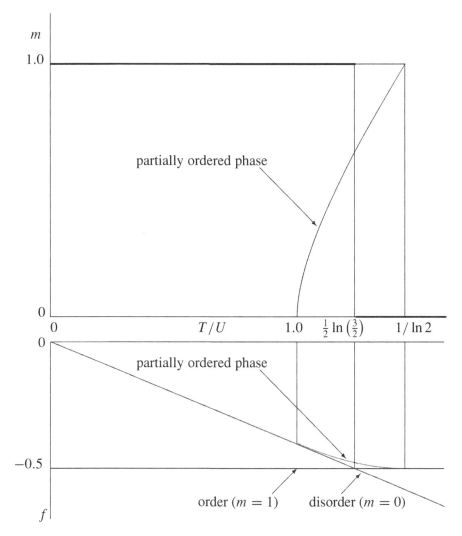

Fig. 8.4. Helmholtz potential $f = F/N$ as a function of temperature T/U.

Exercises

8.1 Prove that $\eta_{k\lambda}$ given by (8.11) satisfies the condition (8.9).

8.2 Derive (8.12) by substituting (8.10) and (8.5) into (8.8).

8.3 Derive (8.104), (8.105), and (8.106).

    [Hint]

    From (8.102),

$$\frac{\eta_{d\sigma}}{\eta_{k\sigma}} = \frac{\epsilon_k^2}{(1 - 2n_{k\sigma})^2 |V|^2}$$

$$= \frac{\epsilon_k^2}{(1 - 4\eta_{k\sigma})|V|^2},$$

$\eta_0(\epsilon)$

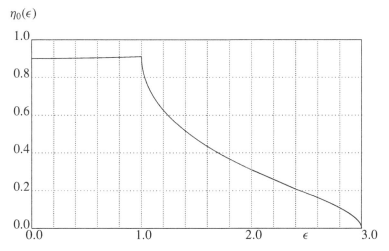

Fig. 8.5. The density of states for a simple cubic lattice is plotted as a function of the energy. The half band width is $6t$, but the dimensionless energy scale 0 to $3 = 6t/2t$ is used.

$1/2t\beta$

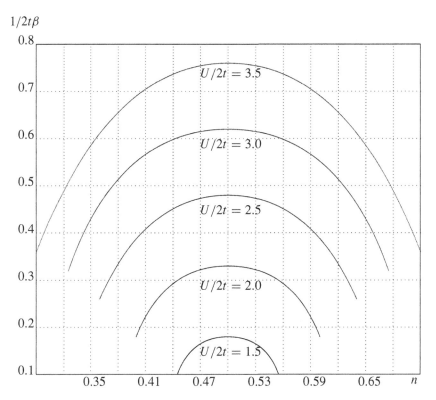

Fig. 8.6. Néel temperature as a function of the electron density n and $U/2t$.

$$\eta_{d\sigma}(1 - 4\eta_{k\sigma})|V|^2 = \epsilon_k^2 \eta_{k\sigma}$$

$$\eta_{d\sigma}|V|^2 = \left(\epsilon_k^2 + 4\eta_{d\sigma}|V|^2\right)\eta_{k\sigma}$$

$$\eta_{k\sigma} = \eta_{d\sigma}|V|^2/\left(\epsilon_k^2 + 4\eta_{d\sigma}|V|^2\right). \tag{E8.1}$$

If this is inserted into (8.102), we see that

$$\epsilon_k - \left(\epsilon_k^2 + 4\eta_{d\sigma}|V|^2\right)^{\frac{1}{2}}(1 - 2n_{k\sigma}) = 0, \tag{E8.2}$$

and hence

$$n_{k\sigma} = \frac{1}{2}\left[1 - \frac{\epsilon_k}{\left(\epsilon_k^2 + 4\eta_{d\sigma}|V|^2\right)^{\frac{1}{2}}}\right]. \tag{E8.3}$$

If this is introduced into (8.101) we immediately find

$$\epsilon_{0\sigma} = 2|V|^2\eta_{d\sigma}^{\frac{1}{2}} \sum_k \left(\epsilon_k^2 + 4\eta_{d\sigma}|V|^2\right)^{-\frac{1}{2}}. \tag{E8.4}$$

8.4 Find the Néel temperature as a function of the electron density n and U.

[**Hint**]
Our problem is to solve (8.160) and (8.161) simultaneously for given values of U and n. These two equations are reproduced in the following:

$$U \int_0^D \frac{1}{\epsilon}\left[\frac{1}{\exp\beta(Un - \mu - \epsilon) + 1} - \frac{1}{\exp\beta(Un - \mu + \epsilon) + 1}\right]\eta_0(\epsilon)d\epsilon = 1; \tag{E8.5}$$

$$\frac{1}{N_0}\int_0^D \left[\frac{1}{\exp\beta(Un - \mu - \epsilon) + 1} + \frac{1}{\exp\beta(Un - \mu + \epsilon) + 1}\right]\eta_0(\epsilon)d\epsilon = 1. \tag{E8.6}$$

In these equations the electron density of states, in the absence of antiferromagnetism, $\eta_0(\epsilon)$, appears. The density of state function is well known and has been calculated by many authors (Morita & Horiguchi, 1971). The graph for the simple cubic lattice is reproduced in Fig. 8.5.

In order to determine the Néel temperature for given values of U and n, one starts with (E8.6), tries a small value of β (a high temperature) and finds the chemical potential μ until the left hand side becomes equal to unity. In order to accomplish this, an interval narrowing method should be employed.

Once the chemical potential μ is determined for the given values of U, n, and β by (E8.6), we return to (E8.5) and evaluate the left hand side integral. If the value of the left hand side is not equal to unity, then we return to (E8.6). We then increase the value of β, i.e., we start at a slightly lower temperature and determine the chemical potential μ again by the interval narrowing method. We return to (E8.5) and see if the left hand side becomes equal to unity.

By repeating these steps, we can finally determine the chemical potential and the Néel temperature for a given set of values of U and n. We then change the electron density n and repeat the same steps as described in the above. Fig. 8.6 is obtained in this fashion.

9

The exact Ising lattice identities

9.1 The basic generating equations

Exact identities which express one Ising spin correlation function as a linear combination of a certain number of Ising spin correlation functions have been known for some time. The identity for the two-dimensional honeycomb lattice was first derived by Fisher (1959) and later it was rederived by several authors (Doman & ter Haar 1962; Callen 1963). A similar identity which holds for the two-dimensional triangular lattice was derived by Stephenson (1964). A general method of deriving such identities, regardless of the dimensionality of the lattice, was presented later by Suzuki (1965).

Let $[f]$ be any function of the honeycomb Ising variables $s_1, s_2, \ldots, s_{N-1}$ (excluding s_0, the origin-site variable in Fig. 9.1), where N is the total number of lattice sites in the system. First of all, the Ising Hamiltonian H is divided into two parts:

$$H = -J \sum_{\langle i,j \rangle} s_i s_j = -J s_0 (s_1 + s_2 + s_3) + H_1, \qquad (9.1)$$

where H_1 is that part of the Hamiltonian not containing s_0.

The canonical thermal average $\langle s_0[f] \rangle$ is constructed as:

$$
\begin{aligned}
Z(\beta) \cdot \langle s_0[f] \rangle \\
&= \mathrm{Tr}_{s_1,\ldots,s_{N-1}} \exp(-\beta H_1)[f] \mathrm{Tr}_{s_0} \exp(+\beta J s_0(s_1 + s_2 + s_3)) s_0 \\
&= \mathrm{Tr}_{s_1,\ldots,s_{N-1}} \exp(-\beta H_1)[f] \mathrm{Tr}_{s_0} \exp(+\beta J s_0(s_1 + s_2 + s_3)) \\
&\quad \times \frac{\mathrm{Tr}_{s_0} \exp(+\beta J s_0(s_1 + s_2 + s_3)) s_0}{\mathrm{Tr}_{s_0} \exp(+\beta J s_0(s_1 + s_2 + s_3))} \\
&= \mathrm{Tr}_{s_0,s_1,\ldots,s_{N-1}} \exp(-\beta H)[f] \frac{\mathrm{Tr}_{s_0} \exp(+\beta J s_0(s_1 + s_2 + s_3)) s_0}{\mathrm{Tr}_{s_0} \exp(+\beta J s_0(s_1 + s_2 + s_3))}, \qquad (9.2)
\end{aligned}
$$

212

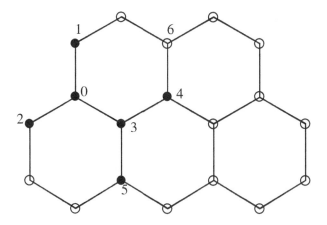

Fig. 9.1. Honeycomb lattice and its "bow tie" cluster made up of the honeycomb lattice sites 0, 1, 2, 3, 4, 5.

where $Z(\beta)$ is the canonical partition function, and hence

$$\langle s_0[f]\rangle = \left\langle [f] \frac{\mathrm{Tr}_{s_0} \exp(+\beta J s_0 (s_1 + s_2 + s_3)) s_0}{\mathrm{Tr}_{s_0} \exp(+\beta J s_0 (s_1 + s_2 + s_3))} \right\rangle. \tag{9.3}$$

Carrying out the Tr_{s_0} summation, we find

$$\frac{\mathrm{Tr}_{s_0} \exp(+\beta J s_0 (s_1 + s_2 + s_3)) s_0}{\mathrm{Tr}_{s_0} \exp(+\beta J s_0 (s_1 + s_2 + s_3))} = \tanh K (s_1 + s_2 + s_3)$$

$$= A(s_1 + s_2 + s_3) + B s_1 s_2 s_3, \tag{9.4}$$

where $K = \beta J$, and

$$A = \tfrac{1}{4}(\tanh 3K + \tanh K), \quad B = \tfrac{1}{4}(\tanh 3K - 3 \tanh K). \tag{9.5}$$

In deriving the second line from the first line in (9.4) use has been made of the facts that $\tanh K (s_1 + s_2 + s_3)$ is expressed as the sum of symmetric, homogeneous and odd functions of s_1, s_2 and s_3, and that $s_i^{2n+1} = s_i$, $s_i^{2n} = 1$, $i = 1, 2, 3$, for an integer n. If the second line agrees with the first line at $s_1 = s_2 = s_3 = 1$ and $s_1 = s_2 = -s_3 = 1$, then the two lines agree identically. Hence,

$$\langle s_0[f]\rangle = A\langle [f](s_1 + s_2 + s_3)\rangle + B\langle [f]s_1 s_2 s_3\rangle, \quad s_0 \notin [f]. \tag{9.6}$$

9.2 Linear identities for odd-number correlations

Using the basic generating equation (9.6) for the Ising correlations on the honeycomb lattice, the detail of the critical behavior of the Ising honeycomb lattice may be analyzed (Munera *et al.*, 1982). Let us look at the "bow tie" cluster (Fig. 9.1) made up of the honeycomb lattice sites 0, 1, 2, 3, 4, 5. Then the following identities are found immediately:

$$y_1 = 3Ay_1 + By_3, \quad [f] = 1, \tag{9.7}$$
$$y_2 = (2A + B)y_1 + Ay_3, \quad [f] = s_1 s_2, \tag{9.8}$$
$$y_3 = A(y_2 + 2y_6) + By_7, \quad [f] = s_4 s_5, \tag{9.9}$$
$$y_4 = A(y_1 + y_4 + y_6) + By_5, \quad [f] = s_1 s_5, \tag{9.10}$$
$$y_5 = A(y_1 + y_5 + y_6) + By_4, \quad [f] = s_1 s_4, \tag{9.11}$$
$$y_7 = A(2y_6 + y_7) + By_2, \quad [f] = s_1 s_2 s_4 s_5, \tag{9.12}$$
$$y_8 = A(2y_4 + y_6) + By_1, \quad [f] = s_1 s_2 s_3 s_4, \tag{9.13}$$

where the various thermal averages are defined as

$$y_1 = \langle s_0 \rangle, \quad y_2 = \langle s_0 s_1 s_2 \rangle, \quad y_3 = \langle s_0 s_4 s_5 \rangle,$$
$$y_4 = \langle s_0 s_1 s_5 \rangle, \quad y_5 = \langle s_0 s_1 s_4 \rangle, \quad y_6 = \langle s_1 s_2 s_4 \rangle,$$
$$y_7 = \langle s_0 s_1 s_2 s_4 s_5 \rangle, \quad y_8 = \langle s_0 s_1 s_2 s_3 s_4 \rangle. \tag{9.14}$$

The set of homogeneous equations (9.7)–(9.13) is exact, and its containing set (9.14) of odd-number correlations exhausts all such (nonequivalent) possibilities defined upon the "bow-tie" cluster of lattice sites 0, 1, 2, 3, 4, 5. Subtracting the y_5 equation from the y_4 equation yields

$$(1 - \tanh K)(y_4 - y_5) = 0, \tag{9.15}$$

where use was made of the coefficient expressions (9.5). Since $1 - \tanh K \neq 0$ (except at $K = \infty$), (9.15) establishes the degeneracy relation

$$y_4 = y_5, \tag{9.16}$$

thereby reducing the number of unknowns. Substituting the degeneracy relation (9.16) into the set of equations (9.7)–(9.13), and after some elementary algebraic manipulations, we obtain the following exact set of six linear inhomogeneous equations in the six ratio quantities $\xi_i = y_i / y_1$, $i = 2, 3, 4, 6, 7, 8$:

$$\xi_3 = B^{-1}(1 - 3A),$$
$$\xi_2 = B + A(2 + \xi_3),$$
$$\xi_7 = (A - B - 1)^{-1}[(A - B)\xi_2 - \xi_3],$$
$$\xi_6 = (2A)^{-1}[(1 - A)\xi_7 - B\xi_2],$$
$$\xi_4 = A(1 - A - B)^{-1}(1 + \xi_6), \quad (\xi_4 = \xi_5),$$
$$\xi_8 = B - A + (1 + A - B)\xi_4. \tag{9.17}$$

By inspection, we find that the system of equations (9.17) is arranged to yield directly exact, closed-form solutions for all the ratio quantities $\xi_i = y_i / y_1$, $i = 2$,

3, 4, 6, 7, 8, i.e., the first of the equations is itself the solution for ξ_3, while every other unknown ratio appearing on the left hand sides of the remaining equations is easily determined from direct substitutions of only those ratio solutions appearing previously in the arrangement. With the known coefficients A and B being analytic functions of temperature for all finite temperatures, the exact solutions for the ratio quantities $\xi_i = y_i/y_1$, $i = 2, 3, 4, 6, 7, 8$, are each found to be similarly analytic, examples of the first few solutions being

$$\xi_3 = \frac{y_3}{y_1} = B^{-1}(1 - 3A), \tag{9.18}$$

$$\xi_2 = \frac{y_2}{y_1} = B + A[B^{-1}(1 - 3A) + 2], \tag{9.19}$$

$$\xi_7 = \frac{y_7}{y_1} = (A - B - 1)^{-1}$$

$$\times \{B^{-1}(1 - 3A)[A(A - B) - 1] + (A - B)(2A + B)\}, \tag{9.20}$$

and so forth. Using (9.5) for A and B, we can choose to rewrite all solutions as functions of K alone. Exact, closed-form solutions for the odd-number correlations y_i, $i = 2, 3, 4, 6, 7, 8$, are now immediately determined by merely multiplying each of the above corresponding ratio solutions $\xi_i = y_i/y_1$ by the spontaneous magnetization y_1 of the honeycomb Ising model. The latter is well known from the literature (Naya, 1954) and is given together with its critical behavior as

$$y_1 = \left(1 - 4\frac{\cosh 3K \cosh^3 K}{\sinh^6 2K}\right)^{\frac{1}{8}} \sim A_1 \epsilon^{\frac{1}{8}}, \quad T \longrightarrow T_c, \tag{9.21}$$

where $\epsilon = (T_c - T)/T_c = (K - K_c)/K_c$ is the fractional deviation of the temperature from its critical value T_c with

$$J/kT_c = K_c = \tfrac{1}{2}\log(2 + \sqrt{3}) = 0.658478\ldots, \tag{9.22}$$

and the critical amplitude A_1 being

$$A_1 = \left[\frac{8\sqrt{3}}{3}\log(2 + \sqrt{3})\right]^{\frac{1}{8}} = 1.253177\ldots. \tag{9.23}$$

It is now seen that all the odd-number correlations possess the same critical temperature T_c and critical exponent $\tfrac{1}{8}$ as the spontaneous magnetization but differing critical amplitues. In order to calculate these critical amplitudes, one uses the value K_c in (9.22) to evaluate first the coefficients A, B in (9.5) at the critical temperature, namely,

$$A_c = -2B_c = 2\sqrt{3}/9 = 0.384900\ldots. \tag{9.24}$$

Now the critical amplitudes of all the odd-number correlations can be found from (9.18)–(9.20) and other similar equations:

$$A_3 = A_1 3(2 - \sqrt{3}) = 1.007363\ldots, \tag{9.25}$$

$$A_2 = A_1\left(\tfrac{5}{3}\sqrt{3} - 2\right) = 1.111257\ldots, \tag{9.26}$$

$$A_7 = A_1(13 - 7\sqrt{3})/(3 - \sqrt{3}) = 0.865443\ldots, \tag{9.27}$$

$$A_6 = A_5(3\sqrt{3} - 5)/(3 - \sqrt{3}) = 0.969336\ldots, \tag{9.28}$$

$$A_4 (= A_5) = A_1 2(7\sqrt{3} - 11)/(5\sqrt{3} - 6) = 1.059310\ldots, \tag{9.29}$$

$$A_8 = A_1(26\sqrt{3} - 39)/3(5\sqrt{3} - 6) = 0.947381\ldots. \tag{9.30}$$

9.3 Star-triangle-type relationships

The star-triangle $(Y - \Delta)$ transformation is due to Onsager (1944) and Wannier (1945), and relates the honeycomb and triangular Ising models by showing that their canonical partition functions, $Z_{h,K}$ and $Z_{t,L}$, differ only by a known multiplicative constant, i.e.,

$$Z_{h,K} = \Delta^{N/2} Z_{t,L}, \tag{9.31}$$

where

$$\Delta^4 = \exp(4L)(\exp(4L) + 3)^2. \tag{9.32}$$

N is the total number of honeycomb lattice sites, and

$$2\cosh 2K = \exp(4L) + 1 \tag{9.33}$$

relates the (dimensionless) interaction parameters K, L of the honeycomb and triangular lattices, respectively. Using the site enumerations in Fig. 9.1, the $Y - \Delta$ transformation is diagrammatically illustrated in Fig. 9.2 and the proof of the $Y - \Delta$ transformation (9.33) is based upon the realization that the trace evaluation

$$\mathrm{Tr}_{s_0} \exp(K s_0(s_1 + s_2 + s_3)) = \Delta \exp(L(s_1 s_2 + s_2 s_3 + s_3 s_1)), \tag{9.34}$$

is similarly valid for each "circle" site appearing within the total trace operation defining $Z_{h,K}$.

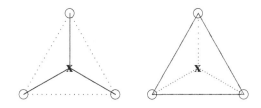

Fig. 9.2. The star-triangle $(Y - \Delta)$ transformation.

Using these correspondence concepts and making reference to Fig. 9.1, the following theorem connecting honeycomb and triangular Ising model correlations may be proved.

Theorem 9.1 *Consider the honeycomb lattice shown in Fig. 9.1 and let* [r] *be any function of Ising variables containing only "circle" sites (or only "**x**" sites). Then*

$$\langle [r] \rangle_{h,K} = \langle [r] \rangle_{t,L}, \qquad (9.35)$$

where $\langle [r] \rangle_{h,K}$ *and* $\langle [r] \rangle_{t,L}$ *denote canonical thermal averages pertaining to the honeycomb and triangular Ising model (dimensionless) Hamiltonian* $H_{h,K}$ *and* $H_{t,L}$, *respectively, with their (dimensionless) interaction parameters* K *and* L *related by* $2 \cosh 2K = \exp(4L) + 1$.

Proof For definiteness, let [r] be any function of Ising variables containing solely "**x**" sites, and introduce the notation $_x\text{Tr}$ to signify the trace operation over the degrees of freedom associated with all the "**x**" sites of the honeycomb lattice, and $_{o,x}\text{Trace}$ to signify the trace operation over the degrees of freedom associated with the totality of honeycomb sites. Then,

$$\langle [r] \rangle_{h,K} = \frac{_{o,x}\text{Tr}[r]\exp(H_{h,K})}{Z_{h,K}} = \frac{_x\text{Tr}[r]\exp(H_{t,L})}{Z_{t,L}} = \langle [r] \rangle_{t,L}, \qquad (9.36)$$

where use has been made of (9.33) and the relationship (9.35) for every "**x**" site of the honeycomb lattice. Noting that a parallel line of argument is easily constructed for the case where [r] is any function of Ising variables comprising solely "circle" sites, this completes the proof of the theorem.

Theorem 9.1 will have a significant consequence in the remainder of the analysis. In general, any honeycomb lattice thermal average associated with a configuration of sites which are in registry with the sites of the triangular lattice can now be equated to the correponding triangular lattice thermal average, and vice versa, in the sense of the $Y - \Delta$ relationships. As an illustration, we can write that

$$\langle s_1 s_2 s_3 \rangle_{h,K} = \langle s_1 s_2 s_3 \rangle_{t,L} = \langle s_1 s_3 s_6 \rangle_{t,L} = \langle s_1 s_3 s_6 \rangle_{h,K}, \qquad (9.37)$$

establishing the interesting degeneracy relation

$$y_3 = \langle s_1 s_2 s_3 \rangle_{h,K} = \langle s_1 s_3 s_6 \rangle_{h,K}, \qquad (9.38)$$

which will be used in the next section. Also, it should be clear that within the context of Fig. 9.1 and Theorem 9.1, all subscripts upon canonical thermal average symbols $\langle \cdots \rangle$ are no longer required and hence will be omitted in the remainder of the analysis; likewise, only the numeric labels of the lattice sites will be placed within the average symbols, e.g., the above degeneracy relation reads

$$y_3 = \langle 123 \rangle = \langle 136 \rangle. \tag{9.39}$$

9.4 Exact solution on the triangular lattice

We can consider the case of a triangular lattice in a similar fashion. In Fig. 9.3, calling site 3 the origin site, sites 1, 2, 6, 7, 8, 9 then become the six nearest-neighbor sites of the origin site 3. By a similar manipulation, as in the case of honeycomb lattice, we obtain

$$
\begin{aligned}
\langle 3[g] \rangle = C \langle (1 + 2 + 6 + 7 + 8 + 9)[g] \rangle + D \langle (126 + 129 + 289 \\
+ 789 + 167 + 678 + 169 + 168 + 128 + 127 + 269 + 279 \\
+ 189 + 689 + 178 + 278 + 267 + 679 + 268 + 179)[g] \rangle \\
+ E \langle (26789 + 16789 + 12789 + 12689 \\
+ 12679 + 12678)[g] \rangle, \quad 3 \notin [g],
\end{aligned} \tag{9.40}
$$

$$
\begin{aligned}
C &= \tfrac{1}{32}(\tanh 6L + 4 \tanh 4L + 5 \tanh 2L), \\
D &= \tfrac{1}{32}(\tanh 6L - 3 \tanh 2L), \\
E &= \tfrac{1}{32}(\tanh 6L - 4 \tanh 4L + 5 \tanh 2L),
\end{aligned} \tag{9.41}
$$

where L is the (dimensionless) interaction parameter of the triangular Ising model.

In order to find exact solutions for odd-number correlations on the triangular lattice, it is more advantageous first to increase the number of honeycomb lattice sites; more specifically, we now add site 6 to the previously considered "bow-tie" cluster of sites, 0, 1, 2, 3, 4, 5 (Fig. 9.1). Using the basic generating equation (9.5), identities are developed starting from the thermal averages

$$y_5 = \langle 036 \rangle, \quad y_{10} = \langle 026 \rangle, \quad y_{11} = \langle 01236 \rangle, \tag{9.42}$$

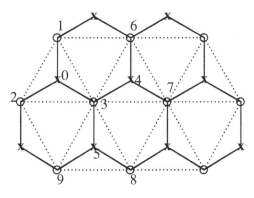

Fig. 9.3. Honeycomb lattice and its decomposition into two interlacing sublattices ("circle" sites and "x" sites, respectively), where each sublattice is triangular.

yielding

$$y_5 = A(y_1 + y_3) + (A + B)y_9, \tag{9.43}$$
$$y_{10} = A(y_1 + 2y_9) + By_3, \tag{9.44}$$
$$y_{11} = A(y_3 + 2y_9) + By_1, \tag{9.45}$$

where $y_9 = \langle 126 \rangle$ and use has been made of the degeneracy relation (9.16) in order to write (9.43). Equations (9.43)–(9.45) may be rewritten in the former ratio notation as

$$\xi_9 = (A + B)^{-1}[\xi_5 - A(1 + \xi_3)], \tag{9.46}$$
$$\xi_{10} = A(1 + 2\xi_9) + B\xi_3, \tag{9.47}$$
$$\xi_{11} = A(\xi_3 + 2\xi_9) + B. \tag{9.48}$$

Since the solutions for ξ_3, ξ_5 were already known, the arranged set of equations (9.46)–(9.48) directly advances the number of exactly known solutions to include ξ_9, ξ_{10}, ξ_{11}. As discussed in the previous section, the latter solutions ξ_9, ξ_{10}, ξ_{11} can now each be multiplied by the honeycomb spontaneous magnetization (9.21) to construct the corresponding exact solutions y_9, y_{10}, y_{11}, having critical amplitudes

$$A_9 = A_1 4(25\sqrt{3} - 43)/(5 - 2\sqrt{3}) = 0.983255\ldots, \tag{9.49}$$
$$A_{10} = A_1(423 - 242\sqrt{3})/3(5 - 2\sqrt{3}) = 1.045392\ldots, \tag{9.50}$$
$$A_{11} = A_1(348 - 199\sqrt{3})/3(5 - 2\sqrt{3}) = 0.903470\ldots, \tag{9.51}$$

where use has been made of expression (9.23) for A_1.

The number of honeycomb lattice sites under inspection is now further augmented so as to include sites 6, 7, 8, 9 (Fig. 9.3). Then, using the basic generating equation (9.6) for the triangular lattice, the following correlation identities are developed:

$$v_1 = 6Cv_1 + 2D(3v_3 + 6v_8 + v_9) + 6Ev_{10},$$
$$v_2 = 2C(v_1 + v_3 + v_8)$$
$$\quad + 2D(2v_1 + v_3 + 4v_8 + v_9 + 2v_{10}) + 2E(v_3 + v_8 + v_{10}),$$
$$v_3 = C(2v_1 + v_3 + 2v_8 + v_9)$$
$$\quad + 4D(2v_1 + v_3 + 2v_8 + v_{10}) + E(v_3 + 2v_8 + v_9 + 2v_{10}),$$
$$v_4 = 2C(v_1 + 2v_8)$$
$$\quad + 2D(2v_1 + 3v_3 + 2v_8 + v_9 + 2v_{10}) + 2E(2v_8 + v_{10}),$$
$$v_5 = 2C(v_3 + v_8 + v_{10})$$
$$\quad + 2D(2v_1 + v_3 + 4v_8 + v_9 + 2v_{10}) + 2E(v_1 + v_3 + v_8),$$
$$v_6 = C(v_3 + 2v_8 + v_9 + 2v_{10})$$
$$\quad + 4D(v_1 + v_3 + 2v_8 + v_{10}) + E(2v_1 + v_3 + 2v_8 + v_9),$$
$$v_7 = 6Cv_{10} + 2d(3v_3 + 6v_8 + v_9) + 6Ev_1,$$
$$v_{11} = 2C(2v_8 + v_{10})$$
$$\quad + 2D(2v_1 + 3v_3 + 2v_8 + v_9 + 2v_{10}) + 2E(v_1 + 2v_8), \tag{9.52}$$

where the triangular lattice thermal averages are diagrammatically represented in Fig. 9.4 and are defined upon the triangular lattice (**x** sites in Fig. 9.3) as

$$v_1 = y_1 = \langle 3 \rangle, \quad v_2 = y_3 = \langle 316 \rangle, \quad v_3 = y_9 = \langle 326 \rangle,$$
$$v_4 = \langle 318 \rangle, \ v_5 = \langle 31267 \rangle, \ v_6 = \langle 31268 \rangle, \ v_7 = \langle 3126789 \rangle,$$
$$v8 = \langle 128 \rangle, \ v_9 = \langle 179 \rangle, \ v_{10} = \langle 12679 \rangle, \ v_{11} = \langle 31278 \rangle. \tag{9.53}$$

Subtracting the v_2-equation from the v_3-equation in (9.52), one obtains

$$v_9 = v_3 + (C - 2D + E)^{-1}(v_3 - v_2), \tag{9.54}$$

which gives, after using the hyperbolic identity $C - 2D + E = -(2A)^{-1}B$ and the $Y - \Delta$ correspondence theorem (9.35), the result

$$y_{12} = v_9 = 2AB^{-1}y_3 + (1 - 2AB^{-1})y_9. \tag{9.55}$$

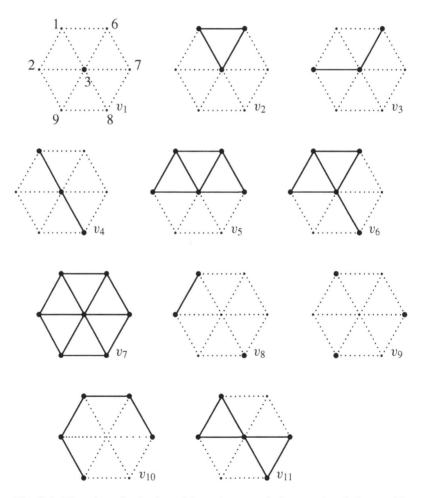

Fig. 9.4. The triangular lattice odd-number correlations $v_i, i = 1, 2, \ldots, 11$.

Since the solutions have already been found for both y_3 and y_9, (9.55) determines the exact solution for y_{12}, whose critical amplitude is evaluated using (9.24), (9.25) and (9.49) to be

$$A_{12} = A_1 4(152\sqrt{3} - 263)/(5 - 2\sqrt{3}) = 0.886821\dots. \qquad (9.56)$$

9.5 Identities for diamond and simple cubic lattices

9.5.1 Diamond lattice

$$\langle 0[f] \rangle = A\langle (1+2+3+4)[f] \rangle + B\langle (123+234+124+134)[f] \rangle, \qquad (9.57)$$

where

$$A = \tfrac{1}{8}(\tanh 4K + 2\tanh 2K),$$
$$B = \tfrac{1}{8}(\tanh 4K - 2\tanh 2K). \qquad (9.58)$$

Derivation of some linear identities for the diamond lattice Ising model correlation functions is left as Exercise 9.3. See also Appendix 5 for details.

9.5.2 Simple cubic lattice

$$\begin{aligned}
\langle 0[f] \rangle = {}& C\langle (1+2+3+4+5+6)[f] \rangle + D\langle (123+124+125 \\
& + 126+134+135+136+145+146+156+234+235 \\
& + 236+245+246+256+345+346+356+456)[f] \rangle \\
& + E\langle (12345+12346+12356+12456 \\
& + 13456+23456)[f] \rangle,
\end{aligned} \qquad (9.59)$$

where

$$C = \tfrac{1}{32}(\tanh 6K + 4\tanh 4K + 5\tanh 2K),$$
$$D = \tfrac{1}{32}(\tanh 6K - 3\tanh 2K),$$
$$E = \tfrac{1}{32}(\tanh 6K - 4\tanh 4K + 5\tanh 2K). \qquad (9.60)$$

These equations for the simple cubic lattice are formally identical with the equations for the triangular lattice because both lattices have the same lattice coordination number, although the correlation functions involved are different.

9.6 Systematic naming of correlation functions on the lattice

In the preceding sections, it was demonstrated that the basic generating equations of the Ising model played an interesting role in finding many exact identities for the Ising model correlation functions. The process of creating new identities can be continued indefinitely while introducing new correlation functions one after another. If the process of creating new identities is terminated after some steps, we find that the number of exact equations is always smaller than the number of independent correlation functions included. It means that the system of exact linear equations is not self-closed for the unknown correlation functions. In the case of the set of equations for the odd-number correlation functions, however, the number of equations is only one less than the number of unknowns. Fortunately, y_1, the smallest odd-number correlation function, is known exactly from the Naya (1954) formulation. When the exact y_1, which is known as a function of temperature, is utilized into the set, the set becomes closed and the honeycomb Ising problem may be solved exactly for the odd-number correlation functions.

The situation is different with the even-number correlation functions. We can again find a set of exact linear identities for even-number correlation functions by terminating the process of creating new equations after some steps. This set will not be self-closed unless a certain number, at least three, of exact even-number correlations are supplied as input information. The method of solution for the honeycomb even-number correlation functions will be available as website information.

Another interesting problem arising from the process of creating new identities is the way in which a newly appearing correlation function is given a numeric name definition. The numeric name definition, so far, has been carried out in an arbitrary way, resulting in some redundancies; a single correlation function was mistakenly given two or more different numeric names, and hence, later, only one of the numbers was retained. This can easily happen if one deals with 80 to 90 such correlation functions (Barry *et al.*, 1982, 1988, 1990, 1991). Since the process of utilizing the generating equation is so unambiguously defined, the process will be executed automatically by computer if a logical definition of naming the correlation function is devised.

Let us consider a group of lattice sites (Fig. 9.5) made up of the central site 3, the nearest-neighbor sites 0, 4, 5, and the second-neighbor sites 1, 2, 6, 7, 8 and 9. The group contains ten lattice sites. Site 0 is chosen as the origin site on the left hand side of the equation below and $[f]$ can be chosen from the remaining nine sites:

$$\langle 0[f] \rangle = A \langle [f](1+2+3) \rangle + B \langle [f]123 \rangle, \quad 0 \notin [f]. \tag{9.61}$$

When $[f]$ is a product of an odd number of Ising variables, identities relating the even-number correlation functions are obtained, and if $[f]$ contains an even number of Ising variables, identities relating only the odd-number correlation functions will result.

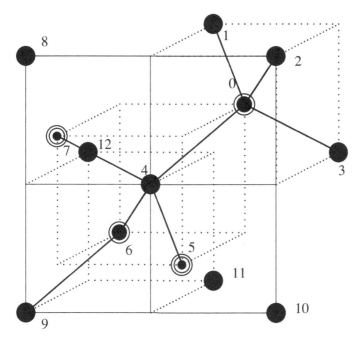

Fig. 9.5. The diamond lattice structure is constructed by starting out with the face centered cubic (fcc) structure. Filled circles (1, 2, 3, 4, 8, 9, 10, 11, 12) are the lattice sites on the fcc lattice. Sites (0, 6) are on the rear side and sites (5, 7) are on the front side of the cube face (2, 4, 8, 9, 10).

9.6.1 Characterization of correlation functions

In this subsection, different ways of characterizing the correlation functions of the two-dimensional lattices are presented, and eventually the most convenient representation will be found. The method presented here can be applied to the correlation functions on the three-dimensional lattices as well.

The terminology *graph* will be used in place of the correlation function in the following analysis.

Site-number representation

A graph may be characterized by its site numbers. For instance,

$$\langle 1459 \rangle \tag{9.62}$$

is an example of the 4-site graph. The representation (9.62) may be called the *site-number representation*. This representation has the advantage of direct appeal to human eyes, and hence it can be used for the purpose of first defining a graph. However, it does not provide any means of identifying or differentiating between two graphs made up of the same number of sites.

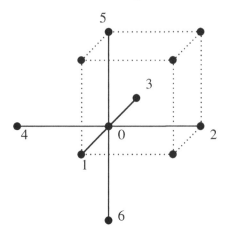

Fig. 9.6. Simple cubic lattice.

Occupation-number representation

One may introduce a representation in which all the occupied sites are indicated by 1's and all the unoccupied sites are represented by 0's. The representation (9.62) is now expressed as

$$\langle 0100110001 \rangle. \tag{9.63}$$

In this representation, all the graphs on the ten sites are represented by ten-digit numbers in the binary system. This by itself, again, does not provide any means of identifying or differentiating between two graphs.

Bond representation

The third way of characterizing a graph may be accomplished by listing all the bond lengths appearing in the graph. In this representation, the graph in (9.62) or (9.63) is characterized in the following way:

$$(123445), \tag{9.64}$$

where 1, 2, 3, etc. stand for, respectively, the nearest-, the second-, the third-, etc. neighbor bonds. These numbers may be called the *bond order number*. The representation (9.64) indicates that there are two fourth-order bonds and one each of the first, second, third and fifth bonds in the graph. Furthermore, the bond order numbers are arranged in increasing order.

Unfortunately, this representation is not a one-to-one representation, because there are several graphs which are represented by the same set of bond order numbers.

For instance, two different four-site graphs, which are defined by the site-number representations ⟨0289⟩ and ⟨0368⟩, are represented by the same bond representation (122345).

Vertex-number representation

A one step more complete characterization of the topology of a graph, again in terms of the bond order numbers, may be introduced as follows. For the sake of definiteness, let us take the same graph. The site-number representation of the first graph of Fig. 9.7 is ⟨0289⟩. One can find four sets of bond order numbers, in which the first set contains three bond order numbers starting from site 0 and ending at sites 2, 8 and 9, respectively. The second set contains three bond order numbers starting from site 2 and ending at sites 0, 8 and 9, and so on. In the actual bond order numbers, these four sets are:

$$[0] = (134), \quad [2] = (125), \quad [8] = (245), \quad [9] = (223), \quad (9.65)$$

where one should note the fact that the bond order numbers in each set are arranged again in increasing order. Now we read those four sets as four three-digit numbers and arrange them in increasing order, i.e.,

$$\langle 0289 \rangle = [(125), (134), (223), (245)]. \quad (9.66)$$

The second graph of Fig. 9.7 is similarly represented as

$$\langle 0368 \rangle = [(122), (134), (235), (245)]. \quad (9.67)$$

The representation defined in this way may be called the *vertex-number representation*, and the numbers (125), (134), (223), (245) are called the *vertex numbers*. One

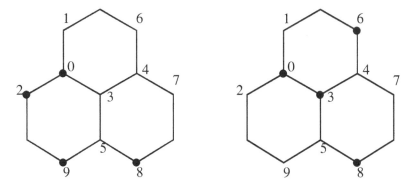

Fig. 9.7. Two graphs, ⟨0289⟩ and ⟨0368⟩, have the same bond order number representation; however, they have different vertex-number representations.

can clearly see that the degeneracy in the bond representations of graphs ⟨0289⟩ and ⟨0368⟩ is lifted in the vertex-number representations.

The vertex-number representation introduced thus far seems to be a complete representation for the purposes of identifying and differentiating between graphs which have different locations and orientations on the lattice. But now a question may arise.

9.6.2 Is the vertex-number representation an over-characterization?

The number of vertex numbers of a given graph is, by definition, equal to the number of vertices, i.e., the number of sites in the graph. In the process of actually establishing exact identities and naming various graphs on the two-dimensional honeycomb lattice, it is found that only the first three vertex numbers of each graph are needed for the purpose of identification and differentiation.

In order to analyze the situation clearly, let us consider two graphs, $(ABCDE)$ and $(ABCDE')$, on the two-dimensional square lattice (Fig. 9.8). Both of them are five-site graphs. The vertex-number representations of these graphs are:

$$(ABCDE) = [(1144), (1335), (1335), (345, 12), (345, 12)], \qquad (9.68)$$
$$(ABCDE') = [(1144), (1335), (1335), (3345), (3345)], \qquad (9.69)$$

where in (9.68) the bond order 12 appears because of the extension of the graph. Both graphs have the same vertex numbers of the first three vertices, A, B, and C, which are colinear. This is because E and E' are located symmetrically with respect to the line joining A, B, and C. Since A, B, and C are colinear the vertex numbers of A, B, and C may be called the colinear vertex numbers. The vertex D is not colinear with respect to A, B, and C. As soon as a noncolinear vertex number (that of D) is added to the sequence of colinear vertex numbers, the distinction between

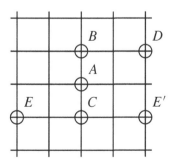

Fig. 9.8. Square lattice.

the two graphs is seen. This does not occur on the honeycomb lattice as long as the region of the lattice is limited to ten sites in Fig. 9.7, i.e., there is no line of symmetry on which there are three colinear sites. The argument may be generalized to the three-dimensional lattices. As soon as a noncoplanar vertex number is added, the degeneracy due to symmetry with respect to a plane is lifted.

In conclusion, it seems to be possible to establish a more efficient way of identifying two graphs. It is a combination of the bond representation and the vertex-number representaion.

A conjecture

(1) Two graphs must have the same bond representation.
(2) Two graphs must have the same pair of the first two vertex numbers.

Instead of a set of noncolinear vertex numbers, the bond representation is included in the criteria for the graph identity.

Some 184 correlation function identities, both the even and odd functions, for the honeycomb lattice are created by computer programs in which the systematic numerical naming is performed by means of the vertex-number representation. These identities are available on the website.

9.6.3 Computer evaluation of the correlation functions

Linear identities for both the even and odd correlation functions on the honeycomb lattice are created by computer programs in which the systematic ordering of the correlation function names is performed by means of the vertex-number representation. A list of 184 identities for the even-correlation functions, together with a FORTRAN program to evaluate the first 50 even-correlation functions, are available on the website.

Exercises
9.1 Establish the following identity:

$$\mathrm{Tr}_{s_0} \exp(K s_0(s_1 + s_2 + s_3)) = \Delta \exp(L(s_1 s_2 + s_2 s_3 + s_3 s_1)), \qquad \text{(E9.1)}$$

where

$$\Delta^4 = \exp(4L)(\exp(4L) + 3)^2, \qquad \text{(E9.2)}$$

and

$$2 \cosh 2K = \exp(4L) + 1. \qquad \text{(E9.3)}$$

[Hint]
If one sets

$$\mathrm{Tr}_{s_0} \exp(K s_0(s_1 + s_2 + s_3))$$
$$= 2\cosh K(s_1 + s_2 + s_3)$$
$$= A + B(s_1 s_2 + s_2 s_3 + s_3 s_1), \tag{E9.4}$$

it can be shown that

$$A = \tfrac{1}{2}(\cosh 3K + 3\cosh K), \quad B = \tfrac{1}{2}(\cosh 3K - \cosh K). \tag{E9.5}$$

Similarly, it can be shown that

$$\exp(L(s_1 s_2 + s_2 s_3 + s_3 s_1)) = C + D(s_1 s_2 + s_2 s_3 + s_3 s_1),$$
$$C = \tfrac{1}{4}(\exp(3L) + 3\exp(-L)), \quad D = \tfrac{1}{4}(\exp(3L) - \exp(-L)). \tag{E9.6}$$

Then,

$$\mathrm{Tr}_{s_0} \exp(K s_0(s_1 + s_2 + s_3))$$
$$= A + \frac{B}{D}[\exp(L(s_1 s_2 + s_2 s_3 + s_3 s_1)) - C]. \tag{E9.7}$$

If we further set

$$B \cdot C = A \cdot D, \tag{E9.8}$$

we finally find

$$\mathrm{Tr}_{s_0} \exp(K s_0(s_1 + s_2 + s_3)) = \Delta \exp(L(s_1 s_2 + s_2 s_3 + s_3 s_1)), \tag{E9.9}$$

where

$$\Delta = \frac{B}{D}. \tag{E9.10}$$

By some simple manipulations, we can show that (E9.8) and (E9.10) are equivalent to

$$2\cosh 2K = \exp(4L) + 1, \tag{E9.11}$$
$$\Delta^4 = \exp(4L)(\exp(4L) + 3)^2, \tag{E9.12}$$

respectively.

9.2 Show that the condition for vanishing of the spontaneous magnetization (9.21) leads to the critical temperature (9.22).
[Hint]
The trigonometric relations employed are:

$$\sinh^2 2K = \cosh^2 2K - 1, \quad 2\cosh^2 K = \cosh 2K + 1,$$
$$\cosh 3K = \cosh K(2\cosh 2K - 1); \tag{E9.13}$$

then,

$$\sinh^6 2K - 4\cosh 3K \cosh^3$$
$$= (\cosh 2K - 1)^3(\cosh 2K + 1)^3 - 4\cosh^4 K(2\cosh 2K - 1)$$
$$= (\cosh 2K - 1)^3(\cosh 2K + 1)^3$$
$$- (\cosh 2K + 1)^2(2\cosh 2K - 1) = 0; \qquad \text{(E9.14)}$$

hence

$$(\cosh 2K - 1)^3(\cosh 2K + 1) - (2\cosh 2K - 1)$$
$$= (\cosh^2 2K - 1)(\cosh^2 2K - 2\cosh 2K + 1)$$
$$- (2\cosh 2K - 1) = 0, \qquad \text{(E9.15)}$$

which leads to

$$\cosh 2K = 2, \quad 2K = \log(2 + \sqrt{3}). \qquad \text{(E9.16)}$$

9.3 As an exercise, derive the first few odd-number correlation function identities for the diamond lattice.

[Hint]

$$y_1 = 4Ay_1 + 4By_3,$$
$$y_2 = 2A(y_1 + y_3) + 2B(y_1 + y_3),$$
$$y_4 = 4Ay_3 + 4By_1, \qquad \text{(E9.17)}$$

where

$$y_1 = \langle 0 \rangle = \langle 1 \rangle = \langle 2 \rangle = \langle 3 \rangle = \langle 4 \rangle,$$
$$y_2 = \langle 012 \rangle = \langle 013 \rangle = \langle 014 \rangle \qquad \text{(E9.18)}$$
$$= \langle 023 \rangle = \langle 024 \rangle = \langle 034 \rangle,$$
$$y_3 = \langle 123 \rangle = \langle 234 \rangle = \langle 124 \rangle = \langle 134 \rangle,$$
$$y_4 = \langle 01234 \rangle. \qquad \text{(E9.19)}$$

See Appendix 5 for more details.

10

Propagation of short range order[†]

10.1 The radial distribution function

One of the main themes of the theories of classical liquids, of which monatomic rare gas liquid played a central role, is the oscillatory behavior of the *radial distribution function*. The cohesion mechanism of such a liquid is explained in terms of the interatomic potential between monatomic atoms which have a relatively simple electronic structure. The widely accepted *interatomic potential*, such as the *Lennard–Jones potential*, has a typical short range repulsive part and a somewhat longer range *van der Waals* attractive part, given by the so-called 6–12 *potential*

$$\phi(r) = \frac{A}{r^{12}} - \frac{B}{r^6}.\tag{10.1}$$

Let us now consider the molecular structure of the *monatomic liquid*. The number of atoms within a thin spherical shell of thickness dr at distance r from the position of an atom located at the origin is given by

$$4\pi\rho g(r)r^2 \mathrm{d}r, \quad \rho = \frac{N}{V},\tag{10.2}$$

where $g(r)$ is called the radial distribution function.

Experimentally, $g(r)$ is given by the Fourier transform of the angular distribution of the diffuse x-ray, or inelastic neutron, scattering intensity by a given liquid. Fig. 10.1 shows the radial distribution function for liquid argon. The position of the first peak is interpreted as roughly corresponding to the distance from the atom at the origin to the minimum of the Lennard–Jones potential. While the behavior of $g(r)$, which goes to zero rapidly at shorter distances, is understood naturally to be due to the repulsive part of the interatomic potential, the oscillatory behavior of the radial distribution function at distances larger than the first peak has not been explained in simple terms.

[†] This chapter is mostly reproduced from Y. Oota, T. Kakiuchi, S. Miyazima & T. Tanaka (1998), 'Propagation of short-range order in the lattice gas', *Physica* A, vol. 250, pp. 103–14, with permission from Elsevier Science.

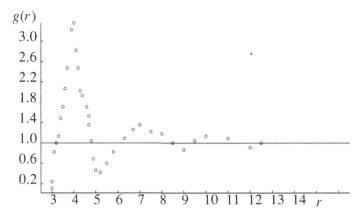

Fig. 10.1. Radial distribution function for liquid argon (schematic only; for more details, see Gingrich, 1945).

In the history of the development of statistical mechanics there was a certain period during which extensive studies of the radial distribution function were performed. The oscillatory behavior of the radial distribution function is certainly due to many-body correlation effects, and at least the effect of three-body correlation must be taken into account.

The general method of approach in the theory of the radial distribution function is to start with an exact equation defining the two-body distribution function, and then differentiate the equation with respect to a parameter. In the case of the formulation by Kirkwood (1946, 1947), the defining equation is differentiated with respect to the so-called coupling parameter. In the case of the formulations by Born & Green (1946) the same defining equation is differentiated with respect to the coordinate of the distribution function. As a result of differentiation, in either formulation, we obtain an integro-differential equation for the two-body distribution function.

Let us, in the following, reproduce the Born–Green equation. To start with, the reduced distribution functions are defined:

$$\int \rho^{(q+1)} d\boldsymbol{r}^{(q+1)} = (N - q)\rho^{(q)},$$

$$\rho^{(N)} = \frac{\exp(-\beta\Phi)}{Z_N},$$

$$Z_N = \frac{1}{N!} \int \exp(-\beta\Phi) d\boldsymbol{r}^{(1)} \cdots d\boldsymbol{r}^{(N)},$$

$$(N - 2)! \rho^{(2)}(\boldsymbol{r}, \boldsymbol{r}') = \frac{1}{Z_N} \int \exp(-\beta\Phi) d\boldsymbol{r}^{(3)} \cdots d\boldsymbol{r}^{(N)},$$

$$\rho^{(2)}(\boldsymbol{r}, \boldsymbol{r}') = \rho^2 g(|\boldsymbol{r} - \boldsymbol{r}'|), \tag{10.3}$$

where $\beta = 1/kT$ and Φ is the sum of the intermolecular potentials for the systems of N atoms,

$$\Phi = \sum_{i,j} \phi(r_i, r_j). \tag{10.4}$$

If the $\rho^{(2)}(r, r')$ equation in (10.3) is differentiated with respect to r', and if we use the relation

$$\nabla_{r'}\Phi = \nabla_{r'}\phi(r, r') + \sum_{k=3}^{N} \nabla_{r'}\phi(r', r_k), \tag{10.5}$$

we find the following equation:

$$-\frac{1}{\beta}\nabla_{r'} \log \rho^{(2)}(r, r') = \nabla_{r'}\phi(r, r') + \int \nabla_{r'}\phi(r', r''')\frac{\rho^{(3)}(r, r', r''')}{\rho^{(2)}(r, r')}dr'''. \tag{10.6}$$

This is an exact equation. In order to find $\rho^{(2)}$ explicitly from this equation, however, some knowledge of $\rho^{(3)}$ is required.

In the theories of the radial distribution function, Kirkwood & Buff (1943) introduced an assumption called the superposition approximation expressed by

$$\rho^3\rho^{(3)}(r, r', r''') = \rho^{(2)}(r, r')\rho^{(2)}(r', r''')\rho^{(2)}(r, r'''). \tag{10.7}$$

This is a plausible approximation which holds asymptotically in the limit of low density.

Born & Green adopted the same approximation in (10.6) and solved the resulting nonlinear integro-differential equation for the two-particle distribution function.

There is another method of approach to the theory of the structure of a monatomic liquid, i.e, the lattice gas theories of liquid. The method is based on the general idea that a continuous space coordinate variable may be approximately represented by a set of discrete values of the variable. The approximation may be improved step by step by changing the mesh size of the space coordinate variables. Hence the structure of a liquid can be described by the occupation numbers of atoms in the lattice if the grid size or the mesh size of the lattice is made sufficiently small.

In this chapter it will be demonstrated that the oscillatory behavior of the radial distribution function is strongly controlled by the repulsive part of the interatomic potential. This will be accomplished by means of examples of the superionic conductors AgI and Ag_2S.

10.2 Lattice structure of the superionic conductor αAgI

There is a group of ionic crystals called the type I superionic conductor of which AgI and Ag_2S are good examples. The ionic crystal AgI, for example, undergoes

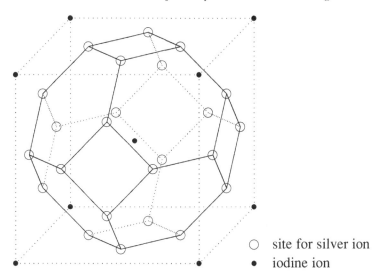

○ site for silver ion
● iodine ion

Fig. 10.2. Tetrahedral sites as vertices of a truncated octahedron. Open circles are the sites over which silver ions can move.

a phase transition from the low temperature, almost nonconducting *β-phase* to the high temperature, superionic conducting *α-phase* associated with a finite latent heat. The change of the electric conductivity at the phase transition is of the order of 10^6, and this is due to a sudden change in the motions of positive Ag ions at the transition.

In superionic conductors αAgI and αAg$_2$S the iodine or sulfur ions form a body centered cubic lattice and silver ions occupy mainly tetrahedrally coordinated positions in this body centered cubic lattice (Boyce *et al.*, 1977) (Fig. 10.2). In the *β-phase*, silver ions and iodine ions are arranged alternately and form a *zinc-blende* structure, and these ions do not move out of their lattice points. Thus the number of available lattice points for silver ions in *α-phase* becomes suddenly six times the actual number of silver ions at the transition point, resulting in some random arrangement of silver ions over the many equivalent lattice sites. The sudden change of the motion of silver ions at the transition point would, therefore, account for a major part of the latent heat. There is usually another source of latent heat such as the change of the vibrational motions; however, inelastic neutron scattering studies (Cava *et al.*, 1977; Hoshino *et al.*, 1977; Wright & Fender, 1977) performed for AgI and Ag$_2$S indicate that there is almost no change in the vibrational contribution at the transition point. For this reason it is assumed that the configurational disorder entropy is the only source of latent heat.

At this moment the connection between the structure of a monatomic liquid as described earlier and the distribution of silver ions in AgI must be mentioned briefly. Since the number of available sites for silver ions is six times the actual number of the silver ions in the lattice, the motion of silver ions is described in almost the same way as the motion of fluid. Hence the distribution of silver ions or the radial distribution function of silver ions must have some resemblance to that of the monatomic liquid.

If the experimentally observed values of the latent heat of transition (Notting, 1963; Beyeler & Strassler, 1979) in these systems is examined more closely, however, one finds that the motions of silver ions, over the tetrahedrally coordinated positions, are not entirely random, and indeed the distribution of silver ions seems to be strongly correlated. The correlated distribution of silver ions is also measured experimentally in the radial distribution functions of silver ions in those systems.

In the following, the disorder entropy of silver ions and the silver–silver pair correlation function in these systems are calculated based on the lattice gas model assuming that the silver ions occupy only the tetrahedral positions in a body centered cubic lattice.

In order to simplify calculations, the electrostatic Coulomb repulsion between silver ions is totally ignored and the overlap repulsion of silver ions is replaced by a rigid sphere repulsion.

Since the distribution of silver ions is limited mainly over the tetrahedral lattice sites, the cluster variation method is the most convenient method of approach. One can start with the mean-field approximation and then move to higher order approximations successively.

10.3 The mean-field approximation

As explained earlier, we assume that silver ions occupy only the tetrahedral sites in the body centered cubic lattice formed by iodine ions (see Fig. 10.2). Since all the tetrahedral sites are equivalent and are surrounded by iodines tetrahedrally, the attractive Coulomb potential of silver ions due to iodines is independent of the distribution of silver ions in the lattice and is strictly a constant. The energy term can be ignored, and hence the minimization of the variational potential is accomplished by the maximization of the entropy S,

$$S = -kG_N. \tag{10.8}$$

In the mean-field approximation we ignore the correlations of silver ions altogether, and hence the entropy S is expressed only by g_1,

$$-S/k = G_{6N} = 6Ng_1, \tag{10.9}$$

where N is the number of silver ions in the lattice, and we include $6N$ lattice sites because there are six times as many available sites for silver ions in αAgI.

Since one treats the motion of silver ions classically, the density matrices are all diagonal. The one-site density matrix has two elements,

$$\rho_1(1) = n, \quad \rho_1(2) = 1 - n, \tag{10.10}$$

where n is the particle density, which is equal to $\frac{1}{6}$ for αAgI and $\frac{1}{3}$ for αAg$_2$S. So the entropy per silver ion is

$$S/k = -6g_1, \quad g_1 = n \log(n) + (1 - n) \log(1 - n), \quad (10.11)$$

$$\text{AgI } n = \tfrac{1}{6}: \quad S = 2.703k \text{ (m-f)}, \quad 1.7440k \text{ (ex)},$$

$$\text{Ag}_2\text{S } n = \tfrac{1}{3}: \quad S = 3.819k \text{ (m-f)}, \quad 1.1185k \text{ (ex)},$$

where 'm-f' refers to the mean-field approximation, and 'ex' refers to the experimental value. The comparison shows clearly that the distribution of silver ions in the lattice is far from totally random, and that silver–silver correlations would be very important. The reason why the discrepancy is greater for αAg$_2$S is because the function g_1 has a maximum at $n = 0.5$ and the density $n = \frac{1}{3}$ is closer to 0.5 than $n = \frac{1}{6}$ to 0.5.

10.4 The pair approximation

The next-order approximation is to take the effect of pair correlations into account. In the pair approximation, however, the range of correlation is equal to the range of interaction, and hence two possibilities may be tested. One is the nearest-neighbor exclusion and the other is up to the second-neighbor exclusion. Here exclusion refers to the strong repulsive interaction when two silver ions come close to each other. Since in this simplified lattice gas model of the motion of silver ions the distance between two ions has only discrete values, a detailed functional form of the repulsive interaction is not important, and hence the repulsive interaction may be replaced by a well defined hard-core radius known from alloy studies. The known value of the hard-core diameter of silver ions is certainly larger than the nearest-neighbor distance between two tetrahedral sites, and could be close to the second-neighbor distance. For this reason it is worthwhile testing two cases of exclusion. The entropy per silver ion is then given by

$$-S/k = 6[g_1 + 2g_2(0, 1) + g_2(1, 2)], \quad (10.12)$$

where $g_2(0, 1)$ and $g_2(1, 2)$ are the cumulant functions for the first- and second-neighbor pairs, respectively (Fig. 10.3). The coefficient 2 in front of $g_2(0, 1)$ is the number of nearest neighbors per site, and $g_2(1, 2)$ appears only once because the number of second-neighbor bonds per lattice site is unity. The two-site cumulant functions are given by

$$g_2(0, 1) = x_1 \log x_1 + 2(n - x_1) \log(n - x_1) + (1 - 2n + x_1) - 2g_1,$$
$$g_2(1, 2) = x_2 \log x_2 + 2(n - x_2) \log(n - x_2) + (1 - 2n + x_2) - 2g_1, \quad (10.13)$$

where x_1 and x_2 are the nearest- and second-nearest-neighbor pair correlation

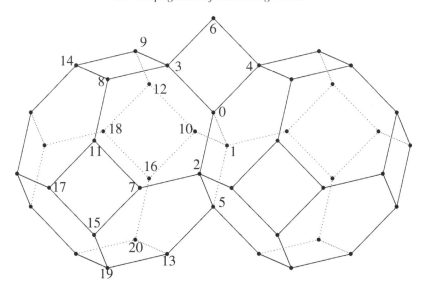

Fig. 10.3. A wide range view of lattice sites over which silver ions can move. The graph contains all the seven pair distances and 23 triangles.

functions, respectively. Here the pair correlation function is defined as the probability that the pair of sites are occupied simultaneously by silver ions.

In the case of the first-neighbor exclusion, we have

$$x_1 = 0, \quad x_2 = n^2, \quad g_2(1, 2) = 0; \tag{10.14}$$

and in the case of the second-neighbor exclusion,

$$x_1 = 0, \quad x_2 = 0. \tag{10.15}$$

These solutions give the following entropy values:

$$\text{AgI } n = \tfrac{1}{6}: \quad 2.301k \ (1) \quad 2.099k \ (2) \quad 1.7440k \ (\text{ex}),$$

$$\text{Ag}_2\text{S } n = \tfrac{1}{3}: \quad 1.726k \ (1) \quad 0.680k \ (2) \quad 1.1185k \ (\text{ex}), \tag{10.16}$$

where (1) and (2) mean the nearest- and second-neighbor exclusions, respectively. These values yield some useful information. The first-neighbor exclusion reduces the entropy substantially compared with the mean-field calculations. The reduction is more pronounced in the case of Ag_2S because of the higher particle density. If we include the effect of the second-neighbor exclusion, the entropy is further reduced. On comparison with the experimental values, we conclude that the second-neighbor exclusion is likely to occur in AgI. In the case of Ag_2S, however, the second-neighbor exclusion is too strong a condition, and only the nearest-neighbor exclusion is likely to be present.

The pair approximation can only be a guide, because the range of correlation is equal to the range of the interaction potential. A more detailed oscillatory behavior of the silver–silver radial distribution function can never be found unless one goes beyond this approximation.

10.5 Higher order correlation functions

The disorder entropy of αAgI will now be calculated in the approximation in which all the correlation functions containing up to the third-neighbor distances are included. The cumulant functions in this approximation may be found in the following form:

$$-S/k = 6[g_1 + 2g_2(0, 1) + g_2(1, 2) + 4g_2(2, 3) + 2g_3(0, 1, 2)$$
$$+ 4g_3(1, 2, 3) + \tfrac{4}{3}g_3(0, 7, 8) + \tfrac{1}{2}g_4(0, 3, 4, 6)$$
$$+ 4g_4(0, 1, 2, 3) + g_4(1, 2, 3, 4) + g_5(0, 1, 2, 3, 4)], \qquad (10.17)$$

where, for example, $g_3(1, 2, 3)$ indicates that the cluster is made up of three sites and their site numbers are 1, 2, and 3 (Fig. 10.3).

The cumulant functions, g, are now expressed in terms of the cluster functions, G. When this is done, the entropy will have the form:

$$-S/k = 6G_5(0, 1, 2, 3, 4) + 3G_4(0, 3, 4, 6) + 8G_3(0, 7, 8)$$
$$- 12G_3(0, 1, 2) - 24G_2(2, 3) + 24G_1. \qquad (10.18)$$

Let us now define the correlation functions appearing in the present formulation (Fig. 10.3):

$$x_1 = \langle n_0 n_1 \rangle = \langle n_1 n_5 \rangle, \quad \text{etc.}, \quad x_2 = \langle n_1 n_2 \rangle = \langle n_3 n_4 \rangle,$$
$$x_3 = \langle n_1 n_3 \rangle = \langle n_2 n_3 \rangle, \quad \text{etc.},$$
$$y_1 = \langle n_0 n_1 n_2 \rangle = \langle n_1 n_2 n_5 \rangle, \quad y_2 = \langle n_0 n_1 n_3 \rangle = \langle n_0 n_1 n_4 \rangle,$$
$$y_3 = \langle n_1 n_2 n_3 \rangle = \langle n_2 n_3 n_4 \rangle, \quad y_4 = \langle n_0 n_7 n_8 \rangle,$$
$$z_1 = \langle n_0 n_1 n_2 n_5 \rangle, \quad z_2 = \langle n_0 n_1 n_2 n_3 \rangle = \langle n_0 n_1 n_2 n_4 \rangle,$$
$$z_3 = \langle n_1 n_2 n_3 n_4 \rangle, \quad u_1 = \langle n_0 n_1 n_2 n_3 n_4 \rangle. \qquad (10.19)$$

The cluster functions are then given in terms of these correlation functions. As a demonstration, all the steps leading to the cluster function G_5 will be worked out. This cluster function is defined by

$$G_5 = \text{tr}\big[\rho_t^{(5)} \log \rho_t^{(5)}\big], \qquad (10.20)$$

where $\rho_t^{(5)}$ is the density matrix pertaining to sites 0, 1, 2, 3, 4 (Fig. 10.3). Since this is a five-site density matrix, there are $2^5 = 32$ elements. These elements are the

Table 10.1. *Coefficients of correlation functions in the four-site density matrix* $G_4(0, 3, 4, 6)$.

Element	Rest	x_1	x_2	x_3	y_1	y_2	y_3	z_1	g
$R(1111)$								1	1
$R(0111)$					1			-1	4
$R(0011)$		1				-2		1	4
$R(0101)$				1		-2		1	2
$R(0001)$	n	-2	-1			3		-1	4
$R(0000)$	$1 - 4n$	4	4			-4		1	1

The column g indicates the multiplicity of each element.

thermal average values of the respective projection operators made up of products of occupation numbers, and they are listed in Table 10.2. In Table 10.2, $R(1, 0, 1, 0, 1)$, for instance, should read

$$R(1, 0, 1, 0, 1) = \langle n_0(1 - n_1)n_2(1 - n_3)n_4 \rangle$$
$$= y_2 - 2z_2 + u_1. \tag{10.21}$$

10.5.1 AgI

If we compare the numerical results of the pair approximation with the experimental value of the entropy for AgI, we can surmise that the second-neighbor exclusion is rather likely to be the case. Proceeding with this assumption we now set all the correlation functions containing the first- and second-nearest neighbors equal to zero:

$$x_1 = x_2 = 0, \quad y_1 = y_2 = y_3 = 0, \quad z_1 = z_2 = z_3 = 0, \quad u_1 = 0. \tag{10.22}$$

When these values are substituted into (10.18) we find

$$\begin{aligned}
-S/k = {}& 24(n - 2x_3)\log(n - 2x_3) + 6(1 - 5n + 4x_3)\log(1 - 5n + 4x_3) \\
& + 8y_4 \log y_4 + 24(x_3 - y_4)\log(x_3 - y_4) \\
& + 24(n - 2x_3 + y_4)\log(n - 2x_3 + y_4) \\
& + 8(1 - 3n + 3x_3 - y_4)\log(1 - 3n + 3x_3 - y_4) \\
& - 48(n - x_3)\log(n - x_3) - 24(1 - 2n + x_3)\log(1 - 2n + x_3) \\
& + 6n \log n + 3(1 - 4n)\log(1 - 4n) \\
& - 12(1 - 3n)\log(1 - 3n) + 24(1 - n)\log(1 - n).
\end{aligned} \tag{10.23}$$

The entropy in this form is maximized with respect to x_3 and y_4. The value of the entropy is then

$$\text{AgI:} \quad S = 1.9785k, \quad 1.7440k \text{ (ex)},$$
$$x_3 = 0.03347, \quad \langle n_1 n_3 \rangle; \quad y_4 = 0.007918, \quad \langle n_4 n_6 n_9 \rangle. \tag{10.24}$$

Table 10.2. *Coefficients of correlation functions in the five-site density matrix* $G_5(0, 1, 2, 3, 4)$.

Element	Rest	x_1	x_2	x_3	y_1	y_2	y_3	z_2	z_3	u_1	g
$R(11111)$										1	1
$R(01111)$									1	−1	1
$R(10111)$								1		−1	4
$R(00111)$							1	−1	−1	1	4
$R(10011)$					1			−2		1	2
$R(10101)$						1		−2		1	4
$R(10001)$			1		−1	−2		3		−1	4
$R(00011)$			1		−1	−2		2	1	−1	2
$R(00101)$				1		−1	−2	2	1	−1	4
$R(00001)$	n	−1	−1	−2	1	2	3	−3	−1	1	4
$R(10000)$	n	−4			2	4		−4		1	1
$R(00000)$	$1 - 5n$	4	2	4	−2	−4	−4	4	1	−1	1

10.5.2 Ag$_2$S

If we again compare the numerical results of the pair approximation with the experimental value of the entropy for Ag$_2$S we can assume that the nearest-neighbor exclusion is likely to be the only case. We then set all the correlation functions containing the nearest-neighbor distance equal to zero:

$$x_1 = 0, \quad y_1 = y_2 = 0, \quad z_1 = z_2 = 0. \tag{10.25}$$

When these values are substituted into (10.18), we find

$$
\begin{aligned}
-S/k = {} & 6z_3 \log z_3 + 24(y_3 - z_3) \log(y_3 - z_3) \\
& + 12(x_2 - 2y_3 + z_3) \log(x_2 - 2y_3 + z_3) \\
& + 24(x_3 - 2y_3 + z_3) \log(x_3 - 2y_3 + z_3) \\
& + 24(n - x_2 - 2x_3 + 3y_3 - z_3) \log(n - x_2 - 2x_3 + 3y_3 - z_3) \\
& + 6(1 - 5n + 2x_2 + 4x_3 - 4y_3 + z_3) \log(1 - 5n + 2x_2 + 4x_3 - 4y_3 + z_3) \\
& - 6x_2 \log x_2 - 12(n - x_2) \log(n - x_2) \\
& + 3(1 - 4n + 2x_2) \log(1 - 4n + 2x_2) + 8y_4 \log y_4 \\
& + 24(x_3 - y_4) \log(x_3 - y_4) + 24(n - 2x_3 + y_4) \log(n - 2x_3 + y_4) \\
& + 8(1 - 3n_3x_3 - y_4) \log(1 - 3n_3x_3 - y_4) \\
& - 12(1 - 3n + x_2) \log(1 - 3n + x_2) - 24x_3 \log x_3 \\
& - 48(n - x_3) \log(n - x_3) - 24(1 - 2n + x_3) \log(1 - 2n + x_3) \\
& + 18n \log n + 24(1 - n) \log(1 - n). \tag{10.26}
\end{aligned}
$$

The entropy in this form is maximized with respect to x_2, x_3, y_3, y_4, and z_3. The value of the entropy is then

$$Ag_2S: \quad S = 1.149k, \quad 1.1185k \text{ (ex)}. \tag{10.27}$$

In summary, there are six times and three times more available lattice sites than the total number of silver ions in AgI and Ag_2S lattices, respectively, and hence a large disorder entropy is expected in each of these superionic systems. If, however, as demonstrated in these subsections, a totally random distribution of silver ions is assumed, the resulting entropy values, i.e., the mean-field values, become much larger than the experimental values. The distribution of silver ions is therefore much more restricted; this is possibly due to the nearest-neighbor exclusion or even the second-neighbor exclusions. This exclusion effect is interpreted as resulting from the relative size of the ionic diameter of silver ions compared with the nearest-neighbor or the second-neighbor distance in the lattice. Based on successively more careful calculations of the disorder entropy by means of cluster variation methods, it has been seen that the entropy value improves substantially compared with the mean-field calculation. The agreement between the five-site approximation and the experimental value is still not satisfactory; however, a qualitative interpretation of the situation is clear.

In order to see the propagation of short range order out to distant lattice points, many more pair correlation functions must be added to the entropy. It turns out, however, that these pair correlation functions are not coupled without intervention, at least, of the three-site correlation functions.

10.6 Oscillatory behavior of the radial distribution function

In order to be able to see the oscillatory behavior of the radial distribution function, three maxima and three minima for example, it is necessary to include the two-site correlation functions up to the seventh-neighbor distance. This may sound quite ambitious and mathematically involved; however, inclusion of three-site correlation functions does not present a problem because they are found almost trivially.

The entropy per lattice site of the following form is now proposed:

$$-S/k = g_1 + \sum_{i=1,7} c_2(i)g_2(i) + \sum_{i=1,23} c_3(i)g_3(i)$$

$$+ g_4(1, 2, 3, 4) + 4g_4(0, 1, 2, 3) + g_5(0, 1, 2, 3, 4). \tag{10.28}$$

The nature of approximation (10.28) is that the entropy contains the pair correlation functions up to the seventh-neighbor pair, all three-site corelation functions containing up to the seventh-neighbor distance, the smallest five-site cumulant function $g_5(01234)$ and its subclusters.

Table 10.3. *Pair correlation function; number of pairs per site.*[a]

Correlation function						c
x_1	(0,1) (1,5) (3,9) (8,11) (10,16) (15,17)	(0,2) (2,5) (4,6) (8,14) (11,17) (15,19)	(0,3) (2,7) (5,13) (9,12) (12,18) (16,18)	(0,4) (3,6) (7,11) (9,14) (13,19) (16,20)	(1,10) (3,8) (7,15) (10,12) (13,20)	2
x_2	(0,5) (7,17) (19,20)	(0,6) (8,9)	(1,2) (10,18)	(3,4) (11,15)	(3,14) (12,16)	1
x_3	(0,7) (1,12) (2,13) (5,10) (7,19) (13,16)	(0,8) (1,13) (2,15) (5,20) (8,17) (17,19)	(0,9) (1,16) (3,11) (6,8) (9,10) (18,20)	(0,10) (2,3) (3,12) (6,9) (9,18)	(1,3) (2,11) (5,7) (7,8) (10,20)	4
x_4	(0,11) (2,19) (7,13)	(0,12) (3,7) (10,13)	(1,9) (3,10) (14,17)	(1,20) (5,15) (14,18)	(2,8) (5,16)	2
x_5	(0,13) (2,6) (4,8) (9,11) (16,19)	(0,14) (2,10) (4,9) (9,16)	(1,6) (2,17) (6,14) (11,19)	(1,7) (3,5) (8,12) (12,20)	(1,18) (4,5) (8,15) (15,20)	4
x_6	(0,15) (2,20) (5,11) (10,14)	(0,16) (3,17) (5,12) (13,17)	(1,8) (3,18) (6,11) (13,18)	(1,19) (4,7) (6,12)	(2,9) (4,10) (7,14)	4
x_7	(0,17)	(0,18)	(0,19)	(0,20)		8

[a] In Table 10.3, (0,1) means that the pair correlation function connects the sites 0 and 1. The last column c is $c_2(i)$ in (10.28).

The newly introduced three-site density matrices, R_3, which appear in the three-site cumulant functions, g_3, have the same structure; and if we denote the three sites by a, b, and c, the matrix elements are given as follows:

$$R_3(1, 1, 1) = \langle n_a n_b n_c \rangle = y,$$
$$R_3(0, 1, 1) = \langle (1 - n_a) n_b n_c \rangle = a_1 - y,$$
$$R_3(1, 0, 1) = \langle n_a (1 - n_b) n_c \rangle = a_2 - y,$$
$$R_3(1, 1, 0) = \langle n_a n_b (1 - n_c) \rangle = a_3 - y,$$

$$R_3(1, 0, 0) = \langle n_a(1 - n_b)(1 - n_c)\rangle = n - a_2 - a_3 + y,$$
$$R_3(0, 1, 0) = \langle (1 - n_a)n_b(1 - n_c)\rangle = n - a_1 - a_3 + y,$$
$$R_3(0, 0, 1) = \langle (1 - n_a)(1 - n_b)n_c\rangle = n - a_1 - a_2 + y,$$
$$R_3(0, 0, 0) = \langle (1 - n_a)(1 - n_b)(1 - n_c)\rangle$$
$$= 1 - 3n + a_1 + a_2 + a_3 - y, \qquad (10.29)$$

where

$$a_1 = \langle n_b n_c\rangle, \;\; a_2 = \langle n_a n_c\rangle, \;\; a_3 = \langle n_a n_b\rangle, \;\; y = \langle n_a n_b n_c\rangle. \qquad (10.30)$$

When the extremum condition on the entropy with respect to the three-site correlation function, y, is imposed, and if y is not included in the five-site cluster (01234), we obtain the following equation:

$$y(n - a_1 - a_2 + y)(n - a_1 - a_3 + y)(n - a_2 - a_3 + y)$$
$$= (a_1 - y)(a_2 - y)(a_3 - y)(1 - 3n + a_1 + a_2 + a_3 - y). \qquad (10.31)$$

In (10.31) it is observed that the fourth-order terms of y cancel out on both sides of the equation, and hence y satisfies a cubic equation. Since the cubic equation is readily solvable (Abramowitz & Stegun, 1972), and hence the triangular correlations are trivially determined.

Table 10.4. *Three-site correlation functions which are zero.*[a]

Correlation function					c
$y_1(112)$	(0,1,2)	(0,3,4)	(0,2,5)	(0,1,5)	2
$y_2(113)$	(0,2,3)	(0,1,3)	(0,1,4)	(0,2,4)	4
	(0,2,7)	(0,3,8)			
$y_3(125)$	(0,1,6)	(0,2,6)	(0,3,5)	(0,4,5)	4
$y_4(134)$	(0,2,8)	(0,3,7)	(2,3,7)		8
$y_5(135)$	(0,1,7)	(1,5,7)	(0,2,10)		8
$y_6(136)$	(0,1,8)	(0,2,9)			8
$y_7(147)$	(0,1,20)	(0,2,19)			8
$y_8(157)$	(0,1,18)	(0,2,17)			8
$y_9(167)$	(0,1,19)	(0,2,20)			16

[a] In Table 10.4, $y_1(112)$ indicates that the three-site correlation function y_1 is made up of three sides of bond lengths 1, 1, 2, and symbol (0,1,2) indicates the three-site correlation function containing site numbers 0, 1, 2. The last column c is $c_3(i)$ in (10.28).

Table 10.5. *Three-site correlation functions which are not zero.*[a]

Correlation function					c
$y_{10}(233)$	(0,5,7)	(0,5,10)	(1,2,3)	(1,2,4)	4
$y_{11}(237)$	(0,5,19) (0,7,17)	(0,5,20) (0,10,18)	(0,6,7)	(0,6,10)	8
$y_{12}(246)$	(0,5,11) (0,11,15)	(0,5,12) (0,12,16)	(0,5,11)	(0,5,12)	8
$y_{13}(255)$	(3,4,5)				2
$y_{14}(277)$	(0,5,17)	(0,5,18)	(0,19,20)		4
$y_{15}(333)$	(0,7,8)	(0,9,10)			4/3
$y_{16}(337)$	(0,7,9) (0,9,18)	(0,7,19) (0,10,20)	(0,8,10)	(0,8,10)	8
$y_{17}(345)$	(0,7,10) (0,12,14)	(0,8,12)	(0,9,11)	(0,11,14)	8
$y_{18}(356)$	(0,7,14) (0,13,15)	(0,8,15) (0,13,16)	(0,9,16)	(0,10,14)	16
$y_{19}(377)$	(0,8,18)	(0,9,17)	(0,17,19)	(0,18,20)	12
$y_{20}(457)$	(0,11,19)	(0,12,20)	(0,14,17)	(0,14,18)	8
$y_{21}(557)$	(1,6,7)	(2,6,10)			8
$y_{22}(567)$	(0,14,15)	(0,14,16)	(0,15,20)	(0,16,19)	16
$y_{23}(777)$	(0,17,20)	(0,18,19)			8/3

[a] In Table 10.5, $y_{10}(233)$ indicates that the three-site correlation function y_{10} is made up of three sides of bond lengths 2, 3, 3, and symbol (0,5,7) indicates the three-site correlation function containing site numbers 0, 5, 7. The last column c is $c_3(i)$ in (10.28).

It should be noted that (10.28) contains $g_2(1)$, $g_3(1)$, ..., $g_3(9)$ which are functions of x_1, y_1, ..., y_9. These correlation functions are identically zero because of the nearest-neighbor exclusion; however, the corresponding cumulant functions are not zero, and hence contribute to the entropy.

By maximizing the entropy with respect to x_2, ..., y_7, y_{10} ..., y_{23}, and z_{1234} we find the following correlation functions:

$$x_2 = 0.05099, \quad x_3 = 0.02290, \quad x_4 = 0.04045,$$
$$x_5 = 0.02341, \quad x_6 = 0.037950, \quad x_7 = 0.02196,$$
$$y_{10} = 0.005146, \quad y_{11} = 0.005843, \quad y_{12} = 0.1498, \quad y_{13} = 0.006306,$$
$$y_{14} = 0.005627, \quad y_{15} = 0.002542, \quad y_{16} = 0.002431, \quad y_{17} = 0.004799,$$
$$y_{18} = 0.004477, \quad y_{19} = 0.002324, \quad y_{20} = 0.004612, \quad y_{21} = 0.002547,$$
$$y_{22} = 0.004299, \quad y_{23} = 0.002220, \quad z_{1234} = 0.0008724. \tag{10.32}$$

The five-site correlation function $w_{01234} = 0$ because it contains nearest-neighbor bonds.

10.7 Summary

A comparison of the results of several calculations with different degrees of approximation is presented here. In all of the calculations, it is assumed that the first-neighbor exclusion, i.e., the probability of occupying the nearest-neighbor sites simultaneously is set strictly to zero ($x_1 = 0$). In the following the values of radial distribution function at distances $\sqrt{2}$, $\sqrt{3}$, etc., up to $\sqrt{7}$ are listed.

(a) Mean-field approximation
   $S = 2.70$; this is an exception in which $g_1 = 1.0000$.
(b) Pair approximation
   $S = 2.301$, $g_1 = 0.0000$, $g_2 = 1.00000$.
(c) Triangles only, with the superposition approximation:
   $S = 2.269$, $g_1 = 0.00000$, $g_2 = 1.8271$, $g_3 = 0.8171$, $g_4 = 1.4955$,
   $g_5 = 0.8339$, $g_6 = 1.3981$, $g_7 = 0.7746$.
(d) Triangles only, without the superposition approximation
   $S = 2.248$, $g_1 = 0.00000$,
   $g_2 = 1.8378$, $g_3 = 0.8134$, $g_4 = 1.4975$, $g_5 = 0.8312$, $g_6 = 1.4001$,
   $g_7 = 0.7732$.
(e) Five-body cluster, no additional triangles:
   $S = 2.1565$, $g_1 = 0.00000$, $g_2 = 1.4165$, $g_3 = 1.2097$.
(f) Five-body cluster, with additional triangles (Fig. 10.4):
   $S = 2.072$, $g_1 = 0.00000$, $g_2 = 1.8357$, $g_3 = 0.8245$, $g_4 = 1.4562$, $g_5 = 0.8427$,
   $g_6 = 1.3661$, $g_7 = 0.7905$.

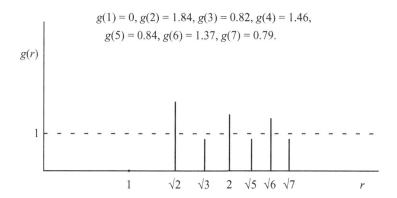

$g(1) = 0, g(2) = 1.84, g(3) = 0.82, g(4) = 1.46,$
$g(5) = 0.84, g(6) = 1.37, g(7) = 0.79.$

Fig. 10.4. The radial distribution function up to the seventh neighbor.

 (i) If judged from the value of the entropy, the three-site approximation without the super-position approximation seems to be slightly better than the one with the superposition approximation. Compare (c) with (d).

 (ii) The five-body calculation, even without additional triangles, gives a better value for the entropy than the triangle only approximation, although the peak of the radial distribution function is not as pronounced as in the simple triangle approximation. Compare (e) with (d).

(iii) The five-body calculation with additional triangles (f) seems to yield a rather satisfactory result both in terms of the oscillations of the radial distribution function and the value of the entropy.

(iv) The oscillation of the radial distribution function seems to be more pronounced compared with that which is observed in the rare gas liquid. This is acceptable because the superionic AgI is primarily a solid crystal, not a liquid.

 (v) The value of the calculated entropy $S = 2.072$ is still far removed from that observed ($S = 1.744$) in AgI; however, this may be due to the fact that we have neglected the effect of either soft-core or the hard-core extending out to the second neighbor. A symptom of this was already noted in the pair approximation in Sec. 10.4. It should be pointed out, however, that the entropy in the five-site cluster with triangles is smaller than the entropy with second-neighbor exclusion in the pair approximation. Whether the five-body correlation should be extended out to further neighbors or the second-neighbor exclusion should be imposed is undecided. This will be left for a future investigation.

A FORTRAN program which will produce the radial distribution function up to the seventh-neighbor distance is available on the website. This program contains a set of two subroutines, MINMAX and MATRIX. This set may be used in order to solve any coupled nonlinear equations which always appear in the application of the cluster variation method.

11

Phase transition of the two-dimensional Ising model

In this chapter a brief excursion through the method of the exact solution of the two-dimensional Ising model will be made. The two-dimensional Ising model on the square lattice is the first model which was solved exactly in the history of the progress of statistical mechanics (Onsager, 1944). The original formulation presented by Onsager, however, is based on an extensive knowledge of quarternion algebra, and hence the mathematical formulation employed is beyond the comprehension of most graduate students. Since Onsager's original formulation, many alternative and more comprehensive methods of solution have been presented. In this chapter the *Pfaffian* formulation, which is thought of as more tractable, will be presented.

11.1 The high temperature series expansion of the partition function

The partition function for an Ising model is expressed as

$$
\begin{aligned}
Z(\beta) &= \sum_{\sigma} \exp\left(K \sum_{(i,j)} \sigma_i \sigma_j\right) \\
&= \sum_{\sigma} \prod_{(i,j)} [\cosh K + (\sigma_i \sigma_j) \sinh K] \\
&= (\cosh K)^{qN/2} \sum_{\sigma} \prod_{(i,j)} [1 + x(\sigma_i \sigma_j)],
\end{aligned}
\tag{11.1}
$$

where N is the total number of sites in the lattice, $K = J/kT$, q is the number of nearest neighbors, $x = \tanh K$, the product is taken over all $qN/2$ nearest-neighbor pairs (i, j), and the summation over σ means

$$
\sum_{\sigma} = \sum_{\sigma_1 = \pm 1} \sum_{\sigma_2 = \pm 1} \cdots \sum_{\sigma_N = \pm 1}.
\tag{11.2}
$$

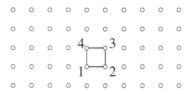

Fig. 11.1. Closed polygons on the square lattice.

When the product is expanded the partition function is expressed as an infinite series in x as follows:

$$Z = (\cosh K)^{qN/2} \sum_{\sigma} \prod_{(i,j)} (1 + x\,\sigma_i\sigma_j)$$

$$= (\cosh K)^{qN/2}$$

$$\cdot \sum_{\sigma} \left[1 + x \sum_{(i,j)}(\sigma_i\sigma_j) + x^2 \sum_{(i,j)}\sum_{(k,l)}(\sigma_i\sigma_j)(\sigma_k\sigma_l) + \cdots \right.$$

$$\left. + x^n \sum_{(p_1,q_1)} \cdots \sum_{(p_n,q_n)} \left(\sigma_{p_1}\sigma_{q_1}\right)\cdots\left(\sigma_{p_n}\sigma_{q_n}\right) + \cdots \right], \qquad (11.3)$$

where the x^n term is the product of n pairs of nearest-neighbor spin variables. The n pairs must each be different from one another.

When the summation, $\sum_{\sigma}$, is performed, all the pair products in which any spin variable appears an odd number of times vanish because of the property

$$\sum_{\sigma_i=\pm 1} \sigma_i^{2m+1} = \sum_{\sigma_i=\pm 1} \sigma_i = 0, \qquad (11.4)$$

and hence the only nonvanishing terms are such that all the pairs in the product can form closed polygons on the square lattice.[†]

In the case of the two-dimensional square lattice, the first nonvanishing term, besides the leading term of unity, in the infinite series is given by

$$x^4 \sum_{\sigma}(\sigma_1\sigma_2)(\sigma_2\sigma_3)(\sigma_3\sigma_4)(\sigma_4\sigma_1) = 2^N x^4. \qquad (11.5)$$

In this way, the partition function is represented as

$$Z = 2^N(\cosh K)^{2N}\left(1 + \sum_{n}\Omega_n x^n\right), \qquad (11.6)$$

where Ω_n is the number of n-sided closed polygons on the lattice.

[†] Here the closed polygon is a closed, one stroke figure which can be drawn on a sheet of paper without lifting the tip of the pen from the paper, going from and returning to the starting point.

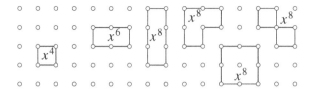

Fig. 11.2. Closed polygons contributing to x^4, x^6, and x^8.

The first few Ω_n are found rather easily:

$$\Omega_4 = N, \quad \Omega_6 = 2N, \quad \Omega_8 = \tfrac{1}{2}N^2 + \tfrac{9}{2}N, \quad \dots . \tag{11.7}$$

Ω_8 is made up of two parts: connected figures and disconnected figures. In Fig. 11.2, there are $7N$ connected figures. A disconnected figure is made up of two disjoint squares. With one square fixed there are $N - 5$ possible positions for the other square, giving a total of $\tfrac{1}{2}N(N - 5)$.

In this way the evaluation of the partition function for the square lattice Ising model is reduced to the counting of the number of closed polygons on the square lattice.

11.2 The Pfaffian for the Ising partition function

11.2.1 Lattice terminals

In the previous section it was shown that the partition function can be expressed as an infinite series (11.6) in terms of $x = \tanh J/kT$. The first few Ω_n can be found rather easily; however, it becomes more and more difficult to find them for large n. This method of calculating successively higher order terms of the partition function is called the *combinatorial formulation* of the Ising problem.

It was first noticed by Hurst & Green (1960) that the combinatorial formulation can be avoided if one uses the Pfaffian formulation, and subsequently the Pfaffian formulation was applied to the two-dimensional Ising model as well as to the two-dimensional dimer problem, and hence the celebrated Onsager result was reformulated (Kasteleyn, 1961, 1963; Montroll *et al.*, 1963; Stephenson, 1964, 1966, 1970). A full acount of the Pfaffian formulation of the two-dimensional Ising model can be found in Green & Hurst (1964) and McCoy & Wu (1973).

Pfaffians were known to mathematicians during the 19th century (Thompson & Tait, 1879); however, since they appeared to be mathematically uninteresting, they were rapidly forgotten.

If we examine the square lattice, on which any number of closed polygons have been drawn, we find that the number of bonds connecting each lattice point to neighboring points of the lattice is either zero, two, or four. Therefore we introduce,

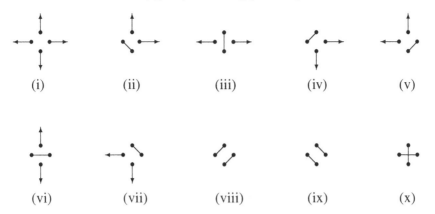

Fig. 11.3. Different modes of connection between terminals.

in the immediate neighborhood of each lattice point, a group of four *terminals*, which can serve as the ends of bonds that are connected to neighboring lattice points.

Four terminals thus created at a lattice point (α, β) are given the names: $E_{\alpha,\beta}$, $N_{\alpha,\beta}$, $W_{\alpha,\beta}$, $S_{\alpha,\beta}$, $\alpha = 1, \ldots, m$, $\beta = 1, \ldots, n$. If the nearest-neighbor sites are connected by the Ising interaction in the horizontal direction, the pair is represented by a bond connecting terminals $E_{\alpha,\beta}$ and $W_{\alpha+1,\beta}$, etc. If the number of bonds connecting a lattice point to its neighbors is two, only two terminals are used in the bond formation. The remaining two terminals may then be connected by a shorter bond as illustrated in Fig. 11.4.

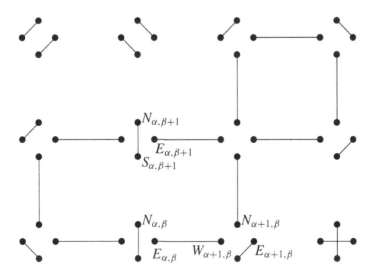

Fig. 11.4. A polygon formed by connecting terminals.

It is worthwhile mentioning that doubly periodic boundary conditions will be imposed in order to eliminate edge effects. This means that the coordinates of lattice points satisfy the conditions

$$(\alpha + m, \beta + n) = (\alpha, \beta), \quad \alpha = 1, \ldots, m, \quad \beta = 1, \ldots, n. \quad (11.8)$$

In this way, all four terminals at each lattice point are used up. This process is called the terminal closing. If the number of bonds connecting one lattice point to its neighbors is zero, there are three different ways to close terminals: two parallel closings and one cross closing (Fig. 11.4).

Once all the terminals in the entire lattice are closed in one way or another, all $4N$ terminals are occupied by dimers in such a way that there are no dimers touching each other and there is no terminal which is not occupied by a dimer. If we could count all the different dimer configurations in the expanded lattice, we should be able to solve the square lattice Ising problem. The dimer problem was first investigated by Kasteleyn (1961) and Temperley & Fisher (1961), who independently solved the dimer problem with the aid of Pfaffians.

11.2.2 The Pfaffian

Let us consider a $2n \times 2n$ antisymmetric matrix A with elements $a(i, j) = -a(j, i)$. The Pfaffian of A, PfA, is roughly half a determinant and is written in the following way:

$$\text{Pf}A = \begin{vmatrix} a(1, 2) & a(1, 3) & \cdots & a(1, 2n) \\ & a(2, 3) & \cdots & a(2, 2n) \\ & & & \vdots \\ & & & a(2n-1, 2n) \end{vmatrix}. \quad (11.9)$$

The precise definition of PfA is

$$\text{Pf}A = \sum_P \delta_P \, a(p_1, p_2) a(p_3, p_4) \cdots a(p_{2n-1}, p_{2n}), \quad (11.10)$$

where the sum is over permutations $(p_1, p_2, \ldots, p_{2n})$ of integers $(1, 2, \ldots, 2n)$ with the restrictions that

$$p_1 < p_2, \; p_3 < p_4, \ldots, \; p_{2n-1} < p_{2n}, \quad (11.11)$$

and

$$p_1 < p_3 < \cdots < p_{2n-1}. \quad (11.12)$$

δ_P, which is called the signature, in (11.10) is $+1$ if the permutation P is even, and -1 if P is odd. For example, when $n = 2$ it is easily verified that

$$\begin{vmatrix} a(1,2) & a(1,3) & a(1,4) \\ & a(2,3) & a(2,4) \\ & & a(3,4) \end{vmatrix} = a(1,2)\,a\,(3,4) - a(1,3)\,a\,(2,4) + a(1,4)\,a\,(2,3).$$

$$(11.13)$$

It can also be verified readily that the Pfaffian is equal to the square root of the antisymmetric determinant (for a general proof see Green & Hurst, 1964):

$$\begin{vmatrix} 0 & a(1,2) & a(1,3) & a(1,4) \\ -a(1,2) & 0 & a(2,3) & a(2,4) \\ -a(1,3) & -a(2,3) & 0 & a(3,4) \\ -a(1,4) & -a(2,4) & -a(3,4) & 0 \end{vmatrix}$$

$$= (a(1,2)\,a\,(3,4) - a(1,3)\,a\,(2,4) + a(1,4)\,a\,(2,3))^2. \quad (11.14)$$

For the sake of demonstration, let us calculate the Ising model partition function for the system made up only of four sites on the square lattice by means of the Pfaffian. The final answer is, of course, trivially known to be:

$$\frac{Z_4(\beta)}{2^4(\cosh K)^8} = 1 + x^4. \quad (11.15)$$

In the above equation, the first term of *unity* is the contribution of the spin configuration in which no site is occupied by the Ising bond and the term of x^4 is the contribution of the square as the smallest polygon formed on the square lattice.

The first question is to see how the first term comes about in the context of the dimer problem. When an Ising lattice site is not occupied, the four terminals of the site must be closed. As explained earlier, there are three ways to close: two parallel closings and one cross closing. The partition function for a single site is represented by

$$Q = \begin{vmatrix} 1 & -1 & 1 \\ & 1 & -1 \\ & & 1 \end{vmatrix} = 1, \quad (11.16)$$

i.e., the contributions of three closings cancel one another.

The Pfaffian for the four-spin system, i.e., the partition function, is

$$Z = \begin{vmatrix}
1 & -1 & 1 & 0 & 0 & -x & 0 & 0 & 0 & 0 & 0 & 0 & 0 & 0 & 0 \\
  & 1 & -1 & 0 & 0 & 0 & 0 & 0 & 0 & -y & 0 & 0 & 0 & 0 & 0 \\
  &   & 1 & 0 & 0 & 0 & 0 & 0 & 0 & 0 & 0 & 0 & 0 & 0 & 0 \\
  &   &   & 0 & 0 & 0 & 0 & 0 & 0 & 0 & 0 & 0 & 0 & 0 & 0 \\
  &   &   &   & 1 & -1 & 1 & 0 & 0 & 0 & 0 & 0 & 0 & 0 & 0 \\
  &   &   &   &   & 1 & -1 & 0 & 0 & 0 & 0 & 0 & 0 & 0 & -y \\
  &   &   &   &   &   & 1 & 0 & 0 & 0 & 0 & 0 & 0 & 0 & 0 \\
  &   &   &   &   &   &   & 0 & 0 & 0 & 0 & 0 & 0 & 0 & 0 \\
  &   &   &   &   &   &   &   & 1 & -1 & 1 & 0 & 0 & -x & 0 \\
  &   &   &   &   &   &   &   &   & 1 & -1 & 0 & 0 & 0 & 0 \\
  &   &   &   &   &   &   &   &   &   & 1 & 0 & 0 & 0 & 0 \\
  &   &   &   &   &   &   &   &   &   &   & 0 & 0 & 0 & 0 \\
  &   &   &   &   &   &   &   &   &   &   &   & 1 & -1 & 1 \\
  &   &   &   &   &   &   &   &   &   &   &   &   & 1 & -1 \\
  &   &   &   &   &   &   &   &   &   &   &   &   &   & 1 \\
\end{vmatrix},$$

$$(11.17)$$

where the Pfaffian Q is arranged along the diagonal, and other nonzero elements are

$$a(1,7) = a(9,15) = -x, \quad a(2,12) = a(6,16) = -y,$$
$$x = \tanh K_1 = \tanh J_1/kT, \quad y = \tanh K_2 = \tanh J_2/kT, \quad (11.18)$$

and x and y have the same value $\tanh K$ for an isotropic square lattice; however, they are represented by different mathematical symbols in order to indicate that x represents a terminal connection between E_i and W_{i+1} in the horizontal direction

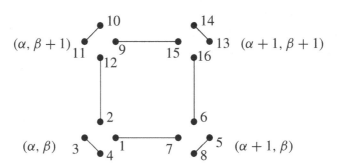

Fig. 11.5. A small system made up of four sites.

with the interaction constant $K_1 = J_1/kT$, and y represents a terminal connection between N_i and S_{i+m} in the vertical direction with $K_2 = J_2/kT$.

The above Pfaffian is evaluated as the square root of the associated antisymmetric determinant to obtain $Z = 1 + x^4$, or simply noting the fact that there are only two nonzero terms when the Pfaffian is expanded, i.e.,

$$Z = Q^4 + a(1, 7)a(2, 12)a(3, 4)a(5, 8)a(6, 16)a(9, 15)a(10, 11)a(13, 14), \tag{11.19}$$

where the sign of the second term is positive according to the signature δ_P in the definition of the Pfaffian (11.10). The Q^4 term is actually the contribution of 3^4 dimer configurations.

11.3 Exact partition function

Let us consider a square lattice of $m \times n = N$ sites with the double cyclid boundary conditions shown in Fig. 11.6. The integer m is the number of lattice sites contained in one ring of lattice sites on the surface of a torus. The partition function for this lattice can be written as

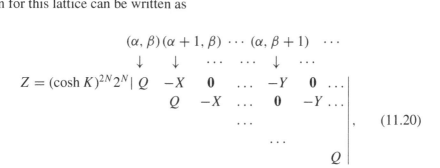

$$\begin{array}{cccccc} & (\alpha, \beta)\,(\alpha+1, \beta) & \cdots & (\alpha, \beta+1) & \cdots \\ & \downarrow \quad\quad \downarrow & \cdots \quad \cdots & \downarrow & \cdots \\ Z = (\cosh K)^{2N} 2^N \Big| \begin{array}{cccccc} Q & -X & 0 & \cdots & -Y & 0 \cdots \\ & Q & -X & \cdots & 0 & -Y \cdots \\ & & & \cdots \\ & & & & \cdots \\ & & & & & Q \end{array} \Big|, \end{array} \tag{11.20}$$

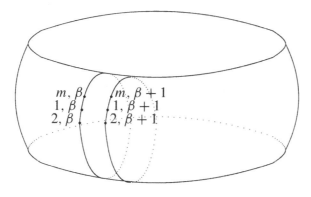

$$m, \beta \quad\quad m, \beta+1$$
$$1, \beta \quad\quad 1, \beta+1$$
$$2, \beta \cdots 2, \beta+1$$

Fig. 11.6. The lattice is wound circularly on a torus to eliminate edge effects.

where

$$X = \begin{bmatrix} 0 & 0 & x & 0 \\ 0 & 0 & 0 & 0 \\ 0 & 0 & 0 & 0 \\ 0 & 0 & 0 & 0 \end{bmatrix}, \qquad Y = \begin{bmatrix} 0 & 0 & 0 & 0 \\ 0 & 0 & 0 & y \\ 0 & 0 & 0 & 0 \\ 0 & 0 & 0 & 0 \end{bmatrix}. \tag{11.21}$$

If the Pfaffian Q appears at (α, β), $(\alpha + 1, \beta)$, etc. positions, the matrix Y appears at $(\alpha, \beta + 1)$, $(\alpha + 1, \beta + 1)$, etc. positions in the above super-Pfaffian, due to the torus boundary condition of the square lattice. The dimension of this Pfaffian is $4 \times N$.

In order to calculate the Pfaffian, it is most convenient to calculate the square root of the determinant of the associated antisymmetric matrix:

$$\frac{Z^2}{(\cosh K)^{4N} 2^{2N}} = |A|, \tag{11.22}$$

given by

$$A = \begin{bmatrix} D & -X & 0 & \cdots & -Y & 0 & \cdots \\ X' & D & -X & \cdots & 0 & -Y & \cdots \\ 0 & X' & D & \cdots & & & \\ \cdots & \cdots & \cdots & \cdots & \cdots & \cdots & \cdots \\ \cdots & \cdots & \cdots & \cdots & \cdots & \cdots & \\ \cdots & Y' & \cdots & \cdots & X' & D & -X \\ \cdots & \cdots & Y' & \cdots & \cdots & X' & D \end{bmatrix}, \tag{11.23}$$

where D is an antisymmetric matrix given by

$$D = \begin{bmatrix} 0 & 1 & -1 & 1 \\ -1 & 0 & 1 & -1 \\ 1 & -1 & 0 & 1 \\ -1 & 1 & -1 & 0 \end{bmatrix}, \tag{11.24}$$

and where X' and Y' are the transposed matrices of X and Y, respectively. The matrix A is also represented by

$$A = D \otimes I - X \otimes \Omega + X' \otimes \Omega^{-1} - Y \otimes \Omega^m + Y' \otimes \Omega^{-m}, \tag{11.25}$$

where Ω is a pseudo-cyclic matrix in N dimensions [Exercises 11.2 and 11.3].

Since the determinant of a matrix is the product of the eigenvalues of that matrix, the strategy one should follow is to find the eigenvalues of the above matrix (11.23).

The eigenvalue problem associated with A can be expressed as

$$\sum_{s=1}^{4N} A_{r,s} \Psi_s^{(q)} = E^{(q)} \Psi_r^{(q)} \qquad r = 1, 2, \ldots, 4N, \tag{11.26}$$

where there are $4N$ eigenvalues $E^{(q)}$, and the associated eigenfunctions $\Psi^{(q)}$, $q = 1, \ldots, 4N$, and

$$A_{r,s} = -A_{s,r}. \tag{11.27}$$

Since many of the elements of matrix A are zero, the foregoing equation may be expressed in terms of only nonvanishing elements of A and four-dimensional vectors $\Psi_{(\alpha,\beta)}^{(q)}$ in the following form

$$D\Psi_{(\alpha,\beta)}^{(q)} - X\Psi_{(\alpha+1,\beta)}^{(q)} + X'\Psi_{(\alpha-1,\beta)}^{(q)}$$
$$-Y\Psi_{(\alpha,\beta+1)}^{(q)} + Y'\Psi_{(\alpha,\beta-1)}^{(q)} = E^{(q)}\Psi_{(\alpha,\beta)}^{(q)},$$
$$\alpha = 1, 2, \ldots, m, \qquad \beta = 1, 2, \ldots, n. \tag{11.28}$$

Due to the pseudo-cyclic nature of the matrix A, $\Psi_{(\alpha,\beta)}^{(q)}$ satisfies the boundary condition

$$\Psi_{(\alpha,\beta+n)}^{(q)} = -\Psi_{(\alpha,\beta)}^{(q)}. \tag{11.29}$$

We can see that (11.28) and (11.29) may both be satisfied by the substitution

$$\Psi_{(\alpha,\beta)}^{(q)} = \omega_p^{[\alpha+(\beta-1)m]}\Psi^{(q)}, \tag{11.30}$$

provided that the following eigenvalue equation is satisfied:

$$\left(D - \omega_p X + \omega_p^{-1} X' - \omega_p^m Y + \omega_p^{-m} Y'\right)\Psi^{(q)} = E^{(q)}\Psi^{(q)}. \tag{11.31}$$

Because of (11.30), ω_p may be any one of the Nth roots of -1:

$$\omega_p = \exp[(2p - 1)\pi i/N], \qquad p = 1, 2, \ldots, N. \tag{11.32}$$

The eigenvalue problem (11.31) shows that there are four eigenvalues $E^{(q)}$ corresponding to each root ω_p, and that these eigenvalues are also eigenvalues of the matrix

$$z\left(\omega_p, \omega_p^m\right) = D - \omega_p X + \omega_p^{-1} X' - \omega_p^m Y + \omega_p^{-m} Y'$$

$$= \begin{bmatrix} 0 & 1 & -1 - \omega_p x & 1 \\ -1 & 0 & 1 & -1 - \omega_p^m y \\ 1 + \omega_p^{-1} x & -1 & 0 & 1 \\ -1 & 1 + \omega_p^{-m} y & -1 & 0 \end{bmatrix}. \tag{11.33}$$

Since only the products of the eigenvalues are required, and the product of the four eigenvalues of z is simply the determinant $|z(\omega_p, \omega_p^m)|$ of the matrix, the product of the $4N$ eigenvalues of A, or the determinant, is given by

$$|A| = \prod_{p=1}^{N} |z(\omega_p, \omega_p^m)|$$

$$= \prod_{p=1}^{N} [(1 + x^2)(1 + y^2) - (\omega_p + \omega_p^{-1})x(1 - y^2) - (\omega_p^m + \omega_p^{-m})y(1 - x^2)].$$

$$(11.34)$$

The partition function can be expressed by

$$\log Z = N \log 2 + N \log(\cosh K_1) + N \log(\cosh K_2) + \tfrac{1}{2} \log |A|,$$

$$\log |A| = \sum_p \log |z(\omega_p, \omega_p^m)|$$

$$= \sum_p \log [(1 + x^2)(1 + y^2)$$

$$- (\omega_p + \omega_p^{-1})x(1 - y^2) - (\omega_p^m + \omega_p^{-m})y(1 - x^2)]. \quad (11.35)$$

If p is represented by j and k in the following way:

$$p = k + (j - 1)m, \quad (p = 1, \ldots, N); \quad (k = 1, \ldots, m; j = 1, \ldots, n),$$

$$(11.36)$$

the summation over p can be represented by the double summation over j and k; and

$$\omega_p + \omega_p^{-1} = 2 \cos \left(\frac{2(j-1)\pi}{m} + \frac{(2k-1)\pi}{N} \right),$$

$$\omega_p^m + \omega_p^{-m} = 2 \cos \left(2(j-1)\pi + \frac{(2k-1)\pi}{n} \right). \quad (11.37)$$

In the limit as n and m go to infinity, the summations are replaced by integrations:

$$\log |A| = \frac{N}{4\pi^2} \int_0^{2\pi} \int_0^{2\pi} \log[(1 + x^2)(1 + y^2)$$

$$- 2x \cos \theta(1 - y^2) - 2y \cos \phi(1 - x^2)] \, d\theta \, d\phi$$

$$= \frac{N}{4\pi^2} \int_0^{2\pi} \int_0^{2\pi} \log(\cosh 2K_1 \cosh 2K_2$$

$$- \sinh 2K_2 \cos \theta - \sinh 2K_1 \cos \phi) \, d\theta \, d\phi. \quad (11.38)$$

This is Onsager's celebrated formula (Onsager, 1944) for the partition function of the rectangular Ising lattice.

Since there is no more advantage in differentiating between the horizontal and vertical interactions, we can consider the isotropic case:

$$x = \tanh J_1/kT = y = \tanh J_2/kT = \tanh K = \tanh J/kT. \quad (11.39)$$

The Helmholtz potential (per lattice site) is now given by

$$-F/kT = \log(2) + \frac{1}{2\pi^2} \int_0^\pi \int_0^\pi \log[\cosh^2 2K$$
$$- \sinh 2K(\cos\theta_1 + \cos\theta_2)] \, d\theta_1 \, d\theta_2. \quad (11.40)$$

The internal energy U is

$$U = -kT^2 \frac{\partial}{\partial T} \frac{F}{kT}$$
$$= J \frac{\partial}{\partial K} \frac{F}{kT}$$
$$= -J \coth 2K$$
$$\times \left[1 + (\sinh^2 2K - 1) \frac{1}{\pi^2} \int_0^\pi \int_0^\pi \frac{d\theta_1 d\theta_2}{\cosh^2 2K - \sinh 2K(\cos\theta_1 + \cos\theta_2)} \right].$$
$$(11.41)$$

The integral in (11.41) is easily evaluated by means of the transformation of the integration variables:

$$\theta_1 = \phi_1 - \phi_2, \quad \theta_2 = \phi_1 + \phi_2, \quad \frac{D(\theta_1, \theta_2)}{D(\phi_1, \phi_2)} = 2,$$

$$\frac{1}{\pi^2} \int_0^\pi \int_0^\pi \frac{d\theta_1 d\theta_2}{\cosh^2 2K - \sinh 2K(\cos\theta_1 + \cos\theta_2)}$$
$$= \frac{2}{\pi^2} \int_0^\pi \int_{-\pi/2}^{\pi/2} \frac{d\phi_1 d\phi_2}{\cosh^2 2K - 2\sinh 2K \cos\phi_1 \cos\phi_2}$$
$$= \frac{1}{\cosh^2 2K} \frac{1}{\pi} \int_{-\pi/2}^{\pi/2} (1 - k_1^2 \cos^2\phi_2)^{-1/2} d\phi_2$$
$$= \frac{1}{\cosh^2 2K} \frac{2}{\pi} K(k_1), \quad (11.42)$$

where $K(k_1)$ is the complete elliptic integral of the first kind and k_1 is the modulus given by[†]

$$K(k_1) = \int_0^{\pi/2} \left(1 - k_1^2 \sin\theta\right)^{-1/2} d\theta,$$

$$k_1 = \frac{2\sinh 2K}{\cosh^2 2K}. \tag{11.43}$$

The internal energy is now given by

$$U = -J \coth 2K \left[1 + \frac{2}{\pi}(2\tanh^2 2K - 1)K(k_1)\right]. \tag{11.44}$$

The specific heat is obtained from the internal energy:

$$C = \frac{2k}{\pi}(K\coth 2K)^2 \left[2K(k_1) - 2E(k_1) \right.$$
$$\left. - 2(1 - \tanh^2 2K)\left(\frac{\pi}{2} + (2\tanh^2 2K - 1)K(k_1)\right)\right], \tag{11.45}$$

where $E(k_1)$ is the complete elliptic integral of the second kind, arising from the differentiation of $K(k_1)$ and defined by

$$E(k_1) = \int_0^{\pi/2} \left(1 - k_1^2 \sin^2\theta\right)^{\frac{1}{2}} d\theta. \tag{11.46}$$

In the neighborhood of $k_1 = 1$ (Abramowitz & Stegun, 1972),

$$K(k_1) \approx \log\left[4\left(1 - k_1^2\right)^{-\frac{1}{2}}\right], \tag{11.47}$$

so from the exact result (11.45) we obtain a logarithmically divergent specific heat. The phase transition temperature is given by

$$k_1 = \frac{2\sinh 2K}{\cosh^2 2K} = 1, \quad \text{i.e., } \sinh 2K_c = 1. \tag{11.48}$$

According to Montroll *et al.* (1963), there is a rather interesting story about the discovery of the spontaneous magnetization of the two-dimensional Ising ferromagnet. It is given by

$$M = (1 - k^2)^{\frac{1}{8}}, \tag{11.49}$$

where

$$k = [\sinh(2J_1/kT)\sinh(2J_2/kT)]^{-1}, \tag{11.50}$$

for an anisotropic Ising interaction, and for the isotropic case

$$M = [1 - (\sinh 2J/kT)^{-4}]^{\frac{1}{8}}. \tag{11.51}$$

[†] If $k_1 > 1$, the above derivation should be modified; however, the final expression is the same except k_1 is replaced by $k_1' = 1/k_1$.

This famous Onsager cryptogram took four years to decipher. It was first exposed to the public on 23 August 1948 on a blackboard at Cornel University at a conference on phase transitions. Lazlo Tissa had just presented a paper entitled 'General theory of phase transitions'. Gregory Wannier opened the discussion with a question concerning the compatibility of the theory with some properties of the Ising model. Onsager continued this discussion and then remarked, incidentally, that the formula for the spontaneous magnetization of the two-dimensional model was just that given by (11.49). To tease a wider audience, the formula was again exhibited during the discussion which followed a paper by Rushbrook at the first postwar IUPAP statistical mechanics meeting in Florence in 1948; it finally appeared in print as a discussion remark in *Nuovo Cimento* (Onsager, 1949). However, Onsager never published his derivation. The puzzle was finally solved by C. N. Yang (1952); Yang's analysis is very complicated. While many derivations of Onsager's thermodynamic formulae exist and are often presented in statistical mechanics courses, no new derivation of the formula appeared for a few years until the Pfaffian method was extensively applied to the Ising combinatorial problem. A paper by Montroll *et al.* (1963) was one of those in which the Pfaffian method is fully employed in order to derive expressions for the partition function and two-site correlation functions for an arbitrary distance between the two sites on the square lattice. As an asymptotic limit of the two-site correlation function Montroll *et al.* derived the spontaneous magnetization as the square root of the correlation function:

$$M^2 = \lim_{m \to \infty} \langle \sigma_{1,1} \sigma_{1,1+m} \rangle. \tag{11.52}$$

They also proved an equivalence between superficially different mathematical expressions for the two-site correlation functions derived by different authors.

11.4 Critical exponents

As was seen in the previous section, the singularity of the specific heat of the Ising model is $-\log|T - T_c|$ and that of the spontaneous magnetization is $(T_c - T)^{\frac{1}{8}}$. These exponents are very important markers which characterize and classify the system. There are other exponents, such as

$$\text{susceptibility:} \quad \chi \sim |T - T_c|^{\gamma} \sim |T - T_c|^{-\frac{7}{4}}, \tag{11.53}$$
$$\text{equation of state:} \quad M \sim H^{-\frac{1}{\delta}} \sim H^{-\frac{1}{15}} \quad \text{at} \quad T = T_c, \tag{11.54}$$
$$\text{correlation function:} \quad \xi \sim |T - T_c|^{-\nu} \sim |T - T_c|^{-1}. \tag{11.55}$$

The important fact here is that these exponents are common for the Ising spin system, the lattice gas, the binary alloy model, and so on. Another fact which should be stressed is that in the case of the Ising spin system, the exponents depend on the dimensionality of the lattice, the magnitude of the spin, and the range of interaction.

There are many important properties of the critical exponents, not only for the Ising model, but also for the gas–liquid critical phenomenon. Readers are referred to Stanley (1971) for an extensive study of the second-order phase transition.

Exercises

11.1 Calculate the partition function of the 6 Ising spin system with the weight x and y for the horizontal and vertical bonds, respectively.

    Answer is $1 + 2x^2y^2 + x^4y^2$. This means that the partition function is the sum of 1, two squares (left and right hand side) and a rectangle.

11.2 Calculate Ω^n for $n = 1, 2, \ldots$, where Ω is given by (E11.1).

11.3 Find the eigenvalues and eigenfunctions of the pseudo-cyclic matrix given by

$$\Omega = \begin{bmatrix} 0 & 1 & 0 & 0 & 0 & \cdots & 0 \\ 0 & 0 & 1 & 0 & 0 & \cdots & 0 \\ 0 & 0 & 0 & 1 & 0 & \cdots & 0 \\ \vdots & & & & \vdots & & \vdots \\ -1 & 0 & 0 & 0 & 0 & \cdots & 0 \end{bmatrix}. \tag{E11.1}$$

[Hint]
$\Omega^N = -1$, and its eigenvalues are $\omega, \omega^2, \ldots, \omega^N$, with the definition

$$\omega = \exp[i(2p - 1)\pi/N], \quad p = 1, 2, \ldots, N. \tag{E11.2}$$

11.4 Find the eigenfunctions and eigenvalues of the n-dimensional pseudo-cyclic matrix

$$M = \begin{bmatrix} a & b & c & & \cdots & z \\ -z & a & b & c & \cdots & 0 \\ \vdots & & & & & \vdots \\ -b & -c & & \cdots & -z & a \end{bmatrix}. \tag{E11.3}$$

[Hint]
The eigenfunctions V_k and eigenvalues w_k of this matrix are

$$V_k = 1, \omega^k, \omega^{2k}, \omega^{3k}, \ldots, \omega^{(n-1)k}, \quad \text{for} \quad k = 1, 2, \ldots, n, \tag{E11.4}$$

and

$$w_k = a + b\omega^k + c\omega^{2k} + d\omega^{3k} + \cdots + z\omega^{(n-1)k}, \quad \text{for} \quad k = 1, 2, \ldots, n. \tag{E11.5}$$

11.5 Derive the relation

$$\frac{d}{dk} K(k) = (1 - k^2)^{-1} E(k) - K(k), \tag{E11.6}$$

where

$$E(k) = \int_0^{\pi/2} (1 - \sin^2\theta)^{\frac{1}{2}} d\theta \tag{E11.7}$$

is the complete elliptic integral of the second kind.

Appendix 1

The gamma function

A1.1 The Stirling formula

The gamma function is defined by

$$\Gamma(z) = \int_0^\infty \exp(-t)t^{z-1}\,dt, \quad z > 1. \tag{A1.1}$$

By the method of integration by parts it is shown immediately that

$$\Gamma(z) = (z-1)\Gamma(z-1), \tag{A1.2}$$

and hence

$$\Gamma(N+1) = N \cdot (N-1)\cdots 2 \cdot 1 = N! \tag{A1.3}$$

We are interested in the asymptotic behavior, as $z \to \infty$, of the function

$$\Gamma(z+1) = \int_0^\infty \exp(-t)t^z\,dt. \tag{A1.4}$$

The power function in (A1.4) is exponentialized, i.e.,

$$t^z = \exp(z \log t), \tag{A1.5}$$

and then the integration variable is transformed:

$$t = z(1+u). \tag{A1.6}$$

The gamma function is now given the form:

$$\Gamma(z+1) = z \exp(z \log z) \int_{-1}^\infty \exp(zf(u))\,du, \tag{A1.7}$$

where

$$f(u) = \log(1+u) - u. \tag{A1.8}$$

In the limit as $z \to \infty$, the major contribution to the integral comes from the region of u where $f(u)$ becomes a maximum:

$$f'(u) = \frac{1}{1+u} - 1 = 0, \quad \text{i.e.,} \quad u = 0, \quad \text{and} \quad f''(0) = -1. \tag{A1.9}$$

Since z is a large positive number, the integrand is appreciable only in the neighborhood of $u = 0$, and hence $f(u)$ can be expanded into an infinite series. At the same time, the lower limit of integration can be extended to $-\infty$ without introducing any appreciable error:

$$f(u) = f(0) - \frac{u^2}{2} + \cdots, \quad f(0) = -1,$$

$$\Gamma(z+1) = z \exp(z \log z - z) \int_{-\infty}^{\infty} \exp\left(-\frac{z}{2}u^2 + \cdots\right) du. \qquad (A1.10)$$

The first term of asymptotic expansion of the gamma function is obtained by retaining only the first two terms of the series expansion of $f(u)$, i.e.,

$$\Gamma(z+1) = z \exp(z \log z - z) \int_{-\infty}^{\infty} \exp\left(-\frac{z}{2}u^2\right) du$$

$$= (2\pi z)^{\frac{1}{2}} \exp(z \log z - z). \qquad (A1.11)$$

Replacing z by N and taking the logarithm yields the *Stirling formula*

$$\log N! = N \log N - N + \tfrac{1}{2} \log(2\pi N). \qquad (A1.12)$$

In order to find more terms of the asymptotic expansion of the gamma function, we return to (A1.7) and introduce the following transformation of the integration variable:

$$\Gamma(z+1) = z \exp(z \log z) \int_{-\infty}^{\infty} \exp(zf(u)) \, du,$$

$$f(u) = f(0) - w^2, \quad f(0) = -1, \qquad (A1.13)$$

$$\Gamma(z+1) = z \exp(z \log z - z) \int_{-\infty}^{\infty} \exp(-zw^2) \frac{du}{dw} \, dw. \qquad (A1.14)$$

This transformation implies that instead of w being defined by an infinite series in terms of u in (A1.13), u must be defined by an infinite series in terms of w in (A1.14). The transformation process is known as the inversion of an infinite series (Morse & Feshbach, 1953):

$$\frac{du}{dw} = \sum_{n=0}^{\infty} a_n w^n,$$

$$\Gamma(z+1) = z \exp(z \log z - z) \left(\frac{\pi a_0^2}{z}\right)^{\frac{1}{2}} \sum_{n=0}^{\infty} \left(\frac{a_{2n}}{a_0}\right) \frac{\Gamma\left(n + \frac{1}{2}\right)}{\Gamma\left(\frac{1}{2}\right)} \left(\frac{1}{z}\right)^n. \qquad (A1.15)$$

The end result, as described in Morse & Feshbach (1953), is

$$\Gamma(z+1) = (2\pi)^{\frac{1}{2}} z^{z+\frac{1}{2}} \exp(-z) \left[1 + \frac{1}{12z} + \frac{1}{288z^2} + \cdots\right]. \qquad (A1.16)$$

The method of the inversion of an infinite series is certainly an elegant way of finding the asymptotic series for the gamma function, and also for similar functions; however, it may be instructive to show that an ordinary and more elementary method of direct expansion would give the same result:

$$\Gamma(z+1)$$

$$= z \exp(z \log z - z) \int_{-\infty}^{\infty} \exp(z[\log(1+u) - u]) \, du$$

$$= z \exp(z \log z - z) \int_{-\infty}^{\infty} \exp\left(-\frac{zu^2}{2} + \frac{zu^3}{3} - \frac{zu^4}{4} + \frac{zu^5}{5} - \frac{zu^6}{6} + \cdots\right) du$$

$$= z \exp(z \log z - z) \int_{-\infty}^{\infty} \exp\left(-\frac{zu^2}{2}\right) \left[1 + \frac{zu^3}{3} - \frac{zu^4}{4} + \frac{zu^5}{5} - \frac{zu^6}{6} + \cdots \right.$$

$$+ \frac{z^2}{2} \left(\frac{u^3}{3} - \frac{u^4}{4} + \frac{u^5}{5} + \cdots\right)^2$$

$$+ \frac{z^3}{6} \left(\frac{u^3}{3} - \frac{u^4}{4} + \cdots\right)^3$$

$$+ \frac{z^4}{24} \left(\frac{u^3}{3} - \frac{u^4}{4} + \cdots\right)^4 + \cdots \bigg] du. \tag{A1.17}$$

Since only the terms of even powers of u will contribute:

$$\Gamma(z+1) = z \exp(z \log z - z) \int_{-\infty}^{\infty} \exp\left(-\frac{zu^2}{2}\right) \left[1 - \frac{zu^4}{4} - \frac{zu^6}{6}\right.$$

$$+ \frac{z^2}{2} \left(\frac{u^6}{9} + \frac{u^8}{16} + \frac{2u^8}{15}\right) - \frac{z^3 u^{10}}{72} + \frac{z^4 u^{12}}{2^3 \cdot 3^5} + \cdots \bigg] du$$

$$= \left(\frac{2\pi}{z}\right)^{\frac{1}{2}} \left(\frac{z}{e}\right)^z \left[1 + \left(-\frac{3}{4} + \frac{15}{18}\right) \frac{1}{z}\right.$$

$$+ \left(-\frac{15}{6} + \frac{47 \cdot 3 \cdot 5 \cdot 7}{2^5 \cdot 3 \cdot 5} - \frac{3 \cdot 5 \cdot 7 \cdot 9}{72} + \frac{3 \cdot 5 \cdot 7 \cdot 9 \cdot 11}{2^3 \cdot 3^5}\right) \frac{1}{z^2}\bigg]$$

$$= \left(\frac{2\pi}{z}\right)^{\frac{1}{2}} \left(\frac{z}{e}\right)^z \left[1 + \frac{1}{12z} + \frac{1}{288z^2} + \cdots\right]. \tag{A1.18}$$

A1.2 Surface area of the N-dimensional sphere

We start with an identity

$$\pi^{N/2} = \int_{-\infty}^{\infty} \int_{-\infty}^{\infty} \cdots \int_{-\infty}^{\infty} \exp\left(-\left(x_1^2 + x_2^2 + \cdots + x_N^2\right)\right) dx_1 dx_2 \cdots dx_N$$

$$= \Omega(N) \int_{0}^{\infty} \exp(-r^2) r^{N-1} dr, \tag{A1.19}$$

where $\Omega(N)$ is the angular factor appearing in the definition of the surface area of the sphere in N-dimensional space, as given by

$$S(2) = 2\pi r, \quad S(3) = 4\pi r^2, \ldots, \quad S(N) = \Omega(N)r^{N-1}. \tag{A1.20}$$

By a transformation of the integration variable,

$$r^2 = t, \quad 2r \, dr = dt, \quad \int_0^\infty \exp(-r^2)r^{N-1} \, dr = \frac{1}{2}\int_0^\infty \exp(-t)t^{\frac{N}{2}-1} \, dt = \frac{1}{2}\Gamma\left(\frac{N}{2}\right),$$

$$\tag{A1.21}$$

and hence

$$\Omega(N) = \frac{2(\pi)^{\frac{N}{2}}}{\Gamma\left(\frac{N}{2}\right)}. \tag{A1.22}$$

Appendix 2

The critical exponent in the tetrahedron approximation

It will be shown that the critical exponent of the spontaneous magnetization is found to be classical even if the approximation is improved beyond the pair approximation. For the sake of illustration, let us examine the Ising model in the tetrahedron approximation (Sec. 6.10). In this case, we can set

$$y_1 = y_2 = y_3 = y_4 = 0, \tag{A2.1}$$

in the formulation of Sec. 6.10, and there appear four correlation functions; x_1, x_2, x_3, and x_4. These correlation functions are classified into two types: the odd and even correlation functions. In the tetrahedron approximation, x_1 and x_3 are the odd correlation functions.

Equations (6.87)–(6.90) are reproduced with the above simplification.

- x_1:

$$0 = \frac{5}{2} \log \frac{1 + x_1}{1 - x_1} - 3 \log \frac{1 + x_2 + 2x_1}{1 + x_2 - 2x_1}$$
$$+ \frac{1}{2} \log \frac{(1 + 6x_2 + x_4) + 4(x_1 + x_3)}{(1 + 6x_2 + x_4) - 4(x_1 + x_3)}$$
$$+ \log \frac{(1 - x_4) + 2(x_1 - x_3)}{(1 - x_4) - 2(x_1 - x_3)}. \tag{A2.2}$$

- x_3:

$$0 = \frac{1}{2} \log \frac{(1 + 6x_2 + x_4) + 4(x_1 + x_3)}{(1 + 6x_2 + x_4) - 4(x_1 + x_3)} - \log \frac{(1 - x_4) + 2(x_1 - x_3)}{(1 - x_4) - 2(x_1 - x_3)}. \tag{A2.3}$$

- x_2:

$$0 = -\frac{3}{2} \log \frac{(1 + x_2)^2 - 4x_1^2}{(1 - x_2)^2} + \frac{3}{4} \log \frac{(1 + 6x_2 + x_4)^2 - 16(x_1 + x_3)^2}{(1 - 2x_2 + x_4)^2} - \frac{1}{2}\beta. \tag{A2.4}$$

- x_4:

$$0 = \frac{1}{8} \log \frac{(1 + 6x_2 + x_4)^2 - 16(x_1 + x_3)^2}{(1 - x_4)^2 - 4(x_1 - x_3)^2} - \frac{3}{8} \log \frac{(1 - x_4)^2 - 4(x_1 - x_3)^2}{(1 - 2x_2 + x_4)^2}. \tag{A2.5}$$

In (A2.2) and (A2.3), the logarithmic terms are expanded for small x_1 and x_3, and

$$0 = 5x_1 \left[1 + \frac{1}{3}x_1^2 \right] - \frac{12x_1}{1+x_2} \left[1 + \frac{4}{3} \left(\frac{x_1}{1+x_2} \right)^2 \right]$$

$$+ \frac{4(x_1+x_3)}{1+6x_2+x_4} \left[1 + \frac{16}{3} \left(\frac{x_1+x_3}{1+6x_2+x_4} \right)^2 \right]$$

$$+ \frac{4(x_1-x_3)}{1-x_4} \left[1 + \frac{4}{3} \left(\frac{x_1-x_3}{1-x_4} \right)^2 \right] ; \qquad (A2.6)$$

$$0 = \frac{4(x_1+x_3)}{1+6x_2+x_4} \left[1 + \frac{16}{3} \left(\frac{x_1+x_3}{1+6x_2+x_4} \right)^2 \right]$$

$$- \frac{4(x_1-x_3)}{1-x_4} \left[1 + \frac{4}{3} \left(\frac{x_1-x_3}{1-x_4} \right)^2 \right]. \qquad (A2.7)$$

As the transition temperature is approached from below, the cubic terms of x_1 and x_3 can be ignored in (A2.7) and (7.23). Then,

$$5x_1 - \frac{12x_1}{1+x_{2c}} + \frac{4(x_1+x_3)}{1+6x_{2c}+x_{4c}} + \frac{4(x_1-x_3)}{1-x_{4c}} = 0, \qquad (A2.8)$$

$$\frac{x_1+x_3}{1+6x_{2c}+x_{4c}} - \frac{x_1-x_3}{1-x_{4c}} = 0, \qquad (A2.9)$$

where x_{2c} and x_{4c} are the values of x_2 and x_4, respectively, at the transition temperature.

Since the above equations are linear and homogeneous with respect to x_1 and x_3, the coefficient determinant must vanish identically in order to have nonzero solutions for x_1 and x_3. Trivial manipulation yields the condition from Sec. 6.10:

$$(3x_{2c} - 1)(5x_{2c} - 1) = 0, \quad x_{2c} = \tfrac{1}{5}. \qquad (A2.10)$$

Those equations, however, give only the limiting value of the ratio x_3/x_1, a sort of point information at the transition temperature.

In order to find the critical exponent for the spontaneous magnetization, equations covering a finite range in the vicinity of the transition point are required. This is the reason why up to the cubic terms are retained in (A2.7) and (7.23). To find the finite range, the logarithmic terms in (A2.4) and (A2.5) are expanded for small deviations ϵ, μ_2, and μ_4 defined by

$$\beta = \beta_c + \epsilon, \qquad (A2.11)$$

and

$$x_2 = x_{2c} + \mu_2, \quad x_4 = x_{4c} + \mu_4. \qquad (A2.12)$$

The following expansions are obtained:

$$\frac{1}{1+6x_2+x_4} = \frac{1}{1+6x_{2c}+x_{4c}} \left[1 - \frac{6\mu_2+\mu_4}{1+6x_{2c}+x_{4c}} \right],$$

$$\frac{1}{1-2x_2+x_4} = \frac{1}{1-2x_{2c}+x_{4c}} \left[1 + \frac{2\mu_2-\mu_4}{1-2x_{2c}+x_{4c}} \right],$$

$$\frac{1}{1+x_4} = \frac{1}{1+x_{4c}}\left[1 - \frac{\mu_4}{1+x_{4c}}\right],$$

$$\frac{1}{1-x_4} = \frac{1}{1-x_{4c}}\left[1 + \frac{\mu_4}{1-x_{4c}}\right]. \tag{A2.13}$$

These expansions are substituted into (A2.7), (7.23), (A2.4), and (A2.5):

$$0 = 5x_1\left[1 + \frac{1}{3}x_1^2\right] - \frac{12x_1}{1+x_{2c}}\left[1 - \frac{\mu_2}{1+x_{2c}} + \frac{4}{3}\left(\frac{x_1}{1+x_{2c}}\right)^2\right]$$

$$+ \frac{4(x_1+x_3)}{1+6x_{2c}+x_{4c}}\left[1 - \frac{6\mu_2+\mu_4}{1+6x_{2c}+x_{4c}} + \frac{16}{3}\left(\frac{x_1+x_3}{1+6x_{2c}+x_{4c}}\right)^2\right]$$

$$+ \frac{4(x_1-x_3)}{1-x_{4c}}\left[1 + \frac{\mu_4}{1-x_{4c}} + \frac{4}{3}\left(\frac{x_1-x_3}{1-x_{4c}}\right)^2\right], \tag{A2.14}$$

$$0 = \frac{4(x_1+x_3)}{1+6x_{2c}+x_{4c}}\left[1 - \frac{6\mu_2+\mu_4}{1+6x_{2c}+x_{4c}} + \frac{16}{3}\left(\frac{x_1+x_3}{1+6x_{2c}+x_{4c}}\right)^2\right]$$

$$- \frac{4(x_1-x_3)}{1-x_{4c}}\left[1 + \frac{\mu_4}{1-x_{4c}} + \frac{4}{3}\left(\frac{x_1-x_3}{1-x_{4c}}\right)^2\right]. \tag{A2.15}$$

It is easily seen that all the first-order terms vanish identically because of (A2.8) and (A2.9), and in particular (A2.15) is reduced to

$$4\left(\frac{x_1+x_3}{1+6x_{2c}+x_{4c}}\right)^2 = 4\left(\frac{x_1-x_3}{1-x_{4c}}\right)^2$$

$$= \frac{\mu_4}{1-x_{4c}} + \frac{6\mu_2+\mu_4}{1+6x_{2c}+x_{4c}}. \tag{A2.16}$$

We can solve these equations for x_1 and x_3 to find:

$$x_1 = \frac{1+3x_{2c}}{2}\left[\frac{\mu_4}{1-x_{4c}} + \frac{6\mu_2+\mu_4}{1+6x_{2c}+x_{4c}}\right]^{\frac{1}{2}}, \tag{A2.17}$$

$$x_3 = \frac{x_{4c}+3x_{2c}}{2}\left[\frac{\mu_4}{1-x_{4c}} + \frac{6\mu_2+\mu_4}{1+6x_{2c}+x_{4c}}\right]^{\frac{1}{2}}. \tag{A2.18}$$

The temperature (ϵ) dependence of μ_2 and μ_4 can be found from (A2.4) and (A2.5). Expanding the logarithmic terms, again for small values of x_1 and x_3, and also since

$\beta = \beta_c + \epsilon$, we obtain:

$$-6\log\frac{1-x_{2c}}{1+x_{2c}} + 3\log\frac{1+6x_{2c}+x_{4c}}{1-2x_{2c}+x_{4c}}$$

$$-12\left(\frac{\mu_2}{1-x_{2c}^2}\right) + 3\left(\frac{6\mu_2+\mu_4}{1+6x_{2c}+x_{4c}} + \frac{2\mu_2-\mu_4}{1-2x_{2c}+x_{4c}}\right)$$

$$+12\frac{x_1^2}{(1+x_{2c})^2} - 12\left(\frac{x_1+x_3}{1+6x_{2c}+x_{4c}}\right)^2 = \beta_c + \epsilon, \tag{A2.19}$$

$$\log\frac{1+6x_{2c}+x_{4c}}{1-x_{4c}} + \frac{6\mu_2+\mu_4}{1+6x_{2c}+x_{4c}} + \frac{\mu_4}{1-x_{4c}}$$

$$-3\log\frac{1-x_{4c}}{1-2x_{2c}+x_{4c}} + \frac{3\mu_4}{1-x_{4c}} - \frac{3(2\mu_2-\mu_4)}{1-2x_{2c}+x_{4c}} = 0. \tag{A2.20}$$

In the above two equations, the zeroth-order terms cancel out because of (A2.4) and (A2.5), if at the transition temperature. From (A2.20), after substituting from (A2.16) and (A2.17), we find

$$-12\left(\frac{\mu_2}{1-x_{2c}^2}\right) + 3\left(\frac{2\mu_2-\mu_4}{1-2x_{2c}+x_{4c}} - \frac{\mu_4}{1-x_{4c}}\right)$$

$$+3\frac{(1+3x_{2c})^2}{(1+x_{2c})^2}\left[\frac{\mu_4}{1-x_{4c}} + \frac{6\mu_2+\mu_4}{1+6x_{2c}+x_{4c}}\right] = \epsilon, \tag{A2.21}$$

$$\frac{6\mu_2+\mu_4}{1+6x_{2c}+x_{4c}} + \frac{4\mu_4}{1-x_{4c}} - \frac{3(2\mu_2-\mu_4)}{1-2x_{2c}+x_{4c}} = 0. \tag{A2.22}$$

These equations are linear with respect to μ_2 and μ_4, and the inhomogeneous terms are simply ϵ and 0. The solutions of these equations are, hence, linear or proportional to ϵ. If these solutions are substituted into (A2.17) and (A2.18) it will be clear that both x_1 and x_3 become proportional to $\epsilon^{1/2}$; the classical value of the critical exponent.

The situation will change if the solution of the four-site approximation is found in terms of the infinite series for the correlation functions. This is treated in detail in Chapter 7.

Appendix 3

Programming organization of the cluster variation method

A3.1 Characteristic matrices

One important feature of the correlation function formulation of the cluster variation method is the fact that all the elements of the reduced density matrices are linear combinations of the correlation functions. In order to determine the thermal properties of the physical system in question the variational potential is minimized with respect to the correlation functions. We can make an observation, therefore, that the coefficients of the linear combinations of correlation functions in various reduced density matrices will completely determine the thermal properties of the physical system. In the two-site approximation the thermal properties should be determined totally by the lattice coordination number (the number of nearest neighbors). By retaining successively three-site, four-site, etc. density matrices in the variational potential the lattice structure is more accurately taken into account. All such detailed information should be built into the coefficients of the linear combinations of the correlation functions in the elements of the density matrices. It will be demonstrated in this section that all the steps of the variational formulation are completely specified in terms of those coefficients.

To demonstrate the ease with which one can work, the tetrahedron-plus-octahedron approximation for the face centered cubic lattice will be chosen as the working system.

In this approximation the entropy per lattice site is given by

$$
\begin{aligned}
-S/kN = {} & g_1(1) + 6g_2(1,2) + 3g_2(1,6) + 8g_3(1,2,3) + 12g_3(1,2,7) \\
& + 2g_4(1,2,3,4) + 3g_4(2,3,5,7) + 12g_3(3,5,6,7) \\
& + 6g_5(1,2,3,5,7) + g_6(1,2,3,5,6,7),
\end{aligned} \tag{A3.1}
$$

where the numerical coefficients of g_2–g_6 represent, respectively per single lattice site, the numbers of nearest-neighbor pairs, equilateral triangles, isosceles triangles, regular tetrahedra, squares, tetrahedra of the type 3567, pyramids, and octahedra.

When the cumulant functions g_1–g_6 are expressed in terms of the cluster functions, G_1–G_6, and an internal energy term is included, the variational potential is found:

$$
\begin{aligned}
F/NkT = {} & -\tfrac{1}{2}\beta x_1 + z_1 G_1 + z_2 G_2(1,2) + z_3 G_3(1,2,3) \\
& + z_4 G_4(1,2,3,4) + z_6 G_6(1,2,3,5,6,7),
\end{aligned} \tag{A3.2}
$$

where $x_1 = \langle s_1 s_2 \rangle$ and $\beta = 12J/kT$, $z_1 = -1$, $z_2 = 6$, $z_3 = -8$, $z_4 = 2$, $z_6 = 1$.

In order to construct the characteristic table for the variational calculations, it is crucially important to lay out all the elements of the reduced density matrices, despite this being a rather trivial procedure.

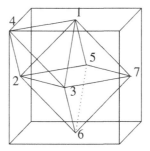

Fig. A3.1. Tetrahedron-plus-octahedron approximation.

The one-site cluster function is

$$G_1 = N_1[R_1(1)\log R_1(1) + R_1(2)\log R_1(2)],$$
$$R_1(1) = (1 + y_1), \quad w_1(1) = 1,$$
$$R_1(2) = (1 - y_1), \quad w_1(2) = 1, \tag{A3.3}$$

where $N_1 = \frac{1}{2}$, $y_1 = \langle s_1 \rangle$ and $R_1(1)$, $R_1(2)$ are the elements of the one-site density matrix. $w_1(1)$ and $w_1(2)$ are the multiplicities of those elements. From these matrix elements the one-site characteristic matrix is defined:

$$AY_1$$

$$\begin{bmatrix} 1 \\ -1 \end{bmatrix}, \tag{A3.4}$$

where $AY_1(1, 1)$ is the coefficient of y_1 in the element $R_1(1)$ and $AY_1(2, 1)$ is the coefficient of y_1 in the element $R_1(2)$. The two-site cluster function in the variational potential is

$$G_2 = \sum_{i=1,2,3} N_2[R_2(i)\log R_2(i)],$$
$$R_2(1) = (1 + x_1 + 2y_1), \quad w_2(1) = 1,$$
$$R_2(1) = (1 + x_1 - 2y_1), \quad w_2(2) = 1,$$
$$R_2(3) = (1 - x_1), \qquad\quad w_2(3) = 2, \tag{A3.5}$$

where $N_2 = \frac{1}{4}$. From these elements of the reduced density matrix, two characteristic matrices are defined:

$$AX_2 \qquad AY_2$$

$$\begin{bmatrix} 1 \\ 1 \\ -1 \end{bmatrix}, \begin{bmatrix} 2 \\ -2 \\ 0 \end{bmatrix}, \tag{A3.6}$$

where $AX_2(1, 1)$ is the coefficient of x_1 in the element $R_2(1)$ and $AX_2(2, 1)$, $AX_2(3, 1)$, $AY_2(1, 1)$, $AY_2(2, 1)$, and $AY_2(3, 1)$ are similarly defined.

The three-site cluster function is

$$G_3 = \sum_{i=1}^{4} N_3[R_3(i) \log R_3(i)],$$

$$
\begin{aligned}
R_3(1) &= (1 + 3x_1 + 3y_1 + y_2), & w_3(1) &= 1, \\
R_3(2) &= (1 + 3x_1 - 3y_1 - y_2), & w_3(2) &= 1, \\
R_3(3) &= (1 - x_1 + y_1 - y_2), & w_3(3) &= 3, \\
R_3(4) &= (1 - x_1 - y_1 + y_2), & w_3(4) &= 3,
\end{aligned}
\tag{A3.7}
$$

where $N_3 = \frac{1}{8}$. From these elements of the reduced density matrix the three-site characteristic matrices are constructed:

$$
\begin{array}{cc}
AX_3 & AY_3
\end{array}
$$

$$
\begin{bmatrix} 3 \\ 3 \\ -1 \\ -1 \end{bmatrix},
\begin{bmatrix} 3 & 1 \\ -3 & -1 \\ 1 & -1 \\ -1 & 1 \end{bmatrix}.
\tag{A3.8}
$$

The four-site cluster function and the reduced density matrix are

$$G_4 = \sum_{i=1}^{5} N_4[R_4(i) \log R_4(i)],$$

$$
\begin{aligned}
R_4(1) &= (1 + 6x_1 + x_3 + 4y_1 + 4y_2), & w_4(1) &= 1, \\
R_4(2) &= (1 + 6x_1 + x_3 - 4y_1 - 4y_2), & w_4(2) &= 1, \\
R_4(3) &= (1 - x_3 + 2y_1 - 2y_2), & w_4(3) &= 4, \\
R_4(4) &= (1 - x_3 - 2y_1 + 2y_2), & w_4(4) &= 4, \\
R_4(5) &= (1 - 2x_1 + x_3), & w_4(5) &= 6,
\end{aligned}
\tag{A3.9}
$$

where $N_4 = \frac{1}{16}$. From these elements the characteristic matrices are found to be

$$
\begin{array}{cc}
AX_4 & AY_4
\end{array}
$$

$$
\begin{bmatrix} 6 & 0 & 1 \\ 6 & 0 & 1 \\ 0 & 0 & -1 \\ 0 & 0 & -1 \\ -2 & 0 & 1 \end{bmatrix},
\begin{bmatrix} 4 & 4 \\ -4 & -4 \\ 2 & -2 \\ -2 & 2 \\ 0 & 0 \end{bmatrix}.
\tag{A3.10}
$$

Finally the six-site cluster function is

$$G_6 = \sum_{i=1}^{10} N_6[R_6(i) \log R_6(i)],$$

$$
\begin{aligned}
R_6(1) &= (S_{12} + L_{12}), & w_6(1) &= 1, \\
R_6(2) &= (S_{12} - L_{12}), & w_6(2) &= 1, \\
R_6(3) &= (S_{34} + L_{34}), & w_6(3) &= 6, \\
R_6(4) &= (S_{34} - L_{34}), & w_6(4) &= 6, \\
R_6(5) &= (S_{56} + L_{56}), & w_6(5) &= 12,
\end{aligned}
$$

$$R_6(6) = (S_{56} - L_{56}), \quad w_6(6) = 12,$$
$$R_6(7) = (S_{78} + L_{78}), \quad w_6(7) = 3,$$
$$R_6(8) = (S_{78} - L_{78}), \quad w_6(8) = 3,$$
$$R_6(9) = (1 - 3x_2 + 3x_4 - x_6), \quad w_6(9) = 8,$$
$$R_6(10) = (1 - 4x_1 + x_2 - x_4 + 4x_5 - x_6), \quad w_6(10) = 12, \tag{A3.11}$$

where $N_6 = \frac{1}{64}$, and

$$S_{12} = 1 + 12x_1 + 3x_2 + 3x_4 + 12x_5 + x_6,$$
$$L_{12} = 6y_1 + 8y_2 + 12y_3 + 6y_4,$$
$$S_{34} = 1 + 4x_1 + x_2 - x_4 - 4x_5 - x_6,$$
$$L_{34} = 4y_1 - 4y_4,$$
$$S_{56} = 1 - x_2 - x_4 + x_6,$$
$$L_{56} = 2y_1 - 4y_3 + 2y_4,$$
$$S_{78} = 1 - 4x_1 + 3x_2 + 3x_4 - 4x_5 + x_6,$$
$$L_{78} = 2y_1 - 8y_2 + 4y_3 + 2y_4. \tag{A3.12}$$

The six-site characteristic matrices are similarly constructed:

$$AX_6 \qquad\qquad\qquad AY_6$$

$$\begin{bmatrix}
12 & 3 & 0 & 3 & 12 & 1 \\
12 & 3 & 0 & 3 & 12 & 1 \\
4 & 1 & 0 & -1 & -1 & -1 \\
4 & 1 & 0 & -1 & -1 & -1 \\
0 & -1 & 0 & -1 & 0 & 1 \\
0 & -1 & 0 & -1 & 0 & 1 \\
-4 & 3 & 0 & 3 & -4 & 1 \\
-4 & 3 & 0 & 3 & -4 & 1 \\
0 & -3 & 0 & 3 & 0 & -1 \\
-4 & 1 & 0 & -1 & 4 & -1
\end{bmatrix},
\begin{bmatrix}
6 & 8 & 12 & 6 \\
-6 & -8 & -12 & -6 \\
4 & 0 & 0 & -4 \\
-4 & 0 & 0 & 4 \\
2 & 0 & -4 & 2 \\
-2 & 0 & 4 & -2 \\
2 & -8 & 4 & 2 \\
-2 & 8 & -4 & -2 \\
0 & 0 & 0 & 0 \\
0 & 0 & 0 & 0
\end{bmatrix}. \tag{A3.13}$$

A3.2 Properties of characteristic matrices

Once the characteristic matrices are constructed, the minimization of the variational potential with respect to each of the correlation functions is straightforward.

x_1-equation (minimization with respect to $x_1 = \langle s_1 s_2 \rangle$)

Note that there is an additional contribution due to the internal energy:

$$\frac{1}{2}\beta = z_2 N_2 \sum_{i=1}^{3} w_2(i) AX_2(i, 1) \log R_2(i) + z_3 N_3 \sum_{i=1}^{4} w_3(i) AX_3(i, 1) \log R_3(i)$$

$$+ z_4 N_4 \sum_{i=1}^{5} w_4(i) AX_4(i, 1) \log R_4(i) + z_6 N_6 \sum_{i=1}^{10} w_6(i) AX_6(i, 1) \log R_6(i). \tag{A3.14}$$

Minimization with respect to $x_2 = \langle s_1 s_6 \rangle$ will yield the following equation (there is no contribution of the four-site density matrix to the x_2-equation):

$$0 = z_6 N_6 \sum_{i=1}^{10} w_6(i) AX_6(i, 2) \log R_6(i). \tag{A3.15}$$

$$x_3 = \langle s_1 s_2 s_3 s_4 \rangle\text{-}equation$$

$$0 = z_4 N_4 \sum_{i=1}^{5} w_4(i) AX_4(i, 3) \log R_3(i) + z_6 N_6 \sum_{i=1}^{10} w_6(i) AX_6(i, 3) \log R_6(i). \tag{A3.16}$$

The equations for x_4, x_5, and x_6, $k = 4, 5, 6$, are:

$$0 = z_6 N_6 \sum_{i=1}^{10} w_6(i) AX_6(i, k) \log R_6(i), \quad k = 4, 5, 6. \tag{A3.17}$$

Minimization with respect to the odd correlation functions, y_1–y_4, will yield similar equations using AY matrices in place of AX matrices.

We note that the utilization of the characteristic matrices makes the minimization of the variational potential an almost trivial procedure by incorporating four factors z_j, $AX_j(i, k)$, or $AY_j(i, k)$, and $w_j(i)$. This free-from-errors procedure is a great advantage in the computer handling of the cluster variation method. Of course, the above equations are equally effective for the algebraic handling of the cluster variation method.

A3.3 Susceptibility determinants

If we require only the critical temperature rather than the low temperature behavior of the odd correlation functions, it will suffice to locate the singularity of the magnetic susceptibility. When an infinitesimally small external magnetic field H is applied, vanishingly small long-range-order correlations x_1, x_3, x_5, and x_8, are created. The equations for these correlation functions are found by expanding the logarithmic terms with respect to those correlation functions up to the first power.

Equation for x_1

$$\beta H = a(1, 1)x_1 + a(1, 2)x_3 + a(1, 3)x_5 + a(1, 4)x_8, \tag{A3.18}$$

where βH represents an external magnetic field divided by the temperature in certain units, and

$$a(1, 1) = 2z_1 N_1 \frac{AY_1(1, 1)AY_1(1, 1)}{R_1(1)} + 2z_2 N_2 \frac{AY_2(1, 1)AY_2(1, 1)}{R_2(1)}$$

$$+ 2z_3 N_3 \left[w_3(1) \frac{A_3(1, 1)A_3(1, 1)}{R_3(1)} + w_3(3) \frac{A_3(3, 1)A_3(3, 1)}{R_3(3)} \right]$$

$$+ 2z_4 N_4 \left[w_4 1 \frac{A_4(1, 1)A_4(1, 1)}{R_4(1)} + w_4(3) \frac{A_4(3, 1)A_4(3, 1)}{R_4(3)} \right] +$$

$$+ 2z_6 N_6 \left[w_6(1) \frac{A_6(1, 1) A_6(1, 1)}{R_6(1)} + w_6(3) \frac{A_6(3, 1) A_6(3, 1)}{R_6(3)} \right.$$

$$\left. + w_6(5) \frac{A_6(5, 1) A_6(5, 1)}{R_6(5)} + w_6(7) \frac{A_6(7, 1) A_6(7, 1)}{R_6(7)} \right],$$

$$a(1, 2) = 2z_3 N_3 \left[w_3(1) \frac{A_3(1, 1) A_3(1, 3)}{R_3(1)} + w_3(3) \frac{A_3(3, 1) A_3(3, 3)}{R_3(3)} \right]$$

$$+ 2z_4 N_4 \left[w_4(1) \frac{A_4(1, 1) A_4(1, 3)}{R_4(1)} + w_4(3) \frac{A_4(3, 1) A_4(3, 3)}{R_4(3)} \right]$$

$$+ 2z_6 N_6 \left[w_6(1) \frac{A_6(1, 1) A_6(1, 3)}{R_6(1)} + w_6(3) \frac{A_6(3, 1) A_6(3, 3)}{R_6(3)} \right.$$

$$\left. + w_6(5) \frac{A_6(5, 1) A_6(5, 3)}{R_6(5)} + w_6(7) \frac{A_6(7, 1) A_6(7, 3)}{R_6(7)} \right],$$

$$a(1, 3) = 2z_6 N_6 \left[w_6(1) \frac{A_6(1, 1) A_6(1, 5)}{R_6(1)} + w_6(3) \frac{A_6(3, 1) A_6(3, 5)}{R_6(3)} \right.$$

$$\left. + w_6(5) \frac{A_6(5, 1) A_6(5, 5)}{R_6(5)} + w_6(7) \frac{A_6(7, 1) A_6(7, 5)}{R_6(7)} \right],$$

$$a(1, 4) = 2z_6 N_6 \left[w_6(1) \frac{A_6(1, 1) A_6(1, 8)}{R_6(1)} + w_6(3) \frac{A_6(3, 1) A_6(3, 8)}{R_6(3)} \right.$$

$$\left. + w_6(5) \frac{A_6(5, 1) A_6(5, 8)}{R_6(5)} + w_6(7) \frac{A_6(7, 1) A_6(7, 8)}{R_6(7)} \right].$$

$$\text{(A3.19)}$$

In these equations, all the long-range-order correlation functions, x_1, x_3, x_5, and x_8 in R_1, R_2, R_3, R_4, and R_6 must be set equal to zero. This same condition should be imposed in all the following equations wherever R_1, R_2, R_3, R_4, and R_6 appear.

Equation for x_3

$$0 = a(2, 1)x_1 + a(2, 2)x_3 + a(2, 3)x_5 + a(2, 4)x_8, \qquad \text{(A3.20)}$$

where

$$a(2, 1) = 2z_3 N_3 \left[w_3(1) \frac{A_3(1, 3) A_3(1, 1)}{R_3(1)} + w_3(3) \frac{A_3(3, 3) A_3(3, 1)}{R_3(3)} \right]$$

$$+ 2z_4 N_4 \left[w_4 1 \frac{A_4(1, 3) A_4(1, 1)}{R_4(1)} + w_4(3) \frac{A_4(3, 3) A_4(3, 1)}{R_4(3)} \right]$$

$$+ 2z_6 N_6 \left[w_6(1) \frac{A_6(1, 3) A_6(1, 1)}{R_6(1)} + w_6(3) \frac{A_6(3, 3) A_6(3, 1)}{R_6(3)} \right.$$

$$\left. + w_6(5) \frac{A_6(5, 3) A_6(5, 1)}{R_6(5)} + w_6(7) \frac{A_6(7, 3) A_6(7, 1)}{R_6(7)} \right],$$

$$a(2,2) = 2z_3N_3\left[w_3(1)\frac{A_3(1,3)A_3(1,3)}{R_3(1)} + w_3(3)\frac{A_3(3,3)A_3(3,3)}{R_3(3)}\right]$$

$$+ 2z_4N_4\left[w_41\frac{A_4(1,3)A_4(1,3)}{R_4(1)} + w_4(3)\frac{A_4(3,3)A_4(3,3)}{R_4(3)}\right]$$

$$+ 2z_6N_6\left[w_6(1)\frac{A_6(1,3)A_6(1,3)}{R_6(1)} + w_6(3)\frac{A_6(3,3)A_6(3,3)}{R_6(3)}\right.$$

$$+ w_6(5)\frac{A_6(5,3)A_6(5,3)}{R_6(5)} + w_6(7)\frac{A_6(7,3)A_6(7,3)}{R_6(7)}\Bigg],$$

$$a(2,3) = 2z_6N_6\left[w_6(1)\frac{A_6(1,3)A_6(1,5)}{R_6(1)} + w_6(3)\frac{A_6(3,3)A_6(3,5)}{R_6(3)}\right.$$

$$+ w_6(5)\frac{A_6(5,3)A_6(5,5)}{R_6(5)} + w_6(7)\frac{A_6(7,3)A_6(7,5)}{R_6(7)}\Bigg],$$

$$a(2,4) = 2z_6N_6\left[w_6(1)\frac{A_6(1,3)A_6(1,8)}{R_6(1)} + w_6(3)\frac{A_6(3,3)A_6(3,8)}{R_6(3)}\right.$$

$$+ w_6(5)\frac{A_6(5,3)A_6(5,8)}{R_6(5)} + w_6(7)\frac{A_6(7,3)A_6(7,8)}{R_6(7)}\Bigg]. \quad \text{(A3.21)}$$

In order to find the paramagnetic susceptibility we consider the following:

Equation for x_5

$$0 = a(3,1)x_1 + a(3,2)x_3 + a(3,3)x_5 + a(3,4)x_8, \quad \text{(A3.22)}$$

where

$$a(3,1) = 2z_6N_6\left[w_6(1)\frac{A_6(1,5)A_6(1,1)}{R_6(1)} + w_6(3)\frac{A_6(3,5)A_6(3,1)}{R_6(3)}\right.$$

$$+ w_6(5)\frac{A_6(5,5)A_6(5,1)}{R_6(5)} + w_6(7)\frac{A_6(7,5)A_6(7,1)}{R_6(7)}\Bigg],$$

$$a(3,2) = 2z_6N_6\left[w_6(1)\frac{A_6(1,5)A_6(1,3)}{R_6(1)} + w_6(3)\frac{A_6(3,5)A_6(3,3)}{R_6(3)}\right.$$

$$+ w_6(5)\frac{A_6(5,5)A_6(5,3)}{R_6(5)} + w_6(7)\frac{A_6(7,5)A_6(7,3)}{R_6(7)}\Bigg],$$

$$a(3,3) = 2z_6N_6\left[w_6(1)\frac{A_6(1,5)A_6(1,5)}{R_6(1)} + w_6(3)\frac{A_6(3,5)A_6(3,5)}{R_6(3)}\right.$$

$$+ w_6(5)\frac{A_6(5,5)A_6(5,5)}{R_6(5)} + w_6(7)\frac{A_6(7,5)A_6(7,5)}{R_6(7)}\Bigg],$$

$$a(3, 4) = 2z_6 N_6 \left[w_6(1) \frac{A_6(1, 5) A_6(1, 8)}{R_6(1)} + w_6(3) \frac{A_6(3, 5) A_6(3, 8)}{R_6(3)} \right.$$

$$\left. + w_6(5) \frac{A_6(5, 5) A_6(5, 8)}{R_6(5)} + w_6(7) \frac{A_6(7, 5) A_6(7, 8)}{R_6(7)} \right].$$

$$\text{(A3.23)}$$

Equation for x_8

$$0 = a(4, 1) x_1 + a(4, 2) x_3 + a(4, 3) x_5 + a(4, 4) x_8, \qquad \text{(A3.24)}$$

where

$$a(4, 1) = 2z_6 N_6 \left[w_6(1) \frac{A_6(1, 8) A_6(1, 1)}{R_6(1)} + w_6(3) \frac{A_6(3, 8) A_6(3, 1)}{R_6(3)} \right.$$

$$\left. + w_6(5) \frac{A_6(5, 8) A_6(5, 1)}{R_6(5)} + w_6(7) \frac{A_6(7, 8) A_6(7, 1)}{R_6(7)} \right],$$

$$a(4, 2) = 2z_6 N_6 \left[w_6(1) \frac{A_6(1, 8) A_6(1, 3)}{R_6(1)} + w_6(3) \frac{A_6(3, 8) A_6(3, 3)}{R_6(3)} \right.$$

$$\left. + w_6(5) \frac{A_6(5, 8) A_6(5, 3)}{R_6(5)} + w_6(7) \frac{A_6(7, 8) A_6(7, 3)}{R_6(7)} \right],$$

$$a(4, 3) = 2z_6 N_6 \left[w_6(1) \frac{A_6(1, 8) A_6(1, 5)}{R_6(1)} + w_6(3) \frac{A_6(3, 8) A_6(3, 5)}{R_6(3)} \right.$$

$$\left. + w_6(5) \frac{A_6(5, 8) A_6(5, 5)}{R_6(5)} + w_6(7) \frac{A_6(7, 8) A_6(7, 5)}{R_6(7)} \right],$$

$$a(4, 4) = 2z_6 N_6 \left[w_6(1) \frac{A_6(1, 8) A_6(1, 8)}{R_6(1)} + w_6(3) \frac{A_6(3, 8) A_6(3, 8)}{R_6(3)} \right.$$

$$\left. + w_6(5) \frac{A_6(5, 8) A_6(5, 8)}{R_6(5)} + w_6(7) \frac{A_6(7, 8) A_6(7, 8)}{R_6(7)} \right]. \quad \text{(A3.25)}$$

Equations (A3.18), (A3.20), (A3.22), and (A3.24) are solved for the induced magnetization, x_1:

$$x_1 = \chi H, \qquad \text{(A3.26)}$$

where

$$\chi = 1 \left/ \left[a(1, 1) - a(2, 1) \frac{\Delta_1}{\Delta_0} + a(3, 1) \frac{\Delta_2}{\Delta_0} - a(4, 1) \frac{\Delta_3}{\Delta_0} \right] \right., \qquad \text{(A3.27)}$$

and

$$\Delta_0 = \begin{vmatrix} a(2,2) & a(2,3) & a(2,4) \\ a(3,2) & a(3,3) & a(3,4) \\ a(4,2) & a(4,3) & a(4,4) \end{vmatrix}, \quad \Delta_1 = \begin{vmatrix} a(1,2) & a(1,3) & a(1,4) \\ a(3,2) & a(3,3) & a(3,4) \\ a(4,2) & a(4,3) & a(4,4) \end{vmatrix},$$

$$\Delta_2 = \begin{vmatrix} a(1,2) & a(1,3) & a(1,4) \\ a(2,2) & a(2,3) & a(2,4) \\ a(4,2) & a(4,3) & a(4,4) \end{vmatrix}, \quad \Delta_3 = \begin{vmatrix} a(1,2) & a(1,3) & a(1,4) \\ a(2,2) & a(2,3) & a(2,4) \\ a(3,2) & a(3,3) & a(3,4) \end{vmatrix}. \quad (A3.28)$$

Appendix 4

A unitary transformation applied to the Hubbard Hamiltonian

As a mathematical exercise, let us examine the diagonalization procedure of the Hubbard Hamiltonian as is given in Sec. 8.5:[†]

$$H = \sum_{0 \le k \le \pi} \{[\epsilon(k) + Un - \mu]a_k^* a_k + [\epsilon(k + \pi) + Un - \mu]a_{k+\pi}^* a_{k+\pi}$$

$$+ UB(a_{k+\pi}^* a_k + a_k^* a_{k+\pi})\},$$

$$\epsilon(k) = -2t(\cos k_x + \cos k_y + \cos k_z), \tag{A4.1}$$

for a simple cubic lattice in which the lattice constant is chosen to be unity.

In order to diagonalize the above given Hamiltonian, the following unitary transformation is introduced:

$$a_k = \cos\theta \alpha_{1k} - \sin\theta \alpha_{2k},$$

$$a_{k+\pi} = \sin\theta \alpha_{1k} + \cos\theta \alpha_{2k}. \tag{A4.2}$$

Then the newly defined operators satisfy the commutation relations

$$\alpha_{1k}\alpha_{1k}^* + \alpha_{1k}^*\alpha_{1k} = 1,$$

$$\alpha_{2k}\alpha_{2k}^* + \alpha_{2k}^*\alpha_{2k} = 1,$$

$$\alpha_{1k}\alpha_{2k}^* + \alpha_{2k}^*\alpha_{1k} = 0. \tag{A4.3}$$

Various terms in the Hamiltonian are, then, transformed into:

$$a_k^* a_k = \cos^2\theta \alpha_{1k}^*\alpha_{1k} + \sin^2\theta \alpha_{2k}^*\alpha_{2k} - \sin\theta\cos\theta(\alpha_{1k}^*\alpha_{2k} + \alpha_{2k}^*\alpha_{1k}), \tag{A4.4}$$

$$a_{k+\pi}^* a_{k+\pi} = \sin^2\theta \alpha_{1k}^*\alpha_{1k} + \cos^2\theta \alpha_{2k}^*\alpha_{2k} + \sin\theta\cos\theta(\alpha_{1k}^*\alpha_{2k} + \alpha_{2k}^*\alpha_{1k}), \tag{A4.5}$$

$$a_{k+\pi}^* a_k + a_k^* a_{k+\pi} = \sin 2\theta(\alpha_{1k}^*\alpha_{1k} - \alpha_{2k}^*\alpha_{2k}) + \cos 2\theta(\alpha_{1k}^*\alpha_{2k} - \alpha_{2k}^*\alpha_{1k}). \tag{A4.6}$$

When these are introduced into (A4.1), we find

$$H = \sum_{0 \le k \le \pi} \{\{[\epsilon(k) + Un - \mu]\cos^2\theta + [\epsilon(k + \pi) + Un - \mu]\sin^2\theta$$

$$+ 2UB\sin\theta\cos\theta\}\alpha_{1k}^*\alpha_{1k}$$

$$+ \{[\epsilon(k) + Un - \mu]\sin^2\theta + [\epsilon(k) + Un - \mu]\cos^2\theta$$

$$- 2UB\sin\theta\cos\theta\}\alpha_{2k}^*\alpha_{2k}$$

$$+ \{[\epsilon(k + \pi) - \epsilon(k)]\sin\theta\cos\theta + UB\cos 2\theta\}(\alpha_{1k}^*\alpha_{2k} - \alpha_{2k}^*\alpha_{1k})\}. \tag{A4.7}$$

[†] It should be noted that the Hamiltonian has been modified in the form of the thermodynamically equivalent Hamiltonian so that it will lead to the same result as obtained in Sec. 8.5.

In order to make the quasiparticle stable, the coefficient of $(\alpha_{1k}^*\alpha_{2k} - \alpha_{2k}^*\alpha_{1k})$ must be set equal to zero; the condition to determine the unknown angle θ is

$$UB\cos 2\theta = \epsilon_{k-}\sin 2\theta, \tag{A4.8}$$

or

$$\tan 2\theta = \frac{UB}{\epsilon_{k-}}, \tag{A4.9}$$

where

$$\epsilon_\pm = \tfrac{1}{2}[\epsilon(k) \pm \epsilon(k+\pi)]. \tag{A4.10}$$

In (A4.8), both B and ϵ_{k-} are negative for less-than-a-half case, and hence $\cos 2\theta$ and $\sin 2\theta$ must have the same sign, and therefore the solutions are

$$\cos 2\theta = \frac{\pm\epsilon_{k-}}{[\epsilon_{k-}^2 + U^2B^2]^{\frac{1}{2}}},$$

$$\sin 2\theta = \frac{\pm UB}{[\epsilon_{k-}^2 + U^2B^2]^{\frac{1}{2}}}, \tag{A4.11}$$

where the double signs must be congruent. The diagonalized Hamiltonian is now given by

$$H = \sum_{0 \le k \le \pi} (E_{1k}\alpha_{1k}^*\alpha_{1k} - E_{2k}\alpha_{2k}^*\alpha_{2k}), \tag{A4.12}$$

where the energy spectra of the quasiparticles are given by

$$E_{1k} = (\epsilon_{k+} + Un - \mu) - [\epsilon_{k-}^2 + U^2B^2]^{\frac{1}{2}},$$

$$E_{1k} = (\epsilon_{k+} + Un - \mu) + [\epsilon_{k-}^2 + U^2B^2]^{\frac{1}{2}}. \tag{A4.13}$$

Because of the form of $\epsilon(k)$ given in (A4.1),

$$\epsilon(k+\pi) = -\epsilon(k), \tag{A4.14}$$

and hence

$$\epsilon_{k-} = \epsilon(k), \quad \text{and} \quad \epsilon_{k+} = 0. \tag{A4.15}$$

Equation (A4.13) shows that there is an energy gap at the center of the band if there is a nonvanishing sublattice magnetization. Then the half-filled lattice becomes an insulator.

The statistical averages of the occupation numbers of quasiparticles are given by

$$\nu_{1k} = \langle \alpha_{1k}^*\alpha_{1k} \rangle = \frac{1}{\exp(\beta E_{1k}) + 1},$$

$$\nu_{2k} = \langle \alpha_{2k}^*\alpha_{2k} \rangle = \frac{1}{\exp(\beta E_{2k}) + 1}. \tag{A4.16}$$

If the equality

$$\sum_{-\pi \le k \le \pi} \langle a_{k+\pi}^* a_k \rangle = \sum_{0 \le k \le \pi} (\langle a_{k+\pi}^* a_k \rangle + \langle a_k^* a_{k+\pi} \rangle) \tag{A4.17}$$

is recognized and we substitute from (A4.6), the sublattice magnetization is found to be

$$A = A(\pi) = \frac{1}{N_0} \sum_{0 \le k \le \pi} \sin 2\theta(\nu_{1k} - \nu_{2k}). \tag{A4.18}$$

When the thermodynamic limit is taken, the summation over the wave vectors is replaced by integrations, and because of (A4.11), with the negative sign, the equation for the sublattice magnetization is given by

$$U \int_0^D \frac{1}{[\epsilon^2 + U^2 A^2]^{\frac{1}{2}}} \left\{ \frac{1}{\exp \beta E_1(\epsilon) + 1} - \frac{1}{\exp \beta E_2(\epsilon) + 1} \right\} \eta_0(\epsilon) d\epsilon = 1, \quad \text{(A4.19)}$$

where

$$E_1(\epsilon) = (Un - \mu) - [\epsilon^2 + U^2 B^2]^{\frac{1}{2}},$$
$$E_2(\epsilon) = (Un - \mu) + [\epsilon^2 + U^2 B^2]^{\frac{1}{2}}; \quad \text{(A4.20)}$$

D is the half band width, $D = 4t$ for a square lattice, $D = 6t$ for a simple cubic lattice, and $\eta_0(\epsilon)$ is the density of states function.

The rest of the formulation is the same as the one by the cluster variation method discussed in Sec. 8.5.

Appendix 5

Exact Ising identities on the diamond lattice

The generating equation for the diamond lattice is

$$\langle 0[f]\rangle = A(\langle 1[f]\rangle + \langle 2[f]\rangle + \langle 3[f]\rangle + \langle 4[f]\rangle)$$
$$+ B(\langle 123[f]\rangle + \langle 124[f]\rangle + \langle 134[f]\rangle + \langle 234[f]\rangle). \qquad (A5.1)$$

A5.1 Definitions of some correlation functions

A5.1.1 Even correlation functions

$x_1 = \langle 01\rangle, \langle 02\rangle, \langle 03\rangle, \langle 04\rangle, \langle 45\rangle, \langle 46\rangle, \langle 47\rangle;$

$x_2 = \langle 05\rangle, \langle 06\rangle, \langle 07\rangle, \langle 12\rangle, \langle 13\rangle, \langle 14\rangle, \langle 23\rangle, \langle 24\rangle, \langle 34\rangle, \langle 56\rangle, \langle 57\rangle, \langle 67\rangle;$

$x_3 = \langle 15\rangle, \langle 17\rangle, \langle 26\rangle, \langle 27\rangle, \langle 35\rangle, \langle 36\rangle;$

$x_4 = \langle 16\rangle, \langle 25\rangle, \langle 37\rangle;$

$x_5 = \langle 0123\rangle, \langle 4567\rangle;$

$x_6 = \langle 0145\rangle, \langle 0147\rangle, \langle 0245\rangle, \langle 0246\rangle, \langle 0346\rangle, \langle 0347\rangle;$

$x_7 = \langle 0146\rangle, \langle 0247\rangle, \langle 0345\rangle;$

$x_8 = \langle 0125\rangle, \langle 0236\rangle, \langle 0137\rangle, \langle 1457\rangle, \langle 2456\rangle, \langle 3467\rangle;$

$x_9 = \langle 0126\rangle, \langle 0127\rangle, \langle 0235\rangle, \langle 0237\rangle, \langle 0135\rangle, \langle 0136\rangle, \langle 1456\rangle, \langle 1467\rangle, \langle 2457\rangle, \langle 2467\rangle,$
$\qquad \langle 3456\rangle, \langle 3457\rangle;$

$x_{10} = \langle 0157\rangle, \langle 0267\rangle, \langle 0356\rangle, \langle 1247\rangle, \langle 1345\rangle, \langle 2346\rangle;$

$x_{11} = \langle 0156\rangle, \langle 0167\rangle, \langle 0256\rangle, \langle 0257\rangle, \langle 0357\rangle, \langle 0367\rangle, \langle 1245\rangle, \langle 1246\rangle, \langle 1346\rangle, \langle 1347\rangle,$
$\qquad \langle 2345\rangle, \langle 2347\rangle;$

$x_{12} = \langle 0567\rangle, \langle 1234\rangle;$

$x_{13} = \langle 1235\rangle, \langle 1236\rangle, \langle 1237\rangle, \langle 1567\rangle, \langle 2567\rangle, \langle 3567\rangle;$

$x_{14} = \langle 1256\rangle, \langle 1257\rangle, \langle 1357\rangle, \langle 1367\rangle, \langle 2356\rangle, \langle 2367\rangle;$

$x_{15} = \langle 1267\rangle, \langle 1356\rangle, \langle 2357\rangle.$

A5.1.2 Odd correlation functions

$y_1 = \langle 0\rangle, \langle 1\rangle, \langle 2\rangle, \langle 3\rangle, \langle 4\rangle, \langle 5\rangle, \langle 6\rangle, \langle 7\rangle;$

$y_2 = \langle 012\rangle, \langle 013\rangle, \langle 014\rangle, \langle 023\rangle, \langle 024\rangle, \langle 034\rangle, \langle 045\rangle, \langle 046\rangle, \langle 047\rangle, \langle 456\rangle, \langle 457\rangle, \langle 467\rangle;$

$y_3 = \langle 015\rangle, \langle 017\rangle, \langle 025\rangle, \langle 026\rangle, \langle 036\rangle, \langle 037\rangle, \langle 145\rangle, \langle 147\rangle, \langle 245\rangle, \langle 246\rangle, \langle 346\rangle, \langle 347\rangle;$

$y_4 = \langle 016\rangle, \langle 027\rangle, \langle 035\rangle, \langle 146\rangle, \langle 247\rangle, \langle 345\rangle;$

$y_5 = \langle 056\rangle, \langle 057\rangle, \langle 067\rangle, \langle 123\rangle, \langle 124\rangle, \langle 134\rangle, \langle 234\rangle, \langle 567\rangle;$

$y_6 = \langle 125\rangle, \langle 137\rangle, \langle 157\rangle, \langle 236\rangle, \langle 256\rangle, \langle 367\rangle;$

$y_7 = \langle 126\rangle, \langle 127\rangle, \langle 135\rangle, \langle 136\rangle, \langle 156\rangle, \langle 167\rangle, \langle 235\rangle, \langle 237\rangle, \langle 257\rangle, \langle 267\rangle, \langle 356\rangle, \langle 357\rangle;$

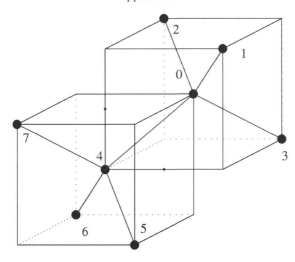

Fig. A5.1. Arrangement of the sites on the diamond lattice showing two intervening tetrahedra.

$y_8 = \langle 01234 \rangle, \langle 04567 \rangle;$
$y_9 = \langle 01235 \rangle, \langle 01236 \rangle, \langle 01237 \rangle, \langle 14567 \rangle, \langle 24567 \rangle, \langle 34567 \rangle;$
$y_{10} = \langle 12567 \rangle, \langle 13567 \rangle, \langle 23567 \rangle, \langle 12356 \rangle, \langle 12357 \rangle, \langle 12367 \rangle;$
$y_{11} = \langle 01567 \rangle, \langle 02567 \rangle, \langle 03567 \rangle, \langle 12345 \rangle, \langle 12346 \rangle, \langle 12347 \rangle;$
$y_{12} = \langle 01267 \rangle, \langle 01356 \rangle, \langle 02357 \rangle, \langle 12467 \rangle, \langle 13456 \rangle, \langle 23457 \rangle;$
$y_{13} = \langle 01256 \rangle, \langle 01257 \rangle, \langle 01357 \rangle, \langle 01367 \rangle, \langle 02356 \rangle, \langle 02367 \rangle, \langle 12456 \rangle, \langle 12457 \rangle,$
$\qquad \langle 13457 \rangle, \langle 13467 \rangle, \langle 23456 \rangle, \langle 23467 \rangle;$
$y_{14} = \langle 01245 \rangle, \langle 01347 \rangle, \langle 01457 \rangle, \langle 02346 \rangle, \langle 02456 \rangle, \langle 03467 \rangle;$
$y_{15} = \langle 01246 \rangle, \langle 01247 \rangle, \langle 01345 \rangle, \langle 01346 \rangle, \langle 02345 \rangle, \langle 02347 \rangle;$
$y_{16} = \langle 01456 \rangle, \langle 01467 \rangle, \langle 02457 \rangle, \langle 02467 \rangle, \langle 03456 \rangle, \langle 03457 \rangle;$
$y_{17} = \langle 0123456 \rangle, \langle 0123457 \rangle, \langle 0123467 \rangle, \langle 0124567 \rangle, \langle 0134567 \rangle, \langle 0234567 \rangle;$
$y_{18} = \langle 0123567 \rangle, \langle 1234567 \rangle.$

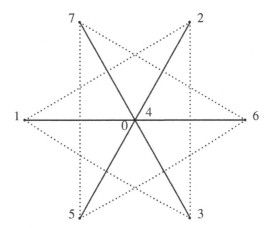

Fig. A5.2. Arrangement of neighboring sites in the diamond lattice looking into the direction of the 0–4 bond.

A5.2 Some of the Ising identities for the odd correlation functions

On the lattice sites of [0], [1], [2], [3], [4], [5], [6], and [7] it is possible to generate the following Ising odd correlation identities:

$$\langle 0 \rangle = y_1, \quad \langle 0[12] \rangle = y_2, \quad \langle 0[45] \rangle = y_2, \quad \langle 0[15] \rangle = y_3,$$
$$\langle 0[16] \rangle = y_4, \quad \langle 0[56] \rangle = y_5, \quad \langle 0[1234] \rangle = y_8,$$
$$\langle 0[4567] \rangle = y_8, \quad \langle 0[1235] \rangle = y_9, \quad \langle 0[1567] \rangle = y_{11},$$
$$\langle 0[1356] \rangle = y_{12}, \quad \langle 0[1256] \rangle = y_{13}, \quad \langle 0[1245] \rangle = y_{14},$$
$$\langle 0[1457] \rangle = y_{14}, \quad \langle 0[1246] \rangle = y_{15}, \quad \langle 0[1456] \rangle = y_{16},$$
$$\langle 0[123456] \rangle = y_{17}, \quad \langle 0[124567] \rangle = y_{17}, \quad \langle 0[123567] \rangle = y_{18}. \tag{A5.2}$$

In the above list, y_2, y_8, y_{14}, and y_{17} appear twice; however, the two definitions of each are nonequivalent with reference to the origin site [0], and hence they give rise to two different identities.

y_6, y_7, and y_{10} are defined in such a way that they do not include the origin site within the eight sites selected in this demonstration, and hence they cannot be used to generate an identity.

The identities thus generated are listed in the following:

$$\langle 0 \rangle = A(\langle 1 \rangle + \langle 2 \rangle + \langle 3 \rangle + \langle 4 \rangle) + B(\langle 123 \rangle + \langle 124 \rangle + \langle 134 \rangle + \langle 234 \rangle),$$
$$y_1 = 4(Ay_1 + By_3); \tag{A5.3}$$
$$\langle 012 \rangle = A(\langle 2 \rangle + \langle 1 \rangle + \langle 123 \rangle + \langle 124 \rangle) + B(\langle 3 \rangle + \langle 4 \rangle + \langle 234 \rangle + \langle 124 \rangle),$$
$$y_2 = 2(A + B)(y_1 + y_3); \tag{A5.4}$$
$$\langle 045 \rangle = A(\langle 145 \rangle + \langle 245 \rangle + \langle 345 \rangle + \langle 5 \rangle) + B(\langle 12345 \rangle + \langle 125 \rangle + \langle 135 \rangle + \langle 235 \rangle),$$
$$y_2 = A(y_1 + 2y_3 + y_4) + B(y_6 + 2y_7 + y_{11}); \tag{A5.5}$$
$$\langle 016 \rangle = A(\langle 6 \rangle + \langle 126 \rangle + \langle 136 \rangle + \langle 146 \rangle) + B(\langle 236 \rangle + \langle 246 \rangle + \langle 346 \rangle + \langle 12346 \rangle),$$
$$y_3 = A(y_1 + y_3 + y_6 + y_7) + B(y_3 + y_4 + y_7 + y_{11}); \tag{A5.6}$$
$$\langle 015 \rangle = A(\langle 5 \rangle + \langle 125 \rangle + \langle 135 \rangle + \langle 145 \rangle) + B(\langle 235 \rangle + \langle 245 \rangle + \langle 345 \rangle + \langle 12345 \rangle),$$
$$y_4 = A(y_1 + y_4 + 2y_7) + B(2y_3 + y_6 + y_{11}); \tag{A5.7}$$
$$\langle 056 \rangle = A(\langle 156 \rangle + \langle 256 \rangle + \langle 356 \rangle + \langle 456 \rangle) + B(\langle 12356 \rangle + \langle 12456 \rangle$$
$$+ \langle 13456 \rangle + \langle 23456 \rangle),$$
$$y_5 = A(y_2 + y_6 + 2y_7) + B(y_{10} + y_{12} + 2y_{13}); \tag{A5.8}$$
$$\langle 01234 \rangle = A(\langle 235 \rangle + \langle 134 \rangle + \langle 124 \rangle + \langle 123 \rangle) + B(\langle 4 \rangle + \langle 3 \rangle + \langle 2 \rangle + \langle 1 \rangle),$$
$$y_8 = 5(A + B)(y_1 + y_5); \tag{A5.9}$$
$$\langle 04567 \rangle = A(\langle 14567 \rangle + \langle 24567 \rangle + \langle 34567 \rangle + \langle 567 \rangle) + B(\langle 1234567 \rangle + \langle 12567 \rangle$$
$$+ \langle 13567 \rangle + \langle 23567 \rangle),$$
$$y_8 = A(y_5 + 3y_9) + B(3y_{10} + y_{18}); \tag{A5.10}$$
$$\langle 01235 \rangle = A(\langle 235 \rangle + \langle 135 \rangle + \langle 125 \rangle + \langle 12345 \rangle) + B(\langle 5 \rangle + \langle 345 \rangle + \langle 245 \rangle + \langle 145 \rangle),$$
$$y_9 = A(y_6 + 2y_7 + y_{11}) + B(y_1 + 2y_3 + y_4); \tag{A5.11}$$
$$\langle 01567 \rangle = A(\langle 567 \rangle + \langle 12567 \rangle + \langle 13567 \rangle + \langle 14567 \rangle) + B(\langle 23567 \rangle + \langle 24567 \rangle$$
$$+ \langle 34567 \rangle + \langle 1234567 \rangle),$$
$$y_{11} = A(y_5 + y_9 + 2y_{10}) + B(2y_9 + y_{10} + y_{18}); \tag{A5.12}$$

$$\langle 01356 \rangle = A(\langle 356 \rangle + \langle 156 \rangle + \langle 12356 \rangle + \langle 13456 \rangle) + B(\langle 256 \rangle + \langle 23456 \rangle$$
$$+ \langle 456 \rangle + \langle 12456 \rangle),$$
$$y_{12} = A(2y_7 + y_{10} + y_{12}) + B(y_2 + y_6 + 2y_{13}); \tag{A5.13}$$
$$\langle 01256 \rangle = A(\langle 256 \rangle + \langle 156 \rangle + \langle 12356 \rangle + \langle 12456 \rangle) + B(\langle 356 \rangle + \langle 456 \rangle$$
$$+ \langle 23456 \rangle + \langle 13456 \rangle),$$
$$y_{13} = A(y_6 + y_7 + y_{10} + y_{13}) + B(y_2 + y_7 + y_{12} + y_{13}); \tag{A5.14}$$
$$\langle 01245 \rangle = A(\langle 245 \rangle + \langle 145 \rangle + \langle 12345 \rangle + \langle 125 \rangle) + B(\langle 345 \rangle + \langle 5 \rangle$$
$$+ \langle 235 \rangle + \langle 135 \rangle),$$
$$y_{14} = A(2y_3 + y_{y_6} + y_{11}) + B(y_1 + y_4 + 2y_7); \tag{A5.15}$$
$$\langle 01457 \rangle = A(\langle 457 \rangle + \langle 12457 \rangle + \langle 13457 \rangle + \langle 157 \rangle) + B(\langle 23457 \rangle + \langle 257 \rangle$$
$$+ \langle 357 \rangle + \langle 12357 \rangle),$$
$$y_{14} = A(y_2 + y_6 + 2y_{13}) + B(2y_7 + y_{10} + y_{12}); \tag{A5.16}$$
$$\langle 01246 \rangle = A(\langle 246 \rangle + \langle 146 \rangle + \langle 12346 \rangle + \langle 126 \rangle) + B(\langle 346 \rangle + \langle 6 \rangle$$
$$+ \langle 236 \rangle + \langle 136 \rangle),$$
$$y_{15} = A(y_3 + y_4 + y_7 + y_{11}) + B(y_1 + y_3 + y_6 + y_7); \tag{A5.17}$$
$$\langle 01456 \rangle = A(\langle 456 \rangle + \langle 12456 \rangle + \langle 13456 \rangle + \langle 156 \rangle) + B(\langle 23456 \rangle + \langle 256 \rangle$$
$$+ \langle 356 \rangle + \langle 12356 \rangle),$$
$$y_{16} = A(y_2 + y_7 + y_{12} + y_{13}) + B(y_6 + y_7 + y_{10} + y_{13}); \tag{A5.18}$$
$$\langle 0123457 \rangle = A(\langle 23456 \rangle + \langle 13456 \rangle + \langle 12456 \rangle + \langle 12356 \rangle) + B(\langle 456 \rangle$$
$$+ \langle 356 \rangle + \langle 256 \rangle + \langle 156 \rangle),$$
$$y_{17} = A(y_{10} + y_{12} + 2y_{13}) + B(y_2 + y_6 + 2y_7); \tag{A5.19}$$
$$\langle 0124567 \rangle = A(\langle 24567 \rangle + \langle 14567 \rangle + \langle 1234567 \rangle + \langle 12567 \rangle) + B(\langle 34567 \rangle$$
$$+ \langle 567 \rangle + \langle 23567 \rangle + \langle 13567 \rangle),$$
$$y_{17} = A(y_{10} + 2y_9 + y_{18}) + B(y_5 + y_9 + 2y_{10}); \tag{A5.20}$$
$$\langle 0123567 \rangle = A(\langle 23567 \rangle + \langle 13567 \rangle + \langle 1234567 \rangle + \langle 12567 \rangle) + B(\langle 567 \rangle$$
$$+ \langle 34567 \rangle + \langle 24567 \rangle + \langle 14567 \rangle),$$
$$y_{18} = A(3y_{10} + y_{18}) + B(y_5 + 3y_9). \tag{A5.21}$$

References

Abramowitz, M. & Stegun, I. A., *Handbook of Mathematical Functions*, Dover Publications, New York, 1972.

Aggarwal, S. K. & Tanaka, T. (1977), *Phys. Rev.* **B16**, 3963.

Anderson, P. W. (1961), *Phys. Rev.* **124**, 41.

Appelbaum, J. A. (1968), *Phys. Rev.* **165**, 632.

Baker, G. A., Jr (1963), *Phys. Rev.* **129**, 99.

Baker, G. A., Jr & Gammel, J. L. (1961), *J. Math. Anal. & Appl.* **2**, 21.

Baker, G. A. Jr, Gammel, J. L. & Wills, J. G. (1961), *J. Math. Anal. & Appl.* **2**, 405.

Bardeen, J, Cooper, L. N. & Schrieffer, J. R. (1957), *Phys. Rev.* **108**, 1175.

Barry, J. H., Múnera, C. H. & Tanaka, T. (1982). *Physica* **113A**, 367.

Barry, J. H., Khatum, M. & Tanaka, T. (1988). *Phys. Rev.* **37B**, 5193.

Beyeler, H. U. & Strassler, S. (1979), *Phys. Rev.* **B 20**, 1980.

Bloch, F. (1930), *Z. Physik* **61**, 206.

Bogoliubov, N. N. (1947), *J. Phys. USSR* **11**, 23.

Bogoliubov, N. N. (1958), *Nuovo Cimento* **7**, 794.

Born, M. & Green, M. S. (1946), *Proc. Roy. Soc.* **A188**, 10.

Bose, S. M. & Tanaka, T. (1968), *Phys. Rev.* **176**, 600.

Bose, S. M., Tanaka, T. & Halow, J. (1973), *Phys. kondens. Materie* **16**, 310.

Boyce, J. B., Hayes, T. M., Stutius, W. & Mikkelson, J. C. (1977), *Phys. Rev. Lett.* **38**, 1362.

Bragg, W. L. & Williams, E. J. (1934), *Proc. Roy. Soc. (London)* **A145**, 699.

Callen, H. B. (1963), *Phys. Lett.* **4**, 161.

Cava, R. J., Reidinger, F. & Wuensch, B. J. (1977), *Solid State Commun.* **24**, 411.

Charap, S. H. & Boyd, E. L. (1964), *Phys. Rev.* **A133**, 811.

Dirac, P. A. M., *Quantum Mechanics*, 4th edn, Oxford University Press, 1958.

Doman, B. G. S. & ter Haar, D. (1962), *Phys. Lett.* **2**, 15.

Domb, C. (1960), *Phil. Mag. Suppl.* **9**, 149.

Domb, C. & Sykes, F. (1956), *Proc. Roy. Soc. (London)* **A235**, 247.

Domb, C. & Sykes, F. (1962), *J. Math. Phys.* **3**, 586.

Dyson, J. F. (1956), *Phys. Rev.* **102**, 1217, 1230.

Fisher, M. E. (1959), *Phys. Rev.* **113**, 969.

Fletcher, N. H., *The Chemical Physics of Ice*, Cambridge University Press, 1970.

Fröhlich, H. (1950), *Proc. Roy. Soc.* **A215**, 291.

Gingrich, N. S. (1945), *Rev. Mod. Phys.* **15**, 90.

Green, H. S. & Hurst, C. A., *Order-Disorder Phenomena*, Wiley Interscience, New York, 1964.

Halow, J., Tanaka, T. & Morita, T. (1968), *Phys. Rev.* **175**, 680.

Haus, J. & Tanaka, T. (1977), *Phys. Rev.* **B16**, 2148.

Hausen, H. (1935), *Phys. Zeits.* **35**, 517.

Heisenberg, W. (1929), *Z. Phys.* **49**, 619.

Heitler, W. & London, F. (1927), *Z. Phys.* **44**, 455.

Hoshino, S., Sakuma, T. & Fujii, Y. (1977), *Solid State Commun.* **22**, 763.

Hurst, C. A. & Green, H. S. (1960), *J. Chem. Phys.* **33**, 1059.

Ising, E. (1925), *Z. Physik.* **31**, 253.

Kasteleyn, P. W. (1961), *Physica*, **27**, 1209.

Kasteleyn, P. W. (1963), *J. Math. Phys.* **4**, 287.

Kikuchi, R. (1951), *Phys. Rev.* **81**, 988.

Kirkwood, J. G. (1946), *J. Chem. Phys.* **14**, 180.

Kirkwood, J. G. (1947), *J. Chem. Phys.* **15**, 72.

Kirkwood, J. G. & Buff, F. P. (1943), *Rev. Mod. Phys.* **15**, 90.

London, F. (1938), *Phys. Rev.* **54**, 947.

McCoy, B. M. & Wu, T. T., *The Two-Dimensional Ising Model*, Harvard University Press, 1973.

Matsubara, T. & Matsuda, H. (1956), *Prog Theoret. Phys.* **16**, 569.

Montroll, E. W., Potts, R. B. & Ward, J. C. (1963), *J. Math. Phys.* **4**, 308.

Morita, T. (1957), *J. Phys. Soc. Japan* **12**, 1060.

Morita, T. (1994), *Prog. Theoret. Phys., Suppl.* **115**, 27.

Morita, T. & Horiguchi, T. (1971), *Table of the Lattice Green's Function for the Cubic Lattices*, Mathematics Division, Department of Applied Science, Faculty of Engineering, Tohoku University, Sendai, Japan.

Morita, T. & Tanaka, T. (1965), *Phys. Rev.*, **138**, A1403.

Morse, P. M. & Feshbach, H., *Methods of Theoretical Physics*, McGraw-Hill, 1953, Pt I, Sec. 4.6.

Mūnera, C. H., Barry, J. H. & Tanaka, T. (1982). *2nd. International Conf. on Solid Films and Surfaces, College Park.*

Nagle, J. F. (1966), *J. Math. Phys.* **7**, 1484.

Naya, S. (1954), *Prog. Theoret. Phys.* **11**, 53.

Notting, J. (1963), *Ber Bunsenges Phys. Chem.* **67**, 172.

Oguchi, T. & Kitatani, H. (1988), *J. Phys. Soc. Japan* **57**, 3973.

Onsager, L. (1944), *Phys. Rev.* **65**, 117.

Onsager, L. (1949), *Nuovo Cimento (Suppl.)* **6**, 261.

Pauling, L., *The Nature of the Chemical Bond*, Cornell University Press, Ithaca, New York, 1960.

Rushbrooke, G. S. & Wakefield, A. J. (1949), *Nuovo Cimento* **2**, 251; suppl. VI, series IX.

Schrödinger, E. (1952), *Statistical Thermodynamics* (Cambridge University Press).

Stephenson, J. (1964), *J. Math. Phys.* **5**, 1009.

Stephenson, J. (1966), *J. Math. Phys.* **7**, 1123.

Stephenson, J. (1970), *J. Math. Phys.* **11**, 413.

Sugiura, Y. (1927), *Z. Phys.* **45**, 484.

Suzuki, M. (1965), *Phys. Lett.* **19**, 267.

Sykes, M. F. (1961), *J. Math. Phys.* **2**, 52.

Tanaka, T., Hirose, T. & Kurati, K. (1994), *Prog. Theoret. Phys., Suppl.* **115**, 41.

Tanaka, T., Katumori, H. & Tosima, S. (1951), *Prog. Theoret. Phys.* **6**, 17.

Tanaka, G. & Kimura, M. (1994), *Prog. Theoret. Phys., Suppl.* **115**, 207.

Tanaka, T. & Libelo, L. F. (1975), *Phys. Rev.* **B12**, 1790.

Tanaka, T. & Morita, T. (1999), *Physica* **A 277**, 555.

Thompson and Tait, *Treatise on Natural Philosophy*, Cambridge University Press, 1879.

Van Vleck, J. H., *The Theory of Electric and Magnetic Susceptibilities*, Oxford University Press, London, 1932.

Wannier, G. H. (1945), *Rev. Mod. Phys.* **17**, 50.

Weiss, P. (1907), *J. Phys.* **6**, 661.

Wentzel, G. (1960), *Phys. Rev.* **120**, 1572.

Whalley, E., Davidson, D. W. & Heath, J. B. R. (1966), *J. Chem. Phys.* **45**, 3976.

Whalley, E., Jones, S. J. & Gold, L. W., *Physics and Chemistry of Ice*, Royal Society of Canada, Ottawa, 1973.

Wright, A. F. & Fender, B. E. F. (1977), *J. Phys.* **C10**, 2261.

Yang, C. N. (1952), *Phys. Rev.* **85**, 809.

.

Bibliography

Anderson, P. W. (1952), *Phys. Rev.* **86**, 694.

Anderson, P. W. (1967), *Phys. Rev.* **164**, 352.

Baker, G. A., Jr (1961, 1964, 1965), *Phys. Rev.* **124**, 768; *ibid.* **136A**, 1376; *Advan. Theoret. Phys.* **1**, 1.

Bethe, H. A. (1935), *Proc. Roy. Soc.* **A150**, 552.

Bragg, W. L. & Williams, E. J. (1935), *Proc. Roy. Soc. (London)* **A151**, 540.

Brown, A. J. & Whalley, E. (1966), *J. Chem. Phys.* **45**, 4360.

Callen, H. B., *Thermodynamics*, John Wiley and Sons, 1960.

Davydov, A. S. *Quantum Mechanics*, 2nd edn, translated, edited and with additions by D. ter Haar, Pergamon Press, 1976.

Emery, V. J. (1987), *Phys. Rev. Lett.* **58**, 2794.

Fermi, Enrico, *Thermodynamics*, Prentice-Hall, 1937; Dover Publications, 1956.

Gibbs, J. W., *Elementary Principles in Statistical Mechanics*, New Haven, 1902.

Green, H. S., *The Molecular Theory of Fluids*, North-Holland Publishing Company, 1952.

Hayes, T. M. & Boyce, J. B. (1979), in Vashishta, P., Mundy, J. N. & Shenoy, G. K. (eds) *Proc. Int. Conf. on Fast Ion Transport in Solid, Electrons and Electrodes*, Lake Geneva, Wisconsin, USA, 21–25 May, 1979, North Holland, New York, 1979, pp. 535, 621.

Heisenberg, W. (1926), *Z. Phys.* **38**, 411.

Herring, C. (1966), in Rado, G. & Suhl, H. (eds), *Magnetism*, Vol. **2B**, Academic Press, New York, p. 1.

Hirsch, J. E. (1985, 1987), *Phys. Rev.* **B 31**, 4403; *Phys. Rev. Lett.* **59**, 228.

Holstein, T. & Primakoff, H. (1940), *Phys. Rev.* **58**, 1908.

Johari, G. P., Lavergre, A. & Whalley, E. (1974), *J. Chem. Phys.* **61**, 4292.

Kasteleyn, P. W., *Fundamental Problems in Statistical Mechanics*, Vol. II (E. G. D. Cohen, ed.), Wiley, New York, 1968.

Kondo, J. (1964), *Progr. Theoret. Phys.* **32**, 37.

Krammers, H. A. & Wannier, G. H. (1941), *Phys. Rev.* **60**, 251, 1230.

Kubo, R. (1952), *Phys. Rev.* **87**, 568.

Kurata, M., Kikuchi, R. & Watari, T. (1953), *J. Chem. Phys.* **21**, 434.

Morita, T. (1972), *J. Math. Phys.* **13**, 115.

Morita, T. & Tanaka, T. (1966), *Phys. Rev.* **145**, 288.

Oota, Y., Kakiuchi, T., Miyajima, S. & Tanaka, T. (1998), *Physica* **A250**, 103.

Penn, D. R. (1966), *Phys. Rev.* **142**, 350.

Pippard, A. B., *The Elements of Classical Thermodynamics*, Cambridge University Press, 1957.

Schrieffer, J. R. & Wolf, P. A. (1966), *Phys. Rev.* **149**, 491.

Schrödinger, E., *Statistical Thermodynamics*, Cambridge University Press, 1948.

Stanley, H. E., *Introduction to Phase Transitions and Critical Phenomena*, Oxford University Press, 1971.

Thompson, C. J., *Mathematical Statistical Mechanics*, The Macmillan Company, New York, 1972.

Thouless, D. J., *The Quantum Mechanics of Many-Body Systems*, Academic Press, 1961.

White, R. M., *Quantum Theory of Magnetism*, McGraw-Hill, 1970.

Wong, P. T. T. & Whalley, E. (1976), *J. Chem. Phys.* **64**, 2349.

Wood, D. C. & Griffiths, J. (1976), *J. Phys.* **A 7**, 409.

Yang, C. N. (1989), *Phys. Rev. Lett.* **63**, 2144.

Zemansky, M. W., *Heat and Thermodynamics*, 5th edn, McGraw-Hill, 1957.

Ziman, J. M., *Elements of Advanced Quantum Theory*, Cambridge University Press, 1980.

Index

adiabatic approximation, 61
adiabatic process, 11, 30, 31, 34, 38
Anderson model, 190, 191, 194
anticommutation relation, 89, 90
antiferromagnetism, 197, 199, 211
a priori probabilities, 52, 64
average value, 69, 70, 71, 77, 87
    statistical, 112, 123 (*see also* correlation function)

BCS Hamiltonian, 175, 178
Bethe (two-site or pair) approximation, 113, 122, 134, 146, 155, 168, 170
Bloch equation, 118
Bloch function, 95, 96
Bloch state, 95, 96
Boltzmann constant, 53, 62, 127
Bose–Einstein condensation, 84, 99, 100
Bose–Einstein statistics, 57, 63, 81, 84, 99
bosons, 57, 63, 81, 84, 99
Boyle–Marriot law, 4, 5
bra, 72
Bragg–Williams approximation, 134

canonical distribution, 63, 114
Carnot cycle, 14
    of an ideal gas, 19
    reverse, 14, 15
charging-up process, 43
chemical potential, 42, 43, 53, 198, 202, 203, 211
Clausius–Clapeyron equation, 44
Clausius' inequality, 22, 24, 26
closed polygon, 247, 264
cluster function, 128, 129, 131, 138, 144, 147, 162, 163, 179, 188
cluster variation method, 127, 130, 131, 136, 140, 143
    correlation function formulation of the, 144, 147, 269
    parent cluster, 130
    simplified (SCVM), 141, 143
    subcluster, 141, 143
    variational potential, 127, 134, 137, 145–7, 149, 167, 179, 191, 200, 207
c-number, 91, 178

coexistence line, 44, 46
combinatorial formulation, 156, 172, 248, 259
commutation relation, 74, 94, 105, 107, 108, 111, 188, 190, 299
commutator, 70, 79
Cooper pair, 99, 179
correlation function, 122, 123, 145, 147, 158, 165, 170, 212, 223
critical exponent, 215, 265
critical temperature, 44, 136, 147, 151, 154, 155, 166, 215, 273
cumulant expansion, 128–30, 134, 137, 144, 145, 172, 185, 188, 206, 235, 237, 240, 269
cumulant function, *see* cumulant expansion
cycle, 12
cyclic engine, 12
cyclic process, 12

density of states, 68, 190, 203, 211, 280
Dirac delta function, 72–4

efficiency of an engine, 18
eigenfunction, 71–3, 95, 102, 103, 106, 110
eigenrepresentation, 74, 75
eigenvalue, 56, 60, 61, 64, 66, 71, 80, 81, 95, 102
electron–phonon interaction, 97, 98
ensemble, 50, 52
    canonical, 50, 59, 62
    grand canonical, 50, 52, 54, 56, 58
    microcanonical, 63, 64
enthalpy, 12, 39
entropy, 24, 26, 28, 36, 38, 44, 53, 55–9, 62, 63, 68, 106, 128, 183, 184, 209, 234–40, 242–5, 269
equal *a priori* probabilities, 52, 64
equation of state, 2, 46
    caloric, 8, 9
    thermal, 3, 8, 9
exact differential, 8, 27–9
exchange integral, 93, 94
expectation value, 70, 80, 184, 185, 187, 200

Fermi–Dirac distribution function, 56, 57
Fermi hole, 91, 92

291

Printed in the United States
By Bookmasters